Operationalizing Expectations and Mapping Challenges of Information Privacy and Data Protection Measures

Indranath Gupta
Editor

Operationalizing Expectations and Mapping Challenges of Information Privacy and Data Protection Measures

A Curated Literature Overview

Springer

Editor
Indranath Gupta
Jindal Global Law School
O.P. Jindal Global University
Sonipat, Haryana, India

ISBN 978-981-99-7370-5 ISBN 978-981-99-7371-2 (eBook)
https://doi.org/10.1007/978-981-99-7371-2

This Springer imprint is published by the registered company Springer Nature Singapore Pte Ltd.
The registered company address is: 152 Beach Road, #21-01/04 Gateway East, Singapore 189721, Singapore

Acknowledgements

The editor is highly grateful to all those who have taken time out of their busy schedule and ensured the timely completion of this project. The editor would like to thank Prof. Paarth Naithani, Jindal Global Law School and the Springer team, who spent several hours on this project. The editor would also like to thank his family and colleagues at O.P. Jindal Global University (JGU) for their constant encouragement. The editor admires the unconditional support of Vice Chancellor Prof. C. Raj Kumar of O.P. Jindal Global University.

Introduction

The general apprehension across jurisdictions is data protection, and privacy measures are at cross purpose with innovation and benefits of the digital economy. While individuals' rights and personal autonomy are of paramount importance, there are important questions of balance that direct to the issue of operationalising expectations surrounding data protection regulatory frameworks. Volume 1 of this two-part volume attempts to address the data protection and privacy issues at a conceptual level alongside dealing with its relations with other laws and the application of consent in different environments. This volume, however, looks at the operational question of implementing the expectations. It positions itself in a context of 'conflict' and 'uneasiness' that often is assigned to the question of technology and privacy. Therefore, it starts with a chapter on these conflicts and contradictions before moving on to articles that effectively implement data protection principles in a tech environment. Often, we do not realise how unknowingly data protection and privacy issues are closer to the home environment. It may be because of the place where we live, because of the way we embrace gadgets and upcoming technology or the time and space that define modern living. The chapter on 'Bringing Privacy and Data Protection Issues Closer to Home' gives us a sense of how privacy and data protection issues are no longer obscure incidents that happen to strangers but are much closer to the general mass of individuals. To harness the potential and benefits of digitisation, optimally adopting technical measures can prevent data breaches from happening. When there is no alternative to technology, there is a need to find technological solutions that preserve the privacy of individuals. For instance, the COVID-19 vaccination process and the recording of such an event by issuance of certificates are irreplaceable for human beings. The entire process is intertwined with questions like the use of personal sensitive data of individuals and their storage in a secure manner. The chapter entitled 'Finding Technical Solutions to Safeguard Privacy of Individuals' provides a snapshot of examples where technical advancements offer effective and practical solutions in implementing data protection measures to protect the privacy of individuals. The final two chapters in this volume portray the extent of risks that individuals face in the tech environment and how often privacy and data protection questions come between technical growth and advancement.

The two volumes in this series reflect that there is no reason to objectivise the issue of privacy as a hindrance to the necessary growth of harnessing the true potential of digitisation. Processes, technical or otherwise, should keep privacy as an essential element that features automatically at the design stage and not as an afterthought to fulfil merely a checklist activity. The contradictions and conflicts in most situations result from including data protection measures protecting privacy as an additional obligation and not as an outcome of best practices that data controllers should adopt. Understandably, the volumes could only capture some of the wide-ranging issues that have emerged and are likely to emerge at the intersection of data protection, privacy, technical advancements, scientific research, and digital data governance. Nonetheless, it is a step in the right direction.

Contents

1 Technology and Privacy: The Conflicts and Contradictions 1
1.1 Introduction by the Editor 1
1.2 Machine Generated Summaries 4
1.2.1 A Comprehensive and Holistic Knowledge Model for Cloud Privacy Protection [1] 4
1.2.2 Privacy in the Shadow of Nanotechnology [18] 8
1.2.3 Security, Privacy and Risks Within Smart Cities: Literature Review and Development of a Smart City Interaction Framework [24] 11
1.2.4 RFID and Privacy: What Consumers Really Want and Fear [41] 14
1.2.5 Accountable Privacy Supporting Services [45] 16
1.2.6 Togetherness and Respect: Ethical Concerns of Privacy in Global Web Societies [66] 20
1.2.7 Informational Privacy, Data Mining, and the Internet [76] 24
1.2.8 Algorithmic Decision-Making Based on Machine Learning from Big Data: Can Transparency Restore Accountability? [77] 24
1.2.9 Disease Management and the Application of Privacy Guidelines [78] 28
1.2.10 Ensuring Privacy Protection in the Era of Big Laparoscopic Video Data: Development and Validation of an Inside Outside Discrimination Algorithm (IODA) [81] 32
References 34

2 General Data Protection Principles and Its Implementation in a Tech-Environment 39
2.1 Introduction by the Editor 39
2.2 Machine Generated Summaries 43

2.2.1 An Analysis of Blockchain and GDPR Under the Data Lifecycle Perspective [120] ... 44
2.2.2 The Role of IS in the Conflicting Interests Regarding GDPR [121] ... 46
2.2.3 Machine Understandable Policies and GDPR Compliance Checking [122] ... 49
2.2.4 Enterprise Architecture Management as a Solution for Addressing General Data Protection Regulation Requirements in a Big Data Context: A Systematic Mapping Study [123] ... 52
2.2.5 On Enabling GDPR Compliance in Business Processes Through Data-Driven Solutions [124] ... 55
2.2.6 Protecting Citizens' Personal Data and Privacy: Joint Effort from GDPR EU Cluster Research Projects [125] ... 59
2.2.7 Modeling Data Protection and Privacy: Application and Experience with GDPR [126] ... 61
2.2.8 Data Governance Through a Multi-DLT Architecture in View of the GDPR [127] ... 65
2.2.9 Modeling Ecosystems of Reference Frameworks for Assurance: A Case on Privacy Impact Assessment Regulation and Guidelines [128] ... 68
2.2.10 Against Explainability Requirements for Ethical Artificial Intelligence in Health Care [129] ... 72
2.2.11 Using Artificial Intelligence to Support Compliance with the General Data Protection Regulation [130] ... 75
2.2.12 An AI Framework to Support Decisions on GDPR Compliance [131] ... 78
2.2.13 Future of Data Analytics in the Era of the General Data Protection Regulation in Europe [132] ... 81
2.2.14 General Data Protection Regulation in Health Clinics [133] ... 84
2.2.15 The GDPR and the Research Exemption: Considerations on the Necessary Safeguards for Research Biobanks [134] ... 86
2.2.16 General Data Protection Regulations (2018) and Clinical Research: Perspectives of Patients and Doctors in an Irish University Teaching Hospital [135] ... 89
2.2.17 GDPR: An Impediment to Research? [136] ... 91
2.2.18 Hypocrisy Around Medical Patient Data: Issues of Access for Biomedical Research, Data Quality, Usefulness for the Purpose and Omics Data as Game Changer [137] ... 95

2.2.19 The European Parliament Proposal for the New EU General Data Protection Regulation May Severely Restrict European Epidemiological Research [138] 100
2.2.20 Clinical Trial Data Transparency in the EU: Is the New Clinical Trials Regulation a Game-Changer? [139] 103
2.2.21 Enablers and Barriers to the Secondary Use of Health Data in Europe: General Data Protection Regulation Perspective [140] 106
2.2.22 The Use of Human Tissue in Epidemiological Research; Ethical and Legal Considerations in Two Biobanks in Belgium [141] 109
2.2.23 Strengthen Electronic Health Records System (EHR-S) Access-Control to Cope with GDPR Explicit Consent [142] 111
2.2.24 Analysis of Solutions for a Blockchain Compliance with GDPR [143] 113
2.2.25 Survey of NCHDs in Ireland to Assess Their Views and Opinions in Relation to Participation in Health Research and the Impact of New Irish Data Protection Regulations [144] 115
2.2.26 Research Under the GDPR—A Level Playing Field for Public and Private Sector Research? [145] 118
2.2.27 A Framework for GDPR Compliance for Small- and Medium-Sized Enterprises [146] 121
2.2.28 Deontic Modality in the GDPR Based Finnish Privacy Notices in the Light of the Transparency Principle [147] 124
References 128

3 Bringing Privacy and Data Protection Issues Closer to Home 137
3.1 Introduction by the Editor 137
3.2 Machine Generated Summaries 140
3.2.1 Privacy Concerns in the Smart Home Context [1] 140
3.2.2 Temporal and Cultural Limits of Privacy in Smartphone App Usage [9] 144
3.2.3 Mobile-Centric Ambient Intelligence in Health- and Homecare—Anticipating Ethical and Legal Challenges [16] 146
3.2.4 Legal Safeguards for Privacy and Data Protection in Ambient Intelligence [19] 149
3.2.5 Techno-Regulation and Intelligent Safeguards [22] 152
3.2.6 Continuous Improvement Process (CIP)-Based Privacy-Preserving Framework for Smart Connected Toys [36] 155

3.2.7 Privacy in Targeted Advertising on Mobile Devices: A Survey [52] 158
3.2.8 On the Privacy of Mental Health Apps [76] 162
References 166

4 Risks, Threats and Way-Outs to Protecting Privacy in Tech-Environment: Finding Technical Solutions to Safeguard Privacy of Individuals 171
4.1 Introduction by the Editor 171
4.2 Machine Generated Summaries 177
4.2.1 Identifying Personal Physiological Data Risks to the Internet of Everything: The Case of Facial Data Breach Risks [307] 178
4.2.2 Challenges in Making Blockchain Privacy Compliant for the Digital World: Some Measures [308] 182
4.2.3 Risk Management and Privacy Violation Detection in the PoSeID-On Data Privacy Platform [309] 187
4.2.4 Privacy Risks of Whole-Slide Image Sharing in Digital Pathology [310] 189
4.2.5 Privacy Preservation in e-health Cloud: Taxonomy, Privacy Requirements, Feasibility Analysis, and Opportunities [311] 191
4.2.6 KDD, Data Mining, and the Challenge for Normative Privacy [312] 194
4.2.7 Privacy Preserving Framework Using Gaussian Mutation Based Firebug Optimization in Cloud Computing [313] 195
4.2.8 PAPIR: Privacy-Aware Personalized Information Retrieval [314] 198
4.2.9 Predictive Privacy: Towards an Applied Ethics of Data Analytics [315] 201
4.2.10 It is not (Only) About Privacy: How Multi-party Computation Redefines Control, Trust, and Risk in Data Sharing [316] 204
4.2.11 Locational Privacy-Preserving Distance Computations with Intersecting Sets of Randomly Labeled Grid Points [317] 207
4.2.12 Consumer Online Privacy Concerns and Responses: A Power-Responsibility Equilibrium Perspective [318] 209
4.2.13 Evaluating a Privacy Requirements Specification Method by Using a Mixed-Method Approach: Results and Lessons Learned [319] 213
4.2.14 Extracting LPL Privacy Policy Purposes from Annotated Web Service Source Code [320] 217

4.2.15 Healthcare Data Breaches: Implications for Digital Forensic Readiness [321] 220
4.2.16 A Blockchain Based Secure and Privacy Aware Medical Data Sharing Using Smart Contract and Encryption Scheme [322] 222
4.2.17 Privacy-Preserving Storage of Sequenced Genomic Data [323] 225
4.2.18 Security and Access of Health Research Data [324] 227
4.2.19 Data Privacy Protection in Scientific Publications: Process Implementation at a Pharmaceutical Company [325] 230
4.2.20 Data Security Crisis in Universities: Identification of Key Factors Affecting Data Breach Incidents [326] 233
4.2.21 Privacy in Pervasive Environments: Next Generation Labeling Protocols [327] 236
4.2.22 Highly Private Blockchain-Based Management System for Digital COVID-19 Certificates [328] 240
4.2.23 A Survey on Security and Privacy Issues in Contact Tracing Application of Covid-19 [329] 245
4.2.24 Realizing Privacy Aspects in Blockchain Networks [330] 249
4.2.25 Information Privacy in Institutional and End-User Tracking and Recording Technologies [331] 251
4.2.26 A Survey of Computational Location Privacy [332] 253
4.2.27 A Survey of RFID Privacy Approaches [333] 256
4.2.28 What it Takes to Boost Internet of Things Privacy beyond Encryption with Unobservable Communication: A Survey And Lessons Learned from the First Implementation of DC-Net [334] 258
4.2.29 Adaptive Key Management-Based Cryptographic Algorithm for Privacy Preservation in Wireless Mobile Adhoc Networks for IoT Applications [335] 262
4.2.30 Enhanced Privacy-Preserving in Student Certificate Management in Blockchain and Interplanetary File System [336] 265
4.2.31 Privacy Enhancing Techniques in the Internet of Things Using Data Anonymisation [337] 269
4.2.32 Addressing Privacy Requirements in System Design: The PriS Method [338] 271
4.2.33 Implications of an Ethic of Privacy for Human-Centred Systems Engineering [339] 274
4.2.34 Privacy-Preserving Process Mining [340] 277
4.2.35 PDGuard: An Architecture for the Control and Secure Processing of Personal Data [341] 281

4.2.36 Using Homomorphic Encryption for Privacy-Preserving Clustering of Intrusion Detection Alerts [342] 284
4.2.37 Privacy-Enhanced BPMN: Enabling Data Privacy Analysis in Business Processes Models [343] 288
4.2.38 P-STORE: Extension of STORE Methodology to Elicit Privacy Requirements [344] 292
4.2.39 Fuzzy-Based Approach to Assess and Prioritize Privacy Risks [345] 296
4.2.40 Privacy Self Management in Enterprise [346] 299
4.2.41 Automated Privacy Negotiations with Preference Uncertainty [347] 301
4.2.42 Embedding a Hiding Function in a Portable Electronic Health Record for Privacy Preservation [348] 305
4.2.43 DeepFake Electrocardiograms Using Generative Adversarial Networks Are the Beginning of the End for Privacy Issues in Medicine [349] 307
4.2.44 Privacy from Promises to Protection [350] 310
References 312

5 Technical Growth vis-à-vis Privacy and Data Protection Issues 331
5.1 Introduction by the Editor 331
5.2 Machine Generated Summaries 336
5.2.1 Evaluating Existing Security and Privacy Requirements for Legal Compliance [1] 336
5.2.2 Integrating Security and Privacy in Software Development [7] 339
5.2.3 An Efficient Clustering-Based Anonymization Scheme for Privacy-Preserving Data Collection in IoT Based Healthcare Services [13] 347
5.2.4 Privacy by Designers: Software Developers' Privacy Mindset [34] 351
5.2.5 Understanding Developers' Privacy and Security Mindsets via Climate Theory [39] 354
5.2.6 Interdisciplinary Perspectives on Privacy Awareness in Lifelogging Technology Development [47] 358
5.2.7 Privacy Preserving Electronic Petitions [50] 360
5.2.8 Multi-level Privacy Analysis of Business Processes: The Pleak Toolset [61] 364
5.2.9 Challenges in Enabling Privacy Self Management [65] 367
5.2.10 Releasing Survey Microdata with Exact Cluster Locations and Additional Privacy Safeguards [68] 370

5.2.11 Visual Privacy Attacks and Defenses in Deep Learning: A Survey [71] ... 372
5.2.12 Information Privacy Assimilation in IT Organizations [92] ... 375
5.2.13 Data Sharing Platforms: Instruments to Inform and Shape Science Policy on Data Sharing? [101] ... 379
5.2.14 Comparative Sensitivity of Social Media Data and Their Acceptable Use in Research [109] ... 382
5.2.15 Comparative Legal Study on Privacy and Personal Data Protection for Robots Equipped with Artificial Intelligence: Looking at Functional and Technological Aspects [113] ... 385
5.2.16 Toward an Understanding of the Antecedents to Health Information Privacy Concern: A Mixed Methods Study [116] ... 387
5.2.17 Security and Privacy Requirements for a Multi-institutional Cancer Research Data Grid: An Interview-Based Study [128] ... 390
5.2.18 Alternatives for Protecting Privacy While Respecting Patient Care and Public Health Needs [129] ... 393
5.2.19 Monitoring Ethical, Legal, and Social Issues in Developing Population Genetic Databases [130] ... 393
5.2.20 Exploring the Need for a Suitable Privacy Framework for mHealth When Managing Chronic Diseases [132] ... 395
5.2.21 So You Want to Build Your Disease's First Online Patient Registry: An Educational Guide for Patient Organizations Based on US and European Experience [138] ... 398
5.2.22 PRISED Tangle: A Privacy-Aware Framework for Smart Healthcare Data Sharing Using IOTA Tangle [151] ... 403
5.2.23 Share and Protect Our Health Data: An Evidence Based Approach to Rare Disease Patients' Perspectives on Data Sharing and Data Protection—Quantitative Survey and Recommendations [163] ... 407
5.2.24 The Italian Neuromuscular Registry: A Coordinated Platform Where Patient Organizations and Clinicians Collaborate for Data Collection and Multiple Usage [167] ... 410
5.2.25 Protection of Visual Privacy in Videos Acquired with RGB Cameras for Active and Assisted Living Applications [173] ... 413

5.2.26 Privacy-Preserving Data Sharing Infrastructures for Medical Research: Systematization and Comparison [180] 416
5.2.27 Health Data Privacy Through Homomorphic Encryption and Distributed Ledger Computing: An Ethical-Legal Qualitative Expert Assessment Study [186] 419
5.2.28 How to Integrate Legal Requirements into a Requirements Engineering Methodology for the Development of Security and Privacy Patterns [191] 422
5.2.29 Pharmacogenetic Research and Data Protection—Challenges and Solutions [196] 426
References 428

About the Editor

Prof. (Dr.) Indranath Gupta is Professor at the Jindal Global Law School and Dean in the Office of Data, Innovation and Technology at O.P. Jindal Global University, India. He is the Director of Jindal Initiative on Research in IP and Competition (JIRICO), and a Senior Fellow at the Jindal Institute of Behavioural Sciences (JIBS). He held the Jean-Monnet Chair- Academic Coordinator for the project 'Multi-dimensional Approaches to the Understanding of the E.U. Data Protection Framework' [2020-23].He has been also invited as a Jean Monnet Visiting Research Fellow at the Centre of Excellence, Jean Monnet Centre for Intellectual Property & Management, Bournemouth University, UK. Prof. Gupta was earlier appointed as the research collaborator by the Università Bocconi, Milan, Italy, for a project funded by the European Commission under the 7th Framework Programme. He participated in a multi-jurisdictional project on Digital Era Competition: A BRICS View, whose report was published by the BRICS Competition Law and Policy Centre.

Professor Gupta holds two LL.M. degrees (taught and research) from the University of Aberdeen, UK, and the University of East Anglia, UK, and a PhD from Brunel University, London. He also taught at the Brunel Law School during this PhD, which looked at the transnational influence of the US Supreme Court's judgement in the Feist case in the framing of the Database Directive in the European Union. Prof. Gupta specialises in Technology Law and aspects of Intellectual Property Law. Over the years, he has offered courses such as Data Protection Law, Information Technology Law,

Music and Copyright, and EU Law and Approaches to New Technology. He has contributed in reputed journals such as the European Intellectual Property Review, *Journal of World Intellectual Property*, *International Journal of Law and Information Technology*, *Journal of Intellectual Property Law & Practice*, *Journal of Data Protection and Privacy*, Economic and Political Weekly, and *European Journal for Law and Technology*. He has worked extensively with Springer on several book projects of all kinds, the latest being an in-process Major Reference Work titled 'Handbook on Originality in Copyright: Cases and Materials', currently available as a Living Edition on SpringerLink.

Chapter 1
Technology and Privacy: The Conflicts and Contradictions

1.1 Introduction by the Editor

There is a casting doubt that advanced form of technologies will breach the boundaries of information privacy. These are technologies that may not have been operationalized at a large scale and have just been trudging along the confines of the labs. What the future beholds and whether large scale applications of these technologies would challenge the present data privacy framework in the existing ecosystem remains to be seen. It is not merely when technologies are used to collect personal data but when they are used or applied to enhance privacy. For their effective implementation, transparency would be a core element with an aim to reduce information asymmetry, thereby helping the data subjects involved to fully ascertain the privacy risks.[1] For instance, there are tools that could protect users' anonymity while initiating financial transactions. For such tools, user privacy is at core, although the data controllers may have other legal liabilities to fulfil.[2] Amongst other related issues, this chapter captures newer privacy concerns[3] in the context of the use of cloud servers, nanotechnology expansions, rise of smart cities, use of data mining techniques and the challenges of implementing algorithmic decision-making.

---

[1] *Spiekermann, Sarah RFID and privacy: what consumers really want and fear. Personal and Ubiquitous Computing (2008)* https://doi.org/10.1007/s00779-008-0215-2.

[2] *Camenisch, Jan; Groß, Thomas; Heydt-Benjamin, Thomas Scott Accountable privacy supporting services. Identity in the Information Society (2009)* https://doi.org/10.1007/s12394-009-0023-5.

[3] *Tavani, Herman T. Informational privacy, data mining, and the Internet. Ethics and Information Technology (1999)* https://doi.org/10.1023/a:1010063528863.

I. Gupta (ed.), *Operationalizing Expectations and Mapping Challenges of Information Privacy and Data Protection Measures*, https://doi.org/10.1007/978-981-99-7371-2_1

Improper storage of personal data not only infringes upon data protection principles[4] but may lead to future data security breach. Often data controllers seek refuge to saving costs. Physical infrastructure used for the storage of personal data, and controlled fully by a data controller is often expensive compared to other forms of storage. Therefore, they select the option of choosing several technologies that provide them an alternative to physical infrastructures. For instance, the use of cloud servers helps in data storage that are not sometimes owned by data controllers involved in the primary collection of personal data. It raises concern about the data protection standards followed by third-party owners of these services, and whether they assuage data privacy issues. The overarching issue is about the extent of control that data subjects would like to exercise over their personal data when matters concerning security, integrity and confidentiality are at stake. Furthermore, data controllers would like to exercise greater controls to prevent incidents of possible data breach. Any data controller owes it to the data subject to provide all necessary details about the rules of disclosure, outsourcing services and the storage norms surrounding the use of personal data. Therefore, it seems necessary to develop the operational norms around cloud storage of personal data.[5]

The use of nanotechnology has the capacity to enhance use and development of existing electronic records. In the process, it has the potential to generate new records. Its use in several sectors raises privacy concerns primarily because of the technology's granular approach. It has the capacity to collect and disseminate sensitive information at a rapid pace with improved efficiency.[6] To see the optimal effect, the technology needs to be a lot closer to the human and to the data that represents the natural person.

Technology seems to have a greater effect on privacy issues when it works in proximity with humans. Smart cities represent an outcome of several connected devices that interact with each other with the final objective of improving the lives of individuals. In the process, data sharing happens with the idea of interoperability as a cornerstone, thereby making the experience of end users seamless. The integration of several technologies and their further integration may pose tangible data security risks to the users. Here, the threat to security is multidimensional in nature and any breach may lead to a significant trust deficit negatively affecting the overall imagination and implementation of smart cities.[7]

---

[4] Article 5, *Regulation (EU) 2016/679 of the European Parliament and of the Council of 27 April 2016 on the protection of natural persons with regard to the processing of personal data and on the free movement of such data, and repealing Directive 95/46/EC (General Data Protection Regulation) (Text with EEA relevance) OJ L 119, 4.5.2016, pp. 1–88.*

[5] *Akremi, Aymen; Rouached, Mohsen A comprehensive and holistic knowledge model for cloud privacy protection. The Journal of Supercomputing (2021)* https://doi.org/10.1007/s11227-020-03594-3.

[6] *Toumey, Chris Privacy in the Shadow of Nanotechnology. NanoEthics (2007)* https://doi.org/10.1007/s11569-007-0023-6.

[7] *Ismagilova, Elvira; Hughes, Laurie; Rana, Nripendra P.; Dwivedi, Yogesh K. Security, Privacy and Risks Within Smart Cities: Literature Review and Development of a Smart City Interaction Framework. Information Systems Frontiers (2020)* https://doi.org/10.1007/s10796-020-10044-1.

We are transiting into a phase where more and more of our decisions are dictated by algorithms.[8] At times, it becomes difficult to locate decisions that are purely taken out of one's free will and decisions that are fully informed and free from all forms of external influence. A greater degree of transparency is required when algorithmic decision-making takes place, and such a threshold of transparency will help to remove opacity and increase accountability amongst data controllers.[9] Algorithmic decision-making may lead to several risks and negatively impact one's privacy. The EU AI Act is an example that lays down the importance of following a risk-based assessment of the use of AI.[10] Article 22 of the GDPR reflects upon instances of automated decision-making and how an individual must be informed at all stages.[11] The overall idea is to bring in a greater degree of transparency and reduce opacity.

With burgeoning technological growth, safeguard of personal sensitive data remains a thorny issue. In the above situations, the nature of personal data involved and processed is likely to be sensitive in nature. There are multitude of additional concerns when data privacy issues are centered around personal sensitive data. The negative repercussions are difficult to fathom and are likely to have a profound impact on a natural person. For instance, health care is one such area where processing of sensitive personal data is likely.[12] On the other hand, health data are extremely useful for the furtherance of medical treatment and for the advancement of technical tools that help in such a treatment. It is imperative to have the privacy conditions embedded in all such situations where sensitive personal data are at risk.[13] The requirement of having a framework in this regard with the responsibilities and accountability structure clearly laid down is the order of the day. One such example is the passage of the Health Insurance Portability and Accountability Act (HIPAA) of 1996. The legal basis for processing health records is crucial where explicit consent is an approachable measure in many jurisdictions.[14]

---

[8] *De Laat, Paul B. Algorithmic Decision-Making Based on Machine Learning from Big Data: Can Transparency Restore Accountability?. Philosophy & Technology (2017)* https://doi.org/10.1007/s13347-017-0293-z.

[9] Ibid.

[10] *OJ L, 2024/1689, 12.7.2024, ELI:* http://data.europa.eu/eli/reg/2024/1689/oj (accessed June, 2025).

[11] *Regulation (EU) 2016/679 of the European Parliament and of the Council of 27 April 2016 on the protection of natural persons with regard to the processing of personal data and on the free movement of such data, and repealing Directive 95/46/EC (General Data Protection Regulation) (Text with EEA relevance) OJ L 119, 4.5.2016, pp. 1–88.*

[12] *Regan, Marybeth Disease Management and the Application of Privacy Guidelines. Disease Management and Health Outcomes (2002)* https://doi.org/10.2165/00115677-200210090-00003.

[13] *Schulze, A.; Tran, D.; Daum, M. T. J.; Kisilenko, A.; Maier-Hein, L.; Speidel, S.; Distler, M.; Weitz, J.; Müller-Stich, B. P.; Bodenstedt, S.; Wagner, M. Ensuring privacy protection in the era of big laparoscopic video data: development and validation of an inside outside discrimination algorithm (IODA). Surgical Endoscopy (2023)* https://doi.org/10.1007/s00464-023-10078-x.

[14] General Data Protection Regulation, Article 9 refers to the application of explicit consent for data that are sensitive in nature.

Privacy in itself is a complex concept. It connects with space, time and culture, and with varied societal and economic dimensions.[15] As a concept, privacy attains shape from several attributes that are not necessarily restricted to that individual, although privacy is an outcome centered around an individual. All these attributes are extremely fluid. For instance, private is determined by the physicality of the space. Your home is a private space. Having a conversation with your friends and family within that confined space is deemed private and further determined by cultural and societal norm. Even in that conversation sometimes individuals are selective in their approach. Every information is not for everyone is a ground rule for many. There is a shift when the rules and understanding pertaining to the physicality of space and how you determine private space leading to privacy of an individual is used to determine informational privacy. Technology with its several advancements speaks a universal language of growth. The challenge is therefore to include the heterogeneity of privacy and private space in the way it functions. The proposal is to bring the human back in the transactions run by technologies.[16]

1.2 Machine Generated Summaries

Disclaimer: The summaries in this chapter were generated from Springer Nature publications using extractive AI auto-summarization: An extraction-based summarizer aims to identify the most important sentences of a text using an algorithm and uses those original sentences to create the auto-summary (unlike generative AI). As the constituted sentences are machine selected, they may not fully reflect the body of the work, so we strongly advise that the original content is read and cited. The auto generated summaries were curated by the editor to meet Springer Nature publication standards. To cite this content, please refer to the original publications.

Machine generated keywords: consumer, provider, mining, algorithm, service, datum mining, traditional, service provider, video, computer, accountability, differ, era, raise, beyond.

1.2.1 A Comprehensive and Holistic Knowledge Model for Cloud Privacy Protection [1]

This is a machine-generated summary of:

Akremi, Aymen; Rouached, Mohsen: A comprehensive and holistic knowledge model for cloud privacy protection [1]

Published in: The Journal of Supercomputing (2021)

[15] *Dodig-Crnkovic, Gordana; Horniak, Virginia Togetherness and respect: ethical concerns of privacy in Global Web Societies. AI and Society (2005)* https://doi.org/10.1007/s00146-005-0024-3.

[16] Ibid.

Link to original: https://doi.org/10.1007/s11227-020-03594-3

If you want to cite the papers, please refer to the original.

For technical reasons we could not place the page where the original quote is coming from.

Abstract-Summary

"Although Cloud computing is gaining popularity by supporting data analysis in an outsourced and cost-effective way, it brings serious privacy issues when sending the original data to Cloud servers."

"The lack of Cloud data governance and the absence of up-to-date dedicated technologies represent serious barriers to satisfy different Cloud stakeholders' privacy needs."

"Several researchers have conducted surveys to understand and target privacy issues in the Cloud."

"We propose a holistic and comprehensive taxonomy for Cloud privacy."

"This study is supported by the results of a systematic literature review, which provides a methodical, structured, and rigorous approach facilitating the understanding of privacy-preserving Cloud strategies and techniques."

"The study's objective is to offer a credible intellectual guide for upcoming researchers in Cloud privacy and identify active privacy research areas to make the most impact."

Introduction

"Despite all these advantages, Cloud computing presents significant privacy issues when storing users' data on Cloud premises."

"Privacy in Cloud computing is a serious concern to users, developers, and Cloud service providers alike, which requires special considerations."

"A holistic and comprehensive study must be conducted to help users increase their trust in the services provided by Clouds service providers and to assist developers in implementing controls that promote a high standard of privacy in Cloud computing services."

"There exist many published surveys on Cloud privacy and security issues, challenges, and solutions [2–10]."

"Although security and privacy are highly correlated, privacy concerns in Cloud computing are complicated, and therefore, they need to be studied separately for better understanding."

"In this mind map paper, we progress beyond the studied surveys and focus on Cloud privacy issues to provide a knowledge model with the different privacy features and study existing efforts within this model."

Cloud Computing and Privacy Issues

"The Cloud service provides proper security to gain user trust, but this will not assure user privacy and never expose individual data without specific and explicit permissions."

"Cloud-service consumers' loss of control represents a severe threat to data integrity, confidentiality, and privacy principles."

"Lack of user control Because consumers' data, applications, and resources are hosted at the Cloud provider's owned premises, they do not directly control their data."

"Cloud providers, therefore, can perform data mining over the users' data, which can lead to privacy issues."

"Legal aspects It is widely considered that data protection and regulatory compliance are among the top security and privacy concerns for Cloud actors."

"Concerning the loss of control, the managerial challenge may consist of developing a comprehensive and effective service-level agreement (SLA) to extend the reach of the in-house security and privacy controls into the Cloud provider organization."

Cloud Privacy Taxonomy Overview

"The Cloud provider must inform the consumer of any access to his private data, where it is stored, and how it is used [11]."

"Privacy assets refer to all private data that could be disclosed to the Cloud and their lifecycle."

"The data usage is the responsibility of the Cloud provider to protect user data when outsourcing the service."

"Privacy metrics have a vital role in improving privacy preservation techniques and verifying the system ability to protect private data [12]."

"Most of the existing privacy metrics are based on security measures since preserving data security will maintain data privacy."

"We identify the privacy risk score which computes the distance between local, global, and social centralities metrics to determine the teachability of a user's private data [13]."

"Both users and Cloud providers may be involved in the data protection process."

Mapping of Privacy Preservation Techniques

"Searching over encrypted data is the most effective technique used to preserve privacy in remote databases such as Cloud storage."

"Asymmetric encryption(ASE): uses public-key encryption to enable the data owner to upload encrypted data to Cloud servers, allowing only authorized users to search and decrypt using the private key [14]."

"To achieve privacy preservation through ASE, the keyword privacy (search keywords), access privacy (returned file as a research result), search pattern privacy (search queries), and query unlikability (information about queries sources) must be protected [15–17] when searching data on untrusted Cloud service."

"According to the generated figure, we observe the following: The extensive use of encryption techniques to protect data privacy in the Cloud."

"Most of the other techniques except for encryption-based approaches are all placed at the user side, which means additional data processing before outsourcing them in the Cloud."

Privacy-Preserving Research Projects in the Clouds/untrusted Environments

"Researchers through RSARC: Trustworthy Computing over Protected Datasets project, funded by the National Science Foundation (NSF), aim to provide community researchers with adequate security and privacy-preserving framework, enabling them to use and compute over their database outsourced in remote environments."

"NSF funded MACS-Modular Approach to Cloud Security project to investigate the efficiency of modular-based approach in preserving security and privacy in the Clouds."

"Big data analysis of DNA through parallel, distributed trusted processing systems that preserve privacy and security is the goal of Enabling Large-Scale, Privacy-Preserving Genomic Computing with a Hardware-Assisted Secure Big-Data Analytics Framework project funded by NSF starting on January 2019."

"The EU Horizon 2020 PRISMACLOUD—PRIvacy and Security MAintaining services in the CLOUD project aims to provide users with end-to-end security and privacy mechanisms tools to control their data."

"The program of H2020 funds CLARUS project to develop a trusted and secure framework allowing users to monitor, audit, and retain their data without losing their privacy."

Open Research Challenges

"Preserving data privacy in the Cloud computing environment is a severe issue that requires special considerations."

"They tackled the problem of conflict between the convenience and accountability of services and privacy protection and considered privacy risk assessment as an essential challenge, which requires an evaluation of the trust level of Cloud service providers and tenants."

"To the end of these open, challenging tasks, we add and highlight some other issues as under: Secure Provenance and Privacy An important issue that arises due to the Cloud nature is secure provenance."

"Although the Cloud environment may be protected against privacy threats, the provenance of the customer data may reveal sensitive information to unauthorized individuals by monitoring the events' sequence."

"Balancing between data provenance and privacy is a significant challenge in Clouds, where physical perimeters are abandoned."

"Only very few studies focused on privacy metrics and evaluations of their impact on Cloud confidence."

Conclusion

"Privacy stands out as a critical issue that prevents the widespread adoption of Clouds."

"The vulnerabilities of Cloud ecosystems can lead to substantial security breaks and significantly hurt user privacy by exposing personal data."

"The global dimension of Cloud computing requires standardized methodologies and technical solutions to enable stakeholders to assess privacy risks and establish adequate protection levels."

"Our aim is to help upcoming researchers identify the open research area of Cloud privacy where they can contribute and address current issues."

"As reported in [2], the privacy protection and data security issues of Cloud computing are not only technical issues but also involve standardization, legal and social aspects, and regulations."

1.2.2 Privacy in the Shadow of Nanotechnology [18]

This is a machine-generated summary of:

Toumey, Chris: Privacy in the Shadow of Nanotechnology [18]
Published in: NanoEthics (2007)
Link to original: https://doi.org/10.1007/s11569-007-0023-6

If you want to cite the papers, please refer to the original.

For technical reasons we could not place the page where the original quote is coming from.

Abstract-Summary

"This sentiment is complicated by the fact that the specific nanotechnologies that might affect privacy are located more in the near future than in the present, so our knowledge of them is more speculative than empirical."

Introduction

"Discussions of the social consequences of nanotech benefit from being anchored in knowledge of the relevant scientific details."

"This is entirely obvious, but it is nevertheless common for serious discussions of nanotech to center upon far-future visions of gray goo, universal assemblers, the end of mortality, and other highly speculative scenarios, with little or no reference to the current state of scientific knowledge."

"Some of the ways that nanotech will affect us are entirely unique to nanotech, but others are closely related to previous experiences of workforce changes, faith in new medical technologies, hope for easy solutions to difficult energy problems, and so on."

"When near-future consequences of nanotechnology constitute a continuation of social changes instigated by earlier technologies, we benefit from situating nanotech in the context of the earlier changes."

"The security forces eventually realize that their sensing system is defective and that Mr. Harger is innocent."

The Question of Privacy

"During the Cold War years, several conditions made threats to privacy seem especially acute, including: mistrust of government; the power of large computers, which were novel at the time; and the rise of the Social Security system in the U.S. Concerns about privacy intensified in the 1990s with the rise of email and the internet, and now most people see technology, especially computers, 'as the main culprit of privacy erosion' [19]."

"Concerns about nanotech-and-privacy can be situated within well-established concerns that numerous large information systems collect data about our personal behavior and values, and they use our data for their own interests, which are not necessarily the same as our interests."

"C. Needham reminds me that in the U.S., the Driver's Personal Privacy Act arose after a stalker obtained someone's personal information from a data base of driver's licenses, and that the Video Privacy Act constituted a reaction to a case of publicizing someone's record of renting movies."

Sensing Technologies

"The secondary relatives were guilty, but the people whose DNA was compared with crime-scene DNA were innocent."

"This is not nanotechnology per se, but one can imagine how nanotech could make this situation even worse by refining DNA identification techniques."

"Nanotechnology can enhance the quality of fingerprints."

"Sometime fingerprints are indistinct because the traditional method for developing latent fingerprints is problematic."

"A new method using gold nanoparticles can greatly enhance latent fingerprints [20, 21]."

"Even if it is stipulated that fingerprint techniques are legitimate law-enforcement tools, nanotechnology offers no guarantee that the fingerprint information it enhances will never be misused or misinterpreted or improperly disseminated."

"Not only can nanotech enhance fingerprints; there is also a procedure to derive 'lifestyle intelligence' from fingerprints."

"That sweat can be recovered in fingerprints, and some metabolites can be identified."

Early Concerns About Nanotech

"The commentaries on nanotech and privacy apparently begin with articles by M. Mehta and J. Weckert in 2002."

"Weckert pursued the issue of privacy in terms of information systems, saying that 'quantum computers would be much more powerful than any available today' (2002:367)."

"For the most part, however he was satisfied that the benefits of nano sensing systems outweighed the risks, that fears of privacy-invading nanotechnologies are exaggerated, and that existing regulatory frameworks are adequate, which is to say that a new order of regulations would constitute over-regulation."

"The most comprehensive treatment of nanotech-and-privacy is a paper by J. van den Hoven and P. Vermaas [22] which urges the reader to understand that, in addition to enabling large spatially-centralized surveillance systems like a panopticon, nanotech will also generate local, transient information."

Privacy in the Shadow of Nanotechnology

"We have three elements: a history of privacy, with threats to privacy and protections against those threats; a technology that can gather and disseminate information about our bodies, our identities and our behavior; and a series of concerns about nanotech threatening privacy."

"To synthesize these three elements, I suggest steering our knowledge and concerns into six considerations that we can observe in the near future: Fears about nanotechnology threatening our privacy are likely to be framed within pre-existing fears of information systems."

"If it is also true that privacy is threatened when the personal medical information of many people is pooled in a large information system, then nanomedicine will be even worse."

"If it can create consumer diagnostic tools, i.e., devices that are portable, affordable, and that do not have to interface with large information systems, then it can be a friend of privacy."

Reflections

"After examining this question in terms of values, policies and technologies, we need to remember that privacy is more than an abstract issue."

"It is, by definition, a very personal question that is likely to touch most people and bother many."

"Then, how will people experience nanotech-and-privacy?"

"There can be ways for nanotech to protect or enhance one's privacy."

"Let there be more like her who can personalize and crystallize threats to privacy, and perhaps show us the way to satisfactory solutions."

"With big thirsty information systems, plenty of eyes at the bottom, molecularly naked patients, and other considerations, it is hard to imagine how multiple developments in nanotech will not intrude further into our privacy [23]."

1.2.3 Security, Privacy and Risks Within Smart Cities: Literature Review and Development of a Smart City Interaction Framework [24]

This is a machine-generated summary of:

Ismagilova, Elvira; Hughes, Laurie; Rana, Nripendra P.; Dwivedi, Yogesh K.: Security, Privacy and Risks Within Smart Cities: Literature Review and Development of a Smart City Interaction Framework [24]

Published in: Information Systems Frontiers (2020)

Link to original: https://doi.org/10.1007/s10796-020-10044-1

If you want to cite the papers, please refer to the original.

For technical reasons we could not place the page where the original quote is coming from.

Abstract-Summary

"An increasing number of studies focus on the security, privacy and risks within smart cities, highlighting the threats relating to information security and challenges for smart city infrastructure in the management and processing of personal data."

"This study analyses many of these challenges, offers a valuable synthesis of the relevant key literature, and develops a smart city interaction framework."

"The study is organised around a number of key themes within smart cities research: privacy and security of mobile devices and services; smart city infrastructure, power systems, healthcare, frameworks, algorithms and protocols to improve security and privacy, operational threats for smart cities, use and adoption of smart services by citizens, use of blockchain and use of social media."

Introduction

"The threats stemming from information security, data privacy and cyber-related factors where unauthorised access to information can cause undesired consequences, highlights the criticality of addressing these issues early within the design and development stage of smart cities (Chatterjee et al. [25]; Elmaghraby and Losavio [26]).”

“This research aims to bridge this gap in the literature by conducting a comprehensive analysis of the many issues and key complexities relating to privacy, security, and risk issues within smart cities.”

“In alignment with the guidance in Pare et al. [27], we conduct a theoretical review of current studies and available literature on privacy, security, and risks within smart cities and develop a smart cities security and privacy framework.”

“The current study aims to address the following questions: What is the current state of knowledge relating to security, privacy, and risk within smart cities?”

Research Methodology

“The study applied the following steps for systematic search of the relevant papers: (1) development of protocol; (2) filtration of research papers by title, keywords and abstract; (3) extraction of data from the selected papers.”

“The search was restricted to peer-reviewed research papers.”

“This database can be useful for searching and locating a substantial proportion of the published peer-reviewed research papers in the area of smart cities research.”

“From the Scopus database, only peer-reviews papers that were published in English were considered.”

“All studies were analysed and processed by the authors then reviewed based on title, keywords and abstract to ensure validity and relevance that the research offered contribution to the smart cities’ discussion in the context of risk, privacy and security.”

“Based on the inclusion criteria, the review of the 99 articles yielded 94 that were selected as offering relevance for this study.”

Literature Review

“The subsections of the article generated discussion around each theme providing insights to research carried under the respective main themes Mobile devices are the backbone of interacting with the smart cities network infrastructure but present new challenges to the security and privacy of users where sensitive data could be vulnerable to attack by third parties.”

“The study by Abosaq [28] analysed the privacy issues faced by smart cities including authentication, access control, confidentiality, trust, data security, policy implementation and secure middleware.”

“Based on the proposed stakeholder model, the study developed a security and privacy framework for secure and privacy-aware service provisioning in smart cities.”

“As smart cities face a number of challenges connected to security and privacy, some studies proposed various frameworks, models and algorithms to improve these issues (Al-Dhubhani et al. [29]; Antonopoulos et al. [30]; Avgerou et al. [31]; Beltran et al. [32]; Burange and Misalkar [33]; Cagliero et al. [34]).”

"Studies have proposed the use of blockchain technologies to solve some of the issues of privacy and security in smart cities (Mora et al. [35]; Noh and Kwon [36]; Ramos and Silva [37])."

Discussion—Smart City Challenges

"These challenges in the context of the security, privacy and risk within smart cities are outlined below: The multidimensional nature of smart city initiatives necessitates the need for interaction with human and social capital using technological solutions as the tool to achieve smart city goals in improving the quality of life for its citizens (Monzon [38])."

"The reality of retrofitting smart cities initiatives effectively integrating the concepts of cities and digital systems, each distinctly highly complex with contingent systems, then binding them together to create environments inherently prone to security vulnerabilities, is a significant challenge and risk to operational capability (Kitchin and Dodge [39]; Townsend [40])."

"The operation of the smart city requires the integration of key technologies such as: IoT, big data, sensors, machine learning and GPS based applications, all of which raise significant threats to the security and integrity of citizen related data."

Conclusion

"This study provides a theoretical review of the smart city literature, focussing on the many threats relating to privacy and security, and how these can impact the operation of smart city processes."

"A number of emergent themes and significant challenges have been discussed to provide a valuable synthesis of the key factors relating to privacy and security threats within the smart cities literature."

"The critical factors relating to technical and security governance of smart cities are key factors relating to operational integrity and citizen trust within the smart city infrastructure."

"Breaches of security that jeopardise data privacy are likely to severely impact citizen engagement and overall trust in smart city systems and services."

"The challenge for smart city designers and planners as innovations such as blockchain based systems and greater use of AI become integral to system architectures, is to maintain humans in the loop to engender the required levels of trust in security and privacy from citizens."

Limitations and Future Research Directions

"Future research should consider how the legal system can be used to solve trust challenges within smart cities."

"It is advised that future research should focus on solving identified challenges of smart cities (trust challenges namely trust challenges, operational and transition challenges, technological challenges and sustainability challenges), which will greatly benefit further smart city initiatives."

"It is important to consider the cost of deploying and operating a blockchain-based system within smart cities."

"This line of thinking can also be extended to other global issues such as pollution and environmental monitoring, whereby a communicating network of smart cities would provide a step-change in the predictive ability of environmental models, with evident benefits for city dwellers, proactive policy development at regional, national and international levels (with regulatory, legal, economic and wider research implications that will inevitably accompany them), with the design and adoption of countermeasures at scale."

1.2.4 RFID and Privacy: What Consumers Really Want and Fear [41]

This is a machine-generated summary of:

Spiekermann, Sarah: RFID and privacy: what consumers really want and fear [41]

Published in: Personal and Ubiquitous Computing (2008)

Link to original: https://doi.org/10.1007/s00779-008-0215-2

If you want to cite the papers, please refer to the original.

For technical reasons we could not place the page where the original quote is coming from.

Abstract-Summary

"This article investigates the conflicting area of user benefits arising through item level radio frequency identification (RFID) tagging and a desire for privacy."

"The third is to let the network access users' RFID tags while adhering to a privacy protocol (network model)."

"The perception and reactions of future users to these three privacy enhancing technologies (PETs) are compared in the present article and an attempt is made to understand the reasoning behind their preferences."

"Enhancing trust through security and privacy 'visibility' as well as PET simplicity may be the road to take for PET engineers in UbiComp."

Introduction

"On the shopfloor, recognizing customers individually and automatically upon arrival, tracking them through the store, observing their interactions with products and offering them personalized advertisements and information services are all activities which can be realized through RFID, but have the potential to be viewed as privacy intrusive [42, 43]."

"One major conclusion drawn by the observers of the focus groups was that participants seemed to unanimously call for RFID tags to be killed at retailer exits."

"Seen these qualitative research results, a question confronted by retailers today is how to treat RFID chips at store exits."

"To answer these questions retailers need to understand how vital the privacy issue really is for their customers and how willing they are to trade their concerns with the technology's benefits; in particular, in the after-sales domain where permanent deactivation of RFID tags would impede any further service potential."

PETs for RFID and User Perceptions

"To User PETs "Network PETs" are based on the idea that RFID tags are unlocked by default and that the network takes the initiative to communicate with a user's tags."

"Hypothesize that for those consumers who appreciate after sales RFID services any PET model, whether that be the user model or the network model is superior to chips being killed: The user model is considered superior to the kill option if people appreciate after-sales RFID services."

"Hypothesize: When confronted with a Network PET users will want to kill RFID tags more readily than when confronted with a User PET."

"Against this background the following hypothesis was formulated: A common set of technology acceptance factors, namely the perceived usefulness and ease of use of RFID, perceived control through the PET and the opinion of others on RFID - will drive users' preference to prefer complex PETs for RFID over a kill approach."

Method to Investigate RFID PET Perceptions

"The independent variables investigated in study ① included the perceived usefulness of RFID after-sales services, the anticipated ease of use of RFID, peer opinion, and perceived control through the PET (in terms of information control through the PET, ease of use of the PET and helplessness despite the PET)."

"The dependent variable used to test hypotheses 1–5 was measured by using one final question asked at the end of the second questionnaire which used an 11-point scale from 1 (totally destroy chips) to 11 (deploy chips with PET) and reads as follows: "Former questions and the film showed that RFID technology can bring about both disadvantages and advantages for consumers."

"The independent variables influencing this assessment in study ① included the perceived usefulness of RFID after-sales services and the anticipated ease of use of RFID generally, peer opinion on RFID and perceived control through the PET."

Results: How Do Users Respond to PETs for RFID?

"In accordance with hypotheses 1 and 2, participants with above average usefulness perceptions of RFID valued both the User and the Agent PET significantly higher than those with low average usefulness ratings."

"On the 11-point scale anchoring the opposing preference for rather killing [2] or rather using a complex PET [12], people appreciating RFID after-sales services in the User Scheme scenario valuated the PET on average at 5.61."

"In the group where participants saw the Agent Scheme, appreciators of RFID valued the complex PET at 4.44 while non-appreciators valued it at 2.26 ($p = 0.002$)."

"These results suggest that the perception of usefulness of RFID after-sales services is an important driver for preferring complex PETs over the kill solution."

"The more helpless users feel despite the User or Agent PET, the more they want to kill RFID tags."

A Qualitative Evaluation of PET Solutions

"A final step in the analysis of PET perception was an attempt to understand why the large majority of participants generally prefer to kill RFID chips at store exits and what drives a smaller portion of users to instead opt for a more complex PET."

"Explanations were given in a free text format (open question) by 175 out of the 208 participants in study ②. We analyzed the reasoning for preferring a complex PET or rather killing tags with the help of a content analysis [44]."

"Each answer typically had one main theme (reason) for why a participant would judge for the User PET or rather favour the killing of RFID tags."

"Out of the 108 (62%) participants who were in favour of killing RFID tags, 70% described some feeling of mistrust in the User PET."

Conclusions on the Acceptance of PET Solutions

"The main finding from the comparative PET study is that complex PETs as they are envisioned today by many UC privacy researchers are highly likely to run into acceptance problems with users."

"Signalling security and trust to users through respective interface design may be very relevant for RFID privacy engineering."

"The results from regression analyses suggest that User Scheme appreciation can be improved by working on the PET itself: Information control provided through the User PET seems to directly influence its appreciation (even though the direction of influence is unclear from the results)."

"If Agent PETs organize users' privacy in a largely autonomous way, then people seem to rely more on the recommendations of peers when deciding not to kill."

"If an Agent PET carried a trust seal and if peer evaluations were available on Agent PETs' performance or Agents provided hints as to the trustworthiness of a respective RFID enabled environment or service rating, then people may be more apt to adopt it."

1.2.5 Accountable Privacy Supporting Services [45]

This is a machine-generated summary of:

Camenisch, Jan; Groß, Thomas; Heydt-Benjamin, Thomas Scott: Accountable privacy supporting services [45]

Published in: Identity in the Information Society (2009)

Link to original: https://doi.org/10.1007/s12394-009-0023-5

If you want to cite the papers, please refer to the original.

For technical reasons we could not place the page where the original quote is coming from.

Abstract-Summary

"Users that do not pay their bill must be held accountable for their behavior."

"This tension between privacy and accountability is fundamental, however a tradeoff is not always required."

"The time capsule together with its related protocols offer support of privacy while retaining strong accountability."

"To existing revocable anonymity systems, our proposed scheme requires less trust in an external authority, while simultaneously making accountability easier (and less costly) to achieve."

Introduction

"However, commercial service providers require accountability."

"By permitting anonymous use of a service provided that identities may be revealed when necessary."

"Users enroll with their service provider anonymously, but give the provider a form of their identity which is encrypted to the court."

"In a dispute, the service provider can ask the court to decrypt this sealed identity to hold the user accountable for improper actions such as non-payment."

"Our protocol can enforce conditions such as a time by which a payment must be made, and can enforce that the service provider learn a user's identity if payment is not made by this time."

"Examine and formalize the problem of accountable privacy enabled service provision."

"In the remainder of this article we examine the background and problem of accountable privacy enabled service provision."

Privacy Supporting Services and Identity Escrow

"Elbonian Public Broadcasting is a privacy preserving service provider that nevertheless can enforce restrictions based on user personal data."

"In identity escrow a user U can encrypt her true identity to the authorities, provide this encrypted data to a service provider SP, and convince SP with a zero-knowledge proof of knowledge that this encrypted data contains a valid user identity that can be opened by the authorities."

"In our example, as a condition of enrollment the user provides an encrypted copy of her complete passport data."

"Even though accountable and privacy-preserving identity management methods for services are already possible with a monolithic and fully trusted third party, we propose to rethink the corresponding primitives to better meet the actual needs of services."

"A malicious service provider holding a verifiable encryption may attempt to betray the user by opening a decryption case at the TTP without good cause. (The TTP alone resolves disputes)."

Time Capsule

"Our scheme provides end-to-end unlinkability of users such that transactions cannot be linked with one another except for when a service provider learns the identity of a misbehaving user."

"When an anonymity revocation is requested by a service provider, RA only knows that it is checking the key for a legitimate transaction, without knowing which transaction or which user."

"The service provider is convinced in zero-knowledge that the time capsule contains the expected piece of data (say, the user's certified identity) Beginning with time ω, the service provider can decrypt the time capsule (say, if after a payment grace period of a month)."

"If the user fulfills a revocation condition before time event ω, the service provider cannot decrypt the time capsule (say, if the user has fulfilled her payment obligation)."

Problem Statement

"In our example scenario a user U obtains a service from a service provider SP."

"This revocation authority does not know the identities of users in the system, but has the ability to take the record of an anonymous service contract and combine it with information from a predetermined time server TS and satisfaction authority SA to determine a key which will then allow SP to unlock the identity sealed into the service agreement."

"By convention, we use verbatim font for principal names, such as user U or service provider SP."

"An interactive interface specification names the inputs and outputs of all principals involved together with the principal's name: This Transaction is between a user U and a service provider SP, where U does not provide any inputs and receives the output tuple outputs."

"If the user has honestly paid her bill, however, this transaction will not succeed, and the service provider learns nothing."

Impossibility Result

"We consider three arguments for necessary conditions for a timed and revocable verifiable encryption."

"We analyze the requirement of a trusted clock, a revocation authority, and the nature of their information flow."

"The system requires at least a trusted clock publishing user-independent immutable time events."

"A trusted clock solely publishing user-independent time events is not sufficient for a system that accepts a revocation condition."

"It therefore is a necessary condition that a trusted entity governs the revocation process and holds imperative decryption information confidential."

Cryptographic Building Blocks

"Allow a verifier to convince a prover without doubt that the verifier knows a secret x without leaking any information about x. The third layer contains three higher cryptographic primitives, which build on top of the basic primitives and realize complex service interfaces: is an advanced form of electronic signatures that enables a signer (a) to certify attribute values and messages as part of the signature itself and (b) to render the signature unlinkable over multiple transaction."

"We specify an abstract interface for integer commitments: for values v_i and a random number r produces an integer commitment C on the values v_i Zero-knowledge proofs of knowledge allow a prover to convince a verifier that the prover knows a secret x without leaking information about x. For instance, a user can prove that she knows an—otherwise secret—credit card number."

Solution Overview

"An example cryptographic implementation of our system using building blocks from anonymous credentials systems such as (Camenisch and Lysyanskaya [46, 47]) and the relatively new committed blind anonymous identity based encryption (Green and Hohenberger [48])."

"We describe here at a high level how these building blocks can implement a system with the requirements and definitions listed in prior sections."

"More details about the properties of the building blocks are explained in the Cryptographic Primitives section."

"The user can revoke the ability of SP to open the time capsule by broadcasting proof of payment on BB."

"If the key is revealed to SP then the capsule can be opened and the user's true identity revealed."

Other Related Works

"A number of authors provided more efficient implementations of these primitives, e.g., group signatures (Ateniese et al. [49]; Boneh et al. [50]; Kiayias and Yung [51]), e-cash (Brands [52]; Camenisch et al. [53]; Frankel et al. [54]), anonymous credentials (Brands [55–57]; Camenisch and Lysyanskaya [46, 47]), traceable signatures (Kiayias et al. [58]), anonymous auctions (Naor et al. [59]), and electronic voting based on blind-signatures (Fujioka et al. [60])."

"All these primitives have in common that some party issues a user some form of certificate that often contains information about the user encoded as attributes."

"There are also some works (Cramer et al. [61]; Brands [62]; Camenisch and Michels [63]; Boudot [64]; Fujisaki and Okamoto [65]) that these authors employ to prove AND, OR and NOT statement about attributes, e.g., "a user has attribute a OR b," basically by showing that some committed value equals a given value OR some other given value."

Conclusion

"We did this by introducing the concept of the time capsule: a cryptographic device that contains the user's real identity, and can only be opened if the user does not fulfill the terms of a predetermined service contract such as a contract to pay a bill on time."

"By leveraging the recently developed anonymous blind committed identity based encryption scheme we are able to form an encryption key for the time capsule that depends on the user's pseudonym within the system, as well as conditions such as time and payment status."

"We have shown that due to the properties of this and other blinding mechanisms, the secret key for decrypting the time capsule can only be derived if the user misbehaves."

"Due to the weaker trust requirements for the TTP it is easier for service providers to perform the anonymity revocation when a user misbehaves, relieving what may be a prohibitive expense for routine transactions."

1.2.6 Togetherness and Respect: Ethical Concerns of Privacy in Global Web Societies [66]

This is a machine-generated summary of:

Dodig-Crnkovic, Gordana; Horniak, Virginia: Togetherness and respect: ethical concerns of privacy in Global Web Societies [66]

Published in: AI and Society (2005)

Link to original: https://doi.org/10.1007/s00146-005-0024-3

If you want to cite the papers, please refer to the original.

For technical reasons we could not place the page where the original quote is coming from.

Abstract-Summary

"Today's computer network technologies are sociologically founded on hunter-gatherer principles; common users may be possible subjects of surveillance and sophisticated internet-based attacks are almost impossible to prevent."

"Information and communication technology, ICT offers the technical possibility of embedded privacy protection."

"This means incorporating options for socially acceptable behaviour in technical systems, and making the basic principles of privacy protection, rights and responsibilities, transparent to the user."

"The current global e-polis already has, by means of different technologies, de facto built-in policies that define the level of user-privacy protection."

Technology and Culture: ICT and a New Renaissance

"Each of the sciences was searching for its own specific and certain truths."

"The post-industrial age has, however, abandoned the rigid mechanical model of a monolithic, deterministically controlled system with "the one right way" and one absolute truth."

"It has embraced the fact that social cohesion through pluralism and polycentrism, cultural diversity, self-organisation and contextual truth is more productive and appropriate for the new epoch."

"Flexibility and fluidity have replaced rigidity and conformance, dynamics have replaced statics."

"In the information–communication era there is a development toward a human-centrism with a potential for a new Renaissance, in which science and the humanities, arts and engineering can reach a new synthesis, through modern computing and communication tools used in global virtual societies (Dodig-Crnkovic [67])."

[Section 2]

"Networking (Gill [68, 69]) at the global level exists in the symbiotic relationship with local resources."

"Gill argues that a rethinking of the development idea in the contemporary globally networked civilization is necessary."

"In the information society, a shift from the techno-centric to a human-centred framework is necessary in consideration of the diversity and the complexity of cross-cultural collaboration."

"Social cohesion in this context results from the ability to participate in the networked society through mutual interaction, exchange of knowledge and sharing of values."

"The relevance of associative networks for a sustainable information and communication society is discussed by Thill [70], while Wagner et al. [71] address the related problem of enhancing e-government in developing countries via virtual communities' knowledge-management."

Why Privacy Matters

"Privacy protects two kinds of basic rights: priority in defining ones own identity."

"Privacy of one's home is a classic example of a private space."

"Under certain special circumstances it is, however, possible for police, for example, to enter your home without your consent, this being strictly regulated by law The following is from Article 8; Right to respect for private and family life of the British Human Rights Act (1998) 1."

"This is enabled through his/her rights to privacy in the sense of the control of one's own personal information."

"There is a different attitude to privacy in Japan because of its specific cultural, linguistic and historical development."

"The culturally developed privacy in individual countries, which is called the rich conception of privacy, is what mainly differentiates the Western world and Japan in this respect."

"The internet is a global technology and each part of the world has its own laws and rights to privacy."

Phenomenology of Cyber Privacy: Many (Inter)faces of Self

"With respect to the difference between the public and the private life of a person leading to a social fraud, some see it as the wearing of different "masks" depending on the current situation in which the person is (Rosen [72])."

"Virtual spaces that many computer users could share and collaborate within, called multi-user dungeons (MUDs) are a new kind of social virtual reality."

"Studies of co-operative work in real-world environments have highlighted the important role of physical space as a resource for negotiating social interaction, promoting peripheral awareness, and sharing artifacts [3]."

"Early studies of social interaction in CVEs stressed the interdependence between virtual and physical space (ibid)."

"On a symbolic level, this problem can be studied in the CVEs, which are virtual worlds shared by users across a computer network."

Privacy as Architecture of Relationships

"According to Mason [73], privacy can be studied through the relationships of four social groups: The first group consists of an individual, I, who has the right to privacy, both to physical privacy and to the protection of personal information."

"The second group consists of all people with whom individual I shares his/her information or private space in return for relationships or services."

"Individuals should acquire information about the second group before beginning a relationship with it."

"The third group does not directly receive the information shared between I and the second group."

"This group has access to the information about I as a result of their professional role."

"The fourth group consists of the rest of the society, the public, who are not in any direct contact with I's private space or information."

State of the Art: Disclosive Ethics

"Brey [74] claims that not just the uses of computer technology, but also other practices that involve computing technology, such as its development and management, require the formulation of policy guidelines."

"It is part of the task of computer ethics to further develop and modify existing moral theory when existing theory is insufficient or inadequate in the light of new demands generated by new practices involving ICT (Brey [74])."

"For Moor, computer ethics is primarily about solving moral problems that arise because there is a policy vacuum about how computer technology should be used."

"Brey claims that a large part of work in computer ethics is about revealing the moral significance of practices that seem to be morally neutral."

"Disclosive computer ethics (Brey [74]) is a multi-level interdisciplinary approach concerned with the moral deciphering of embedded values and norms in computer systems, applications and practices."

Togetherness and Respect: Legitimacy by Design

"Whitworth and de Moor [75] claim that legitimate interaction increases social well-being, and they analyse the ways in which societies traditionally establish legitimacy, and how the development of socio-technical systems changes previously established patterns of behaviour."

"As a remedy, the following three social communication "rights" are proposed: the right to block personal data access, the right to not interact, and the right to return e-mail to its sender."

"This can be an intrusion upon the consumer's right to privacy because, as a rule, the purchaser is not informed of the presence of the tag (ibid)."

"This includes both technical and ethics research into the development of protocols and policies that effectively balance privacy rights with Internet security."

"A system based on digital certificates and a public key infrastructure, which provides persons and services with information about the location of the user but gives the user fine-grain access-control over who is to be informed of his/her location."

Conclusion

"Growing research interest in societal issues such as work and organisational cultures, creativity and innovation, cooperation and participation, and culture and communication among AI and information technology communities shows a sign of hope for future human centred perspectives of IT research and applications."

"ICT supports and promotes the formation of new global virtual communities that are new socio-technological phenomena typical of our time."

"What are the basic principles behind successful virtual community environments?"

"According to Whitworth there are two such principles: Virtual community systems must match the processes of human–human interaction."

"The rights and the ownership must be clearly defined (this can actually be included under the first principle for well-defined human interactions within social organizations)."

"Trust and privacy trade-offs are normal constituents of human social, political and economic interactions, and they consequently must be incorporated in the ICT sphere developed on the principles of human-centrism."

1.2.7 Informational Privacy, Data Mining, and the Internet [76]

This is a machine-generated summary of:

Tavani, Herman T.: Informational privacy, data mining, and the Internet [76]
Published in: Ethics and Information Technology (1999)
Link to original: https://doi.org/10.1023/a:1010063528863

If you want to cite the papers, please refer to the original.

For technical reasons we could not place the page where the original quote is coming from.

Abstract-Summary

"Privacy concerns involving data mining are examined in terms of four questions: (1) What exactly is data mining? (2) How does data mining raise concerns for personal privacy? (3) How do privacy concerns raised by data mining differ from those concerns introduced by 'traditional' information-retrieval techniques in computer databases? (4) How do privacy concerns raised by mining personal data from the Internet differ from those concerns introduced by mining such data from 'data warehouses?'"

1.2.8 Algorithmic Decision-Making Based on Machine Learning from Big Data: Can Transparency Restore Accountability? [77]

This is a machine-generated summary of:

de Laat, Paul B.: Algorithmic Decision-Making Based on Machine Learning from Big Data: Can Transparency Restore Accountability? [77]
Published in: Philosophy and Technology (2017)
Link to original: https://doi.org/10.1007/s13347-017-0293-z

If you want to cite the papers, please refer to the original.

For technical reasons we could not place the page where the original quote is coming from.

Abstract-Summary

"Decision-making assisted by algorithms developed by machine learning is increasingly determining our lives."

"Several objections to full transparency are examined: the loss of privacy when datasets become public, the perverse effects of disclosure of the very algorithms themselves ("gaming the system" in particular), the potential loss of companies' competitive edge, and the limited gains in answerability to be expected since sophisticated algorithms usually are inherently opaque."

"It is argued that algorithmic decisions preferably should become more understandable; to that effect, the models of machine learning to be employed should either be interpreted ex post or be interpretable by design ex ante."

Introduction

"It is this opacity that I submit to be the fifth obstacle to proper accountability concerning computerized systems—in particular, those involving algorithmic decision-making."

"As long as opacity reigns, the discussion about algorithmic decision-making can hardly get started."

"With every step towards disclosure, accountability for the functioning of algorithmic systems would seem to come nearer to being restored."

"Total transparency seems to be the perfect recipe for restoring accountability for algorithmic systems."

"I explore what accounting amounts to when disclosure reigns, for the various phases of algorithmic decision-making."

"Four main types of argument are to be found: (1) Privacy: sensitive data could easily leak into the open; (2) Perverse effects: transparency is an invitation to game the system, making it effectively worthless; (3) Competition: disclosure could hurt companies' competitive edge; and (4) Inherent opacity: disclosed algorithms are often difficult to interpret, so not much insight is gained."

Ranking Algorithms

"Reviews from users involved are solicited for the purpose."

"It would mean full transparency concerning all phases of decision-making: the institutions involved are open about how data have been obtained, make these data available to anyone interested, disclose the ranking algorithm in use (down to the very specifics), and specify whether decision-making (i.e., the actual rankings that become public) proceeds in semi-automated or fully automated fashion."

"The data here are reviews, ratings, or digs from customers and users."

"Another type of gaming is an issue here: constituents may try and manipulate the reviewing system."

"The organizations involved have to put energy into maintaining the integrity of their reviewing system, and build in several checks and controls."

"As argued above, the ranking algorithms involved just function better when they are out in the open."

"Why use this dimension in the algorithm and not another?"

Predictive Algorithms

"Machine learning is very much interrelated all through the stages of data collection, model construction, and model use."

"Any account of how an algorithm has been constructed, cannot do without an account of how datasets have been used in the process (say, as concerns possibly biased data)."

"Accounting for machine learning models can only make sense if all phases are taken into account."

"For all of these defects in training data, one has to find remedies to be applied in the subsequent phase of model construction (see below: "discrimination-aware" modeling)."

"The data available are used as training material for machine learning."

"Training on such an imbalanced dataset may produce a model that over fits to the majority of data representing bona-fide transactions."

"If such data are used inadvertently for model construction, chances are that the biases involved will be built straight into it."

Objections

"The first one was the privacy objection that data could easily leak into the open."

"If full transparency also pertains to the very datasets involved in machine learning, one can easily imagine scenarios in which these, after having been queried by interested individuals or groups, are simply diverted from the intended purposes and leak away to be used for other purposes."

"Prudence suggests that the datasets involved are only to be made available upon request to intermediate parties that have oversight authority of a kind."

"The public at large is not to be entrusted with the power to scrutinize datasets as used in machine learning (although the entries of individuals should of course be open for personal inspection)."

"The restraint stops right there; oversight boards should have full authority to summon up whole databases, whether from public or private parties, for purposes of, say, auditing the machine learning involved."

Perverse Effects

"Most prominent is the argument about "gaming the system": as soon as a model is in the open, interested parties may be able to detect the proxies involved and subsequently evade them."

"Gaming, then, refers to behavior that intends to evade the proxies concerned: omitting the mention of one's occupation, suspending charity donations, getting rid of one's credit cards, paying for the airplane by other means, leaving knives at home, or asking a non-Muslim colleague from elsewhere to take over."

"Intermediate oversight authorities are to retain full rights of transparency as far as the model and its proxies are concerned, otherwise all accountability is gone."

"A third option, obviously, always remains available: prevent the issue of stigmatization from arising at all and restrict transparency concerning the algorithms in use to oversight bodies alone, so far as they exist."

Competitive Edge

"A third type of argument against total transparency of the algorithmic process is an argument about property: companies that develop and maintain algorithms may consider them to be their intellectual property."

"One has to take into account that many public services (such as collecting taxes, border control, and airport security) employ software bought on the market, or outsource the whole process to companies."

"This should also be possible for machine learning algorithms in transparent settings."

"As far as public interests (broadly construed) are concerned, society no longer accepts that algorithms remain completely secret, and therefore unaccounted for."

"Oversight boards, tied to secrecy, may inspect and control the various algorithms the public is subjected to and dependent on."

"After this first step, private companies may possibly be pressed at a later stage to conform to the same regime of transparency and open up their algorithms to oversight boards as well."

Inherent Opacity

"A final objection against transparency, especially of the very algorithms themselves, is to point out that algorithmic end-products of machine learning are often difficult to interpret, even by experts."

"Classifiers, or decision trees, have always been easily interpretable."

"The final classifier is obtained by the summation of all classifiers that have been produced, weighted by how well they actually performed in classifying the (weighted) data points."

"All classifiers obtained vote on the outcome for each data point; so effectively, the majority vote of all classifiers together decides."

"In a nutshell, modern classifiers no longer employ just one single tree for classifying fresh data, which would allow easy interpretation."

"The end model obtained displays all those weights, but cannot be interpreted as to how much the various input variables contribute to the outcome."

Conclusions

"The default position I started from was full transparency of the whole algorithmic cycle all through to the public at large."

"As long as this stance continues to be held, the only feasible alternative to complete opacity is, again, transparency limited to intermediate parties involved in oversight (concerning algorithmic decision-making for public purposes, to begin with)."

"We should normally satisfy ourselves with oversight bodies enjoying full transparency; they are to perform the task of calling algorithmic systems to account in our stead."

"These are sobering conclusions about the limited amount of transparency in algorithmic decision-making that is feasible nowadays, given the state of the art in machine learning and the current political climate."

"It would make sense for algorithms (and their outcomes) to be opened up for perusal to the public at large; accountability for them no longer needs to be implemented behind closed doors only."

1.2.9 Disease Management and the Application of Privacy Guidelines [78]

This is a machine-generated summary of:

Regan, Marybeth: Disease Management and the Application of Privacy Guidelines [78]

Published in: Disease Management and Health Outcomes (2002)

Link to original: https://doi.org/10.2165/00115677-200210090-00003

If you want to cite the papers, please refer to the original.

For technical reasons we could not place the page where the original quote is coming from.

Abstract-Summary

"The Health Insurance Portability and Accountability Act (HIPAA) of 1996 was enacted to provide improved portability of health benefits and greater accountability in the area of healthcare fraud."

"The legislation is divided into five titles with title II addressing the prevention of healthcare fraud and abuse, and the promotion of administrative simplification and privacy."

"This legislation was in response to requested improvements including greater privacy of health information."

"Through the lack of definition, disease management entities were tasked with making assumptions as to the interpretation of the legislation."

"To provide more specificity, the Disease Management Association of America (DMAA) had repeatedly met with the Department of Health and Human Services to understand the intent and purpose behind the legislation."

"At the time of this article, the Department of Health and Human Services are collecting comments on the legislation and have already provided some updated information."

"The DMAA provided input to the Department of Health and Human Services regarding the application of HIPAA to disease management."

"The article addresses the new legislation and addresses how a disease management organization can apply the new legislation to its operation."

1. **Background**

"Healthcare providers could access information real-time regarding eligibility, payment, and clinical information."

"Many countries have strengthened their privacy legislation."

"It is a significant step to providing privacy and security, and a standard code for transactions."

"This paper delves into the specifics of privacy legislation and the ability of HIPAA to be applied to disease management methodologies."

2. **Privacy Issues**

"The European Commission took a major step forward in defining the rights of individuals with regard to their personal information, including health data."

"There are exceptions to this including when: processing of health data is necessary to carry out rights of organizations under employment law the individual is physically or legally incapable of giving their consent the data are expressly made public by the individual processing of data is required for the purposes of preventative medicine, medical diagnosis, the provision of care or treatment or the management of healthcare services, and where those data are processed by a health professional subject under national law or rules established by national competent bodies to the obligation of professional secrecy."

"The legislation in the US was enacted after the EU Directive was ordained due to the continuing concern with regard to the privacy of health data."

3. **Health Insurance Portability and Accountability Act (HIPAA): A Response to Healthcare Reform**

"In December 2000, the final regulations were released establishing federal privacy protections for personal health information as required by the HIPAA."

"The final regulations improve on the proposed rule by strengthening several key protections, including extending protections to personal medical records in all forms (including paper records and oral communications), providing for written consent for routine use and disclosure of health records, protecting against unauthorized use of medical records for employment purposes, and ensuring that providers have all the information necessary to appropriately treat their patients."

"In response to more than 50,000 comments submitted by the public, the final regulation strengthens patient protection and control over their health information by extending coverage to personal medical records in all forms including paper records and oral communications."

4. **Assessing the Impact of HIPAA**

"HIPAA contains five sections, or 'titles' of requirements and standards, which apply to virtually every provider, payor and clearinghouse in the US."

"Title II focuses on preventing healthcare fraud and abuse, while title III pertains to tax-related provisions and medical savings accounts."

"The impact of title I on improving health insurance access, portability, and renewability of coverage has been widely debated."

"The cost of implementing these regulations is estimated by the Blue Cross/Blue Shield Association [79]. Estimates of benefit are difficult to find with regard to the impact of title III (tax-related provisions), title IV (application and enforcement of group health plan requirements), and title V (revenue offsets)."

"Of all the components of HIPAA, the fraud and abuse provisions in title II combined with greatly increased federal funding for the Office of Inspector General will likely have the greatest impact on providers and payors."

5. **Definition of Disease Management**

"Disease management: supports the physician or practitioner/patient relationship and plan of care emphasizes prevention of exacerbations and complications utilizing evidence-based practice guidelines and patient empowerment strategies evaluates clinical, humanistic, and economic outcomes on an ongoing basis with the goal of improving overall health."

"Disease management components include: population identification processes evidence-based practice guidelines collaborative practice models to include physician and support-service providers patient self-management education (may include primary prevention, behavior modification programs, and compliance/surveillance) process and outcome measurement, evaluation, and management routine reporting/ feedback loop (may include communication with patient, physician, health plan and ancillary providers, and practice profiling)."

"Programs consisting of fewer components are disease management support services."

6. **HIPAA and Disease Management**

"The rules elevated disease management to the area of 'protected activities', including it in the definition of 'treatment' (covered through the HIPAA legislation) thereby shielding it from many hurdles of the rule—the requirements for obtaining patient consent before health plans and providers can share protected health information with disease management programs."

"Disease management organizations may receive identifiable patient data directly from a health plan via the 'treatment' exception because of a passage buried deep within the preamble: 'We (HHS) note that this rule permits a covered entity (which would include a health plan) to disclose protected health information to any person for treatment purposes, without specific authorization from the patient.'"

7. **Concerns**

"For disease management companies to be able to fulfill their mission in the management of a patient's disease and comorbidities, access to the patient's records regarding all treatment is necessary."

"If a patient's hospital records, or a specialty physician's records, are not provided to the disease management company, the treatment of the patient can be compromised."

"Unless HHS specifically clarifies that healthcare providers do not require such authorization or consent, providers will almost certainly begin to refuse to provide PHI to disease managers, thereby crippling the very programs that have been advocated as the best solutions to the nation's health quality and cost needs' [80]. Healthcare providers are 'any organization or person who furnishes, bills, or is paid for healthcare in the normal course of business', according to the legislation."

"The concern is that disease management organizations could increase their liability and regulatory exposure, if considered a healthcare provider, in addition to the increased cost of the compliance with the HIPAA regulations."

8. **Conclusions**

"Most countries, in addition to the US, have addressed consumers' concerns by strengthening these rules."

"The ability of disease management companies to further the agenda of coordinated healthcare interventions and communications for populations with conditions has been both strengthened and weakened through HIPAA guidelines by the nonspecific nature of the legislation."

"With the consistent application of HIPAA guidelines by the many players in the industry, disease management companies can move forward while awaiting further clarification of HIPAA."

"Organizations that begin HIPAA awareness, assessment, and planning now will be in the best position to manage the HIPAA process and the HIPAA compliance costs."

1.2.10 Ensuring Privacy Protection in the Era of Big Laparoscopic Video Data: Development and Validation of an Inside Outside Discrimination Algorithm (IODA) [81]

This is a machine-generated summary of:

Schulze, A.; Tran, D.; Daum, M. T. J.; Kisilenko, A.; Maier-Hein, L.; Speidel, S.; Distler, M.; Weitz, J.; Müller-Stich, B. P.; Bodenstedt, S.; Wagner, M.: Ensuring privacy protection in the era of big laparoscopic video data: development and validation of an inside outside discrimination algorithm (IODA) [81]

Published in: Surgical Endoscopy (2023)

Link to original: https://doi.org/10.1007/s00464-023-10078-x

If you want to cite the papers, please refer to the original.

For technical reasons we could not place the page where the original quote is coming from.

Abstract-Summary

"The purpose of this study was to ensure data privacy in video recordings of laparoscopic surgery by censoring extra abdominal parts."

"An inside-outside-discrimination algorithm (IODA) was developed to ensure privacy protection while maximizing the remaining video data."

"The data set for algorithm training and testing contained a total of 100 laparoscopic surgery videos of 23 different operations with a total video length of 207 h (124 min ± 100 min per video) resulting in 18,507,217 frames (185,965 ± 149,718 frames per video)."

"Each video frame was tagged either as abdominal cavity, trocar, operation site, outside for cleaning, or translucent trocar."

"The distribution of annotated classes were abdominal cavity 81.39%, trocar 1.39%, outside operation site 16.07%, outside for cleaning 1.08%, and translucent trocar 0.07%."

"Algorithm training on binary or all five classes showed similar excellent results for classifying outside frames with a mean F1-score of 0.96 ± 0.01 and 0.97 ± 0.01, sensitivity of 0.97 ± 0.02 and 0.0.97 ± 0.01, and a false positive rate of 0.99 ± 0.01 and 0.99 ± 0.01, respectively."

"IODA is able to discriminate between inside and outside with a high certainty."

Background

"Surgical procedure videos, for example in laparoscopy, can still not readily be used for AI development, because outside the abdomen people could be recognizable (e.g., skin of the patient with identifying tattoo, faces of personnel)."

"Perfectly anonymized videos could be used and shared without the consent of the patient, because the General Data Protection Regulation does not apply to anonymous information (GDPR art."

"IODA helps to ensure data privacy in video recordings of laparoscopic surgery by censoring extraabdominal parts that may identify patients or personnel while also maximizing the remaining video data."

"By anonymizing laparoscopic videos based on inside and outside scenes, they can be safely used for big data analysis, such as surgical AI, quality management or educational purposes while immensely reducing the effort of data collection."

Materials and Methods

"The aim of using an LSTM is that usually in a laparoscopic video each class (e.g., inside and outside) appears for an extended time and usually not as an isolated frame."

"Going from inside the abdominal cavity to outside and back in such a short time frame is very unlikely."

"The data set contains a total of 100 laparoscopic surgery videos with 23 different operation types distributed over four categories: upper gastrointestinal, cholecystectomy, colorectal, and miscellaneous."

"Statistical analysis of the results is done via F1-score, sensitivity, and specificity of the outside class."

"Sensitivity of the outside class indicates whether all outside frames are detected as outside, and is therefore important in terms of privacy protection."

"Specificity of the outside class on the other hand shows how much of the inside frames are misclassified as outside."

Results

"A total of 9,374 frames deviate from the initial annotation, 5,052 frames are annotated as "Inside" and predicted as "Outside," 4,322 frames are annotated as "Outside" and predicted as "Inside."

"In the multiclass experiment, the predictions for "abdominal cavity" match the annotation in 601,161 out of 605,771 (99.24%) frames."

"For "Outside for cleaning" 8,061 frames are annotated of which 1,036 (12.85%) frames are predicted as "Outside for cleaning," while 6,242 (77.43%) frames are predicted as "Outside."

"For "Outside" 119,541 frames are annotated of which 114,887 (96.10%) frames are predicted as "Outside."

"For "translucent trocar" 477 frames are annotated of which none are predicted as such, 206 (43.19%) frames are predicted as "trocar" and 196 (41.09%) frames as "abdominal cavity."

Discussion

"A striking feature of the analysis of discrepancies between manual annotation and algorithm is that the transition frames between inside and outside are especially critical and prone to errors."

"After analyzing the misclassified frames, especially in the multiclass experiment, it becomes apparent that the algorithm struggles with transitional areas between the camera view of the abdominal cavity and the trocar: 46.2% of frames predicted as "abdominal cavity," 40% as "trocar."

"To apply our results to a practical example in case of the binary experiment: for a specificity of 99.2%, in an 1 h laparoscopic video the time span of 21 s is at risk to be falsely predicted as "Inside" while the camera view shows an outside part, while also taking into account that the video consists of 17.2% outside frames."

Conclusion

"Our Inside-Outside-Discrimination-Algorithm IODA allows for privacy protection when recording laparoscopic video data."

"Implementing this kind of deep learning into surgical video data sets holds the potential to immensely improve the quality and especially the quantity of available video data for secondary use."

"The next step will be a prospective evaluation within a real time setting in the operating room."

References

1. Akremi, Aymen; Rouached, Mohsen A comprehensive and holistic knowledge model for cloud privacy protection. The Journal of Supercomputing (2021). https://doi.org/10.1007/s11227-020-03594-3
2. Sun P (2020) Security and privacy protection in cloud computing: discussions and challenges. J Netw Comput Appl 160:102642
3. Sun PJ (2019) Privacy protection and data security in cloud computing: a survey, challenges, and solutions. IEEE Access 7:147420–147452
4. Ghorbel A, Ghorbel M, Jmaiel M (2017) Privacy in cloud computing environments: a survey and research challenges. J Supercomput 73(6):2763–2800
5. Werner J, Merkle Westphall C, Westphall C (2017) Cloud identity management: a survey on privacy strategies. Comput Netw 122:29–42

6. Ari AAA, Ngangmo OK, Titouna C, Thiare O, Kolyang Mohamadou A, Gueroui AM (2019) Enabling privacy and security in cloud of things: architecture, applications, security & privacy challenges. Appl Comput Inform. https://doi.org/10.1016/j.aci.2019.11.005
7. Singh N, Singh A (2017) Data privacy protection mechanisms in cloud. Data Sci Eng 3:11
8. Gholami A, Laure E (2016) Security and privacy of sensitive data in cloud computing: a survey of recent developments. Comput Sci Inf Technol 5:01
9. Liu Y, Sun Y, Ryoo J, Rizvi S, Vasilakos A (2015) A survey of security and privacy challenges in cloud computing: solutions and future directions. J Comput Sci Eng 9:119–133
10. Domingo-Ferrer J, Farràs O, Ribes-González J, Sánchez D (2019) Privacy-preserving cloud computing on sensitive data: a survey of methods, products and challenges. Comput Commun 140–141:38–60
11. Svantesson D, Clarke R (2010) Privacy and consumer risks in cloud computing. Comput Law Secur Rev 26(4):391–397
12. Trapero R, Luna J, Suri N (2016) Quantifiably trusting the cloud: putting metrics to work. IEEE Secur Privacy 14(3):73–77
13. Alemany J, del Val E, Alberola J, García-Fornes A (2018) Estimation of privacy risk through centrality metrics. Future Gener Comput Syst 82:63–76
14. Boneh D, Crescenzo GDi, Ostrovsky R, Persiano G (2004) Public key encryption with keyword search. In: International Conference on the Theory and Applications of Cryptographic Techniques. Springer, pp 506–522
15. Xia Z, Wang X, Sun X, Wang Q (2016) A secure and dynamic multi-keyword ranked search scheme over encrypted cloud data. IEEE Trans Parallel Distrib Syst 27(2):340–352
16. Sun W, Lou W, Hou YT, Li H (2014) Privacy-preserving keyword search over encrypted data in cloud computing. In: Secure Cloud Computing. Springer, pp 189–212
17. Sun W, Wang B, Cao N, Li M, Lou W, Hou YT, Li H (2013) Privacy-preserving multi-keyword text search in the cloud supporting similarity-based ranking. In: Proceedings of the 8th ACM SIGSAC Symposium on Information, Computer and Communications Security. ACM, pp 71–82
18. Toumey, Chris Privacy in the Shadow of Nanotechnology. NanoEthics (2007). https://doi.org/10.1007/s11569-007-0023-6
19. Li J (2003) Lessons from the last five years. Scarecrow, Lanham MD
20. Holden C (2007). Nanofinger. Science 315:1773
21. Sametband M, Shweky I, Banin U, Mandler D, Almog J (2007) Application of nanoparticles for the enhancement of latent fingerprints. Chem Commun 2007:1142–1144
22. van den Hoven J, Vermaas P (2007) Nanotechnology and privacy. J Med Philos 32:283–297
23. Toumey C (2007) Plenty of eyes at the bottom. Nat Nanotechnol 2(4):192–193
24. Ismagilova, Elvira; Hughes, Laurie; Rana, Nripendra P.; Dwivedi, Yogesh K. Security, Privacy and Risks Within Smart Cities: Literature Review and Development of a Smart City Interaction Framework. Information Systems Frontiers (2020). https://doi.org/10.1007/s10796-020-10044-1
25. Chatterjee, S., Kar, A. K., Dwivedi, Y. K., & Kizgin, H. (2019). Prevention of cybercrimes in smart cities of India: from a citizen's perspective. Information Technology & People, 32(5), 1153–1183.
26. Elmaghraby, A. S., & Losavio, M. M. (2014). Cyber security challenges in smart cities: Safety, security and privacy. Journal of Advanced Research, 5(4), 491–497. https://doi.org/10.1016/j.jare.2014.02.006
27. Pare, G., Trudel, M. C., Jaana, M., & Kitsiou, S. (2015). Synthesizing information systems knowledge: A typology of literature reviews. Information & Management, 52(2), 183–199
28. Abosaq, N. H. (2019). Impact of privacy issues on smart city services in a model smart city. International Journal of Advanced Computer Science and Applications, 10(2), 177–185
29. Al-Dhubhani, R., Mehmood, R., Katib, I., & Algarni, A. (2018). Location privacy in smart cities era https://doi.org/10.1007/978-3-319-94180-6_14

30. Antonopoulos, K., Petropoulos, C., Antonopoulos, C. P., & Voros, N. S. (2017). Security data management process and its impact on smart cities' wireless sensor networks. Paper presented at the South-East Europe Design Automation, Computer Engineering, Computer Networks and Social Media Conference, SEEDA-CECNSM 2017, https://doi.org/10.23919/SEEDA-CECNSM.2017.8088238
31. Avgerou, A., Nastou, P. E., Nastouli, D., Pardalos, P. M., & Stamatiou, Y. C. (2016). On the deployment of citizens' privacy preserving collective intelligent ebusiness models in smart cities. International Journal of Security and its Applications, 10(2), 171–184. https://doi.org/10.14257/ijsia.2016.10.2.16
32. Beltran, V., Martinez, J. A., & Skarmeta, A. F. (2017). User-centric access control for efficient security in smart cities. Paper presented at the GIoTS 2017 - Global Internet of Things Summit, Proceedings, https://doi.org/10.1109/GIOTS.2017.8016287
33. Burange, A. W., & Misalkar, H. D. (2015). Review of internet of things in development of smart cities with data management & privacy. Paper presented at the Conference Proceeding – 2015 International Conference on Advances in Computer Engineering and Applications, ICACEA 2015, 189–195. https://doi.org/10.1109/ICACEA.2015.7164693
34. Cagliero, L., Cerquitelli, T., Chiusano, S., Garino, P., Nardone, M., Pralio, B., & Venturini, L. (2015). Monitoring the citizens' perception on urban security in smart city environments. Paper presented at the Proceedings - International Conference on Data Engineering, 2015-June 112–116. https://doi.org/10.1109/ICDEW.2015.7129559
35. Mora, O. B., Rivera, R., Larios, V. M., Beltran-Ramirez, J. R., Maciel, R., & Ochoa, A. (2019). A use case in cybersecurity based in blockchain to deal with the security and privacy of citizens and smart cities cyberinfrastructures. Paper presented at the 2018 IEEE International Smart Cities Conference, ISC2 2018, https://doi.org/10.1109/ISC2.2018.8656694
36. Noh, J., & Kwon, H. (2019). A study on smart city security policy based on blockchain in 5G age. Paper presented at the 2019 International Conference on Platform Technology and Service, PlatCon 2019 - Proceedings, https://doi.org/10.1109/PlatCon.2019.8669406
37. Ramos, L. F. M., & Silva, J. M. C. (2019). Privacy and data protection concerns regarding the use of blockchains in smart cities. Paper presented at the ACM International Conference Proceeding Series, Part F148155 342–347. https://doi.org/10.1145/3326365.3326410
38. Monzon, A. (2015). Smart cities concept and challenges: Bases for the assessment of smart city projects. In 2015 international conference on smart cities and green ICT systems (SMARTGREENS) (pp. 1–11)
39. Kitchin, R., & Dodge, M. (2011). Code/Space: Software and Everyday Life. Cambridge: MIT Press
40. Townsend, A. (2013). Smart Cities: Big data, civic hackers, and the quest for a new utopia. New York: W.W. Norton & Co
41. Spiekermann, Sarah RFID and privacy: what consumers really want and fear. Personal and Ubiquitous Computing (2008). https://doi.org/10.1007/s00779-008-0215-2
42. Smith H, Milberg J, Burke, J (1996) Information privacy: measuring individuals' concerns about organizational practices. MIS Q 20(2):167–196
43. Berthold O, Guenther O, Spiekermann S (2005) RFID Verbraucherängste und Verbraucherschutz. Wirtschaftsinformatik 47(6):422–430
44. Kassarjian HH (1977) Content analysis in consumer research. J Consumer Res 4(1):8–18
45. Camenisch, Jan; Groß, Thomas; Heydt-Benjamin, Thomas Scott Accountable privacy supporting services. Identity in the Information Society (2009). https://doi.org/10.1007/s12394-009-0023-5
46. Camenisch J, Lysyanskaya A. Efficient non-transferable anonymous multi-show credential system with optional anonymity revocation. In: Pfitzmann B, editor. Advances in cryptology—EUROCRYPT 2001. Lecture notes in computer science. New York: Springer; 2001. vol. 2045, p. 93–118
47. Camenisch J, Lysyanskaya A. Signature schemes and anonymous credentials from bilinear maps. In: Franklin MK, editor. Advances in cryptology—CRYPTO 2004. Lecture notes in computer science. New York: Springer; 2004. vol. 3152, p. 56–72

48. Green M, Hohenberger S. Blind identity-based encryption and simulatable oblivious transfer. In: ASIACRYPT; 2007. p. 265–282
49. Ateniese G, Camenisch J, Joye M, Tsudik G. A practical and provably secure coalition-resistant group signature scheme. In: Bellare M, editor. Advances in cryptology—CRYPTO 2000. Lecture notes in computer science. New York: Springer; 2000. vol. 1880, p. 255–270
50. Boneh D, Boyen X, Shacham H. Short group signatures. In: Franklin MK, editor. Advances in cryptology—CRYPTO 2004. Lecture Notes in Computer Science. New York: Springer; 2004. vol. 3152, p. 41–55
51. Kiayias A, Yung M. Secure scalable group signature with dynamic joins and separable authorities. IJSN 2006;1(1/2):24–45
52. Brands S. An efficient off-line electronic cash system based on the representation problem. Technical Report CS-R9323, CWI; Apr. 1993
53. Camenisch J, Hohenberger S, Lysyanskaya A. Compact E-cash. In: Cramer R, editor. Advances in cryptology—Eurocrypt 2005. Lecture notes in computer science. New York: Springer; 2005. vol. 3494, p. 302–321
54. Frankel Y, Tsiounis Y, Yung M. Fair off-line e-cash made easy. In: Kim K, Matsumoto T, editors. Advances in cryptology—ASIACRYPT '98. Lecture notes in computer science. New York: Springer; 1998. vol. 1514, p. 257–270
55. Brands S. Restrictive blinding of secret-key certificates. Technical Report CS-R9509, CWI; Sept. 1995a
56. Brands S. Secret-key certificates. Technical Report CS-R9510, CWI, Sept. 1995b
57. Brands S. Rethinking public key infrastructures and digital certificates: building in privacy. Cambridge: MIT; 2000
58. Kiayias A, Yung M, Tsiounis Y. Traceable signatures. In: Cachin C, Camenisch J, editors. Advances in cryptology—EUROCRYPT 2004. Lecture notes in computer science. New York: Springer; 2004. vol. 3027, p. 571–589
59. Naor M, Pinkas B, Sumner R. Privacy preserving auctions and mechanism design. In: Proc. 1st ACM conference on electronic commerce. 1999
60. Fujioka A, Okamoto T, Ohta K. A practical secret voting scheme for large scale elections. In: Seberry J, Zheng Y, editors. ASIACRYPT. Lecture notes in computer science. New York: Springer; 1992. vol. 718, p. 244–251
61. Cramer R, Damgård I, Schoenmakers B. Proofs of partial knowledge and simplified design of witness hiding protocols. In: Desmedt YG, editor. Advances in cryptology—CRYPTO '94. Lecture notes in computer science. Springer; 1994. vol. 839, p. 174–187
62. Brands S. Rapid demonstration of linear relations connected by Boolean operators. In: Fumy W, editor. Advances in cryptology—EUROCRYPT '97. Lecture Notes in Computer Science. New York: Springer; 1997. vol. 1233, p. 318–333
63. Camenisch J, Michels M. Proving in zero-knowledge that a number n is the product of two safe primes. Technical Report RS-98-29, BRICS, Departement of Computer Science, University of Aarhus; Nov. 1998
64. Boudot F. Efficient proofs that a committed number lies in an interval. In: Preneel B, editor. Advances in cryptology—EUROCRYPT 2000. Lecture Notes in Computer Science. New York: Springer; 2000. vol. 1807, p. 431–444
65. Fujisaki E, Okamoto T. Statistical zero knowledge protocols to prove modular polynomial relations. In: Kaliski B, editor. Advances in cryptology—CRYPTO '97. Lecture notes in computer science. New York: Springer; 1997. vol. 1294, p. 16–30
66. Dodig-Crnkovic, Gordana; Horniak, Virginia Togetherness and respect: ethical concerns of privacy in Global Web Societies. AI & SOCIETY (2005). https://doi.org/10.1007/s00146-005-0024-3
67. Dodig-Crnkovic G (2003) Shifting the paradigm of the philosophy of science: the philosophy of information and a new renaissance. Minds Mach: Special Issue on the Philosophy of Information 13(4)
68. Gill KS (1997) Knowledge networking and social cohesion in the Information Society, a study for the European Commission

69. Gill KS (2002) Knowledge networking in cross-cultural settings. AI Soc 16:252–277
70. Thill G (1994) The relevance of associative networks for a sustainable information and communication society. AI Soc 6(1):70–77
71. Wagner C, Cheung K, Lee F, Ip R (2003) Enhancing e-government in developing countries: managing knowledge through virtual communities. Electron J Inform Syst Dev Ctries 14(4):1–20
72. Rosen J (2000) Why privacy matters. Wilson Quarta 24(4)
73. Mason RO (2004) A tapestry of privacy, a meta-discussion. (http://www.cyberethics.cbi.msstate.edu/mason2/)
74. Brey P (2000) Method in computer ethics: towards a multi-level interdisciplinary approach. Ethics Inform Technol 2(3):1–5
75. Whitworth B, de Moor A (2003) Legitimate by design: towards trusted virtual community environments. Behav Inform Technol 22(1):31–51
76. Tavani, Herman T. Informational privacy, data mining, and the Internet. Ethics and Information Technology (1999). https://doi.org/10.1023/a:1010063528863
77. de Laat, Paul B. Algorithmic Decision-Making Based on Machine Learning from Big Data: Can Transparency Restore Accountability?. Philosophy & Technology (2017). https://doi.org/10.1007/s13347-017-0293-z
78. Regan, Marybeth Disease Management and the Application of Privacy Guidelines. Disease Management and Health Outcomes (2002). https://doi.org/10.2165/00115677-200210090-00003
79. Fox A. Cost value estimated at: cost value estimated, as reported in the Workgroup for Electronic Data Interchange, the HIPAA Security Task Force. HIPAA Implementation Considerations. Blue Cross and Blue Shield Association, Feb. 19, 2001. Reston (VA)
80. Villagra V. DMAA to Feds: 'Practice What You Preach' in privacy rules. Dis Manage News 2001; 6 (22): 1
81. Schulze, A.; Tran, D.; Daum, M. T. J.; Kisilenko, A.; Maier-Hein, L.; Speidel, S.; Distler, M.; Weitz, J.; Müller-Stich, B. P.; Bodenstedt, S.; Wagner, M. Ensuring privacy protection in the era of big laparoscopic video data: development and validation of an inside outside discrimination algorithm (IODA). Surgical Endoscopy (2023). https://doi.org/10.1007/s00464-023-10078-x

Chapter 2
General Data Protection Principles and Its Implementation in a Tech-Environment

2.1 Introduction by the Editor

There are certain foundational rules to data processing. They are the data protection principles. The assigned responsibilities under GDPR follow these principles. Article 5 of the GDPR reflects upon these principles that create the framework of personal data processing.[1] While lawful, fair and transparent processing are the crucial pillars, it is equally important to have a specific, explicit and legitimate purpose of data processing. A specific purpose determines the scope of processing. Therefore, it designates any processing outside such scope as incompatible. A well-defined purpose establishes relevant, adequate and necessary data processing. That further leads to the data minimisation principle. The idea of data minimisation also restricts the time of storage and gives rise to the storage limitation principle. By way of the accuracy principle, data subjects do have the right to keep personal data updated. More importantly, a data controller should provide such an opportunity to a data subject to operationalise this right. The data controllers have the responsibility to protect the personal data from unauthorised access, destruction and damage and therefore, must adopt measures that help operationalise overall integrity and confidentiality of the process involving data processing.

There are several advanced technologies and operations that struggle fulfilling the principles discussed above. Blockchain due to operational reasons may struggle in fulfilling the storage limitation principle. It may also struggle with the requirement

[1] *Regulation (EU) 2016/679 of the European Parliament and of the Council of 27 April 2016 on the protection of natural persons with regard to the processing of personal data and on the free movement of such data, and repealing Directive 95/46/EC (General Data Protection Regulation) (Text with EEA relevance) OJ L 119, 4.5.2016, pp. 1–88.*

I. Gupta (ed.), *Operationalizing Expectations and Mapping Challenges of Information Privacy and Data Protection Measures*, https://doi.org/10.1007/978-981-99-7371-2_2

of keeping personal details up to date,[2] thereby needing to include some mutability option in the process.[3]

The designing of an internal governance structure helps in complying with the GDPR requirements, especially when there are several facets to the operation of a business.[4] One of the ways is to develop an automated solution mechanism that takes into account the GDPR requirements, so that business practices can be mapped against a specific compliance matrix. The automated compliance checking mechanism takes into consideration the legal basis of processing personal data and the regulatory obligations mentioned under GDPR.[5] The automated tool then validate business practice measures and gives a go ahead to data collection.[6]

To tackle the processing of Big Data and to address the privacy concerns, embedded privacy protection measures are recommended. Identifying the specific purpose behind processing of personal data is a challenge that Big Data processing needs to overcome. Big Data that originates from different data sources that are collected for different reasons will infringe one or several of the data protection principles.[7] Solution mechanism imbibing data minimisation techniques enabled at the design and default stage of the followed process is proposed for businesses running complex operations.[8] At the centre of this mechanism, runs the idea of security, data subjects' rights, and the purpose of processing of personal data connected with the business process that requires access to such data.[9] These steps and many more, like

[2] *Freund, Gislaine Parra; Fagundes, Priscila Basto; de Macedo, Douglas Dyllon Jeronimo An Analysis of Blockchain and GDPR under the Data Lifecycle Perspective. Mobile Networks and Applications (2020)* https://doi.org/10.1007/s11036-020-01646-9.

[3] *Godyn, Mateusz; Kedziora, Michal; Ren, Yingying; Liu, Yongxin; Song, Houbing Herbert Analysis of solutions for a blockchain compliance with GDPR. Scientific Reports (2022)* https://doi.org/10.1038/s41598-022-19341-y.

[4] *Jakobi, Timo; von Grafenstein, Maximilian; Legner, Christine; Labadie, Clément; Mertens, Peter; Öksüz, Ayten; Stevens, Gunnar The Role of IS in the Conflicting Interests Regarding GDPR. Business & Information Systems Engineering (2020)* https://doi.org/10.1007/s12599-020-00633-4.

[5] *Bonatti, Piero A.; Kirrane, Sabrina; Petrova, Iliana M.; Sauro, Luigi Machine Understandable Policies and GDPR Compliance Checking. KI - Künstliche Intelligenz (2020)* https://doi.org/10.1007/s13218-020-00677-4.

[6] Ibid.

[7] *Georgiadis, Georgios; Poels, Geert Enterprise architecture management as a solution for addressing general data protection regulation requirements in a big data context: a systematic mapping study. Information Systems and e-Business Management (2021)* https://doi.org/10.1007/s10257-020-00500-5.

[8] *Zaman, Rashid; Hassani, Marwan On Enabling GDPR Compliance in Business Processes Through Data-Driven Solutions. SN Computer Science (2020)* https://doi.org/10.1007/s42979-020-00215-x.

[9] Ibid.

improved understanding of clear and plain language[10] will effectively increase transparency and regain trust of data subjects. Like, giving back control to data subjects through the use of the adopted technology will increase transparency.[11]

Many of these embedded systems remove the difficult task of interpreting the GDPR requirements and map the expectations of the law, which are anyways difficult for all kinds of data controllers to understand.[12] Effective adoption of GDPR depends on a multitude of things, including the complexity of business processes and nature of personal data used.[13] It also depends on the nature of data controllers in question.[14] These embedded systems help to make GDPR requirements operational and can help in processing situations where impact assessment exercises are essential.[15] However, these systems will involve several stakeholders and a joint operation is required to fully contemplate the overarching criticalities.[16]

There are several operational challenges to overcome where artificial intelligence (AI), either as a tool or a product, plays a pivotal role in processing personal information or is used to help data controllers remain GDPR compliant. For instance, the use of AI in healthcare and the inherent opacity surrounding its use raises doubt.[17] There are challenges to meet the requirement of informed consent primarily because of the way decisions are made and the fog surrounding the steps followed in this regard.[18] The general use of AI to meet the GDPR expectations has been a point of contention.

---

[10] *Rydzewska-Siemiątkowska, Joanna Deontic Modality in the GDPR Based Finnish Privacy Notices in the Light of the Transparency Principle. International Journal for the Semiotics of Law—Revue internationale de Sémiotique juridique (2022)* https://doi.org/10.1007/s11196-022-09939-w.

[11] *Zichichi, Mirko; Ferretti, Stefano; D'Angelo, Gabriele; Rodríguez-Doncel, Víctor Data governance through a multi-DLT architecture in view of the GDPR. Cluster Computing (2022)* https://doi.org/10.1007/s10586-022-03691-3; *de Carvalho Junior, Marcelo Antonio; Bandiera-Paiva, Paulo Strengthen Electronic Health Records System (EHR-S) Access-Control to Cope with GDPR Explicit Consent. Journal of Medical Systems (2020)* https://doi.org/10.1007/s10916-020-01631-5.

[12] *De Carvalho, Renata M.; Del Prete, Camillo; Martin, Yod Samuel; Araujo Rivero, Rosa M.; Önen, Melek; Schiavo, Francesco Paolo; Rumín, Ángel Cuevas; Mouratidis, Haralambos; Yelmo, Juan C.; Koukovini, Maria N. Protecting Citizens' Personal Data and Privacy: Joint Effort from GDPR EU Cluster Research Projects. SN Computer Science (2020)* https://doi.org/10.1007/s42979-020-00218-8.

[13] Lopes, Isabel Maria; Guarda, Teresa; Oliveira, Pedro General Data Protection Regulation in Health Clinics. Journal of Medical Systems (2020) https://doi.org/10.1007/s10916-020-1521-0.

[14] *Brodin, Martin A Framework for GDPR Compliance for Small- and Medium-Sized Enterprises. European Journal for Security Research (2019)* https://doi.org/10.1007/s41125-019-00042-z.

[15] *Ruiz, Alejandra; Martin, Yod-Samuel; Martinez, Jabier; Quintans, Jacobo; Mockly, Guillaume; Gyrard, Amelie; Crepax, Tommaso Modeling ecosystems of reference frameworks for assurance: a case on privacy impact assessment regulation and guidelines. Software and Systems Modeling (2022)* https://doi.org/10.1007/s10270-022-01061-6.

[16] *Torre, Damiano; Alferez, Mauricio; Soltana, Ghanem; Sabetzadeh, Mehrdad; Briand, Lionel Modeling data protection and privacy: application and experience with GDPR. Software and Systems Modeling (2021)* https://doi.org/10.1007/s10270-021-00935-5.

[17] *Kawamleh, Suzanne Against explainability requirements for ethical artificial intelligence in health care. AI and Ethics (2022)* https://doi.org/10.1007/s43681-022-00212-1.

[18] Ibid.

In this regard, the argument is that a rule-based AI driven approach would be better equipped to reason the decisions of data processing to the data subjects.[19] Further, owing to its operational reasons AI can identify risks better. However, depending on its use, AI also runs the risk of making incorrect assessments, thereby increasing the liability of data controllers.[20] AI techniques have been sought to assess security risks and to prevent unlawful disclosure of personal data.[21] In the process, data controllers can measure the level of compliance to the GDPR requirements. The utility of Big Data in the healthcare sector is difficult to undermine. A lot of positive health-related decision-making can happen through Big Data analysis.[22] The challenge is of course to balance the rights that data subjects can exercise vis-à-vis the access required to such data for further analysis.[23]

Article 89(2) of GDPR allows for certain derogation of data subjects' rights when processing of personal data is an outcome of scientific research.[24] These are the rights mentioned under Articles 15, 16, 18 and 21. Of course, these derogations are not without adequate safeguards mentioned under Article 89(1).[25] However, the implementation of this provision could be a challenge in a research intensive area and may construe practices followed in such research as unethical.[26] There are concerns about critical research getting affected as a result of data privacy requirements, thereby having a far reaching impact on the overall well-being of humans.[27] The question of access to data remains a major question and seen in direct conflict with larger societal developments.[28] Access to data is a question that extends to

---

[19] *Kingston, John Using artificial intelligence to support compliance with the general data protection regulation. Artificial Intelligence and Law (2017)* https://doi.org/10.1007/s10506-017-9206-9.

[20] Ibid.

[21] *Lorè, Filippo; Basile, Pierpaolo; Appice, Annalisa; de Gemmis, Marco; Malerba, Donato; Semeraro, Giovanni An AI framework to support decisions on GDPR compliance. Journal of Intelligent Information Systems (2023)* https://doi.org/10.1007/s10844-023-00782-4.

[22] *Kolasa, Katarzyna; Ken Redekop, W.; Berler, Alexander; Zah, Vladimir; Asche, Carl V. Future of Data Analytics in the Era of the General Data Protection Regulation in Europe. PharmacoEconomics (2020)* https://doi.org/10.1007/s40273-020-00927-1.

[23] Ibid.

[24] *Regulation (EU) 2016/679 of the European Parliament and of the Council of 27 April 2016 on the protection of natural persons with regard to the processing of personal data and on the free movement of such data, and repealing Directive 95/46/EC (General Data Protection Regulation) (Text with EEA relevance) OJ L 119, 4.5.2016, pp. 1–88.*

[25] Ibid.

[26] *Staunton, Ciara; Slokenberga, Santa; Mascalzoni, Deborah The GDPR and the research exemption: considerations on the necessary safeguards for research biobanks. European Journal of Human Genetics (2019)* https://doi.org/10.1038/s41431-019-0386-5.

[27] *Tantoso, Erwin; Wong, Wing-Cheong; Tay, Wei Hong; Lee, Joanne; Sinha, Swati; Eisenhaber, Birgit; Eisenhaber, Frank Hypocrisy Around Medical Patient Data: Issues of Access for Biomedical Research, Data Quality, Usefulness for the Purpose and Omics Data as Game Changer. Asian Bioethics Review (2019)* https://doi.org/10.1007/s41649-019-00085-3.

[28] Ibid; See generally, *Zemła-Pacud, Żaneta; Lenarczyk, Gabriela Clinical Trial Data Transparency in the EU: Is the New Clinical Trials Regulation a Game-Changer?. IIC—International Review of Intellectual Property and Competition Law (2023)* https://doi.org/10.1007/s40319-023-01329-4.

different kinds of data controllers, including the legal basis that they would adopt to process personal data.[29] The concern with future critical research engagements emerged even before the enactment of GDPR.[30] Further, Member States in the EU may give a different interpretation to Article 89(2) leading to additional challenges in scientific research.[31] Secondary use of health-related data is a point of concern with major challenges apprehended in data transfer both within the EU and for cross border transfers.[32] There are challenges about having varied privacy perceptions where stakeholders give different meanings to the idea of privacy and to the legal basis of processing. Measures adopted to inform the data subjects are critical towards the correct implementation of GDPR provisions.[33] Adequate information goes a long way in increasing awareness amongst data subjects and also gives an opportunity to understand the possible risks.[34]

2.2 Machine Generated Summaries

Disclaimer: The summaries in this chapter were generated from Springer Nature publications using extractive AI auto-summarization: An extraction-based summarizer aims to identify the most important sentences of a text using an algorithm and uses those original sentences to create the auto-summary (unlike generative AI). As the constituted sentences are machine selected, they may not fully reflect the body of the work, so we strongly advise that the original content is read and cited. The auto generated summaries were curated by the editor to meet Springer Nature publication standards. To cite this content, please refer to the original publications.

---

[29] *Quinn, Paul Research under the GDPR—a level playing field for public and private sector research?. Life Sciences, Society and Policy (2021)* https://doi.org/10.1186/s40504-021-00111-z.

[30] *Nyrén, Olof; Stenbeck, Magnus; Grönberg, Henrik The European Parliament proposal for the new EU General Data Protection Regulation may severely restrict European epidemiological research. European Journal of Epidemiology (2014)* https://doi.org/10.1007/s10654-014-9909-0.

[31] *Clarke, Niamh; Vale, Gillian; Reeves, Emer P.; Kirwan, Mary; Smith, David; Farrell, Michael; Hurl, Gerard; McElvaney, Noel G. GDPR: an impediment to research?. Irish Journal of Medical Science (2019)* 10.1007/s11845-019-01980-2; See generally, *Wallace, Rachel; Greene, Elaine Survey of NCHDs in Ireland to assess their views and opinions in relation to participation in health research and the impact of new Irish data protection regulations. Irish Journal of Medical Science (2020)* https://doi.org/10.1007/s11845-020-02185-8.

[32] *Vukovic, Jakov; Ivankovic, Damir; Habl, Claudia; Dimnjakovic, Jelena Enablers and barriers to the secondary use of health data in Europe: general data protection regulation perspective. Archives of Public Health (2022)* https://doi.org/10.1186/s13690-022-00866-7.

[33] *Davey, Matthew G.; O'Donnell, John P.M.; Maher, Elizabeth; McMenamin, Cliona; McAnena, Peter F.; Kerin, Michael J.; Miller, Nicola; Lowery, Aoife J. General data protection regulations (2018) and clinical research: perspectives of patients and doctors in an Irish university teaching hospital. Irish Journal of Medical Science (2021)* https://doi.org/10.1007/s11845-021-02789-8.

[34] *Truyers, Carla; Kellen, Eliane; Arbyn, Marc; Trommelmans, Leen; Nys, Herman; Hensen, Karen; Aertgeerts, Bert; Bartholomeeusen, Stefaan; Hansson, Mats; Buntinx, Frank The use of human tissue in epidemiological research; ethical and legal considerations in two biobanks in Belgium. Medicine, Health Care and Philosophy (2009)* https://doi.org/10.1007/s11019-009-9230-y.

Machine generated keywords: gdpr, gdpr compliance, health, compliance, scientific research, controller, clinical, scientific, automate, organisation, article, directive, blockchain, consent, business.

2.2.1 An Analysis of Blockchain and GDPR Under the Data Lifecycle Perspective [82]

This is a machine-generated summary of:

Freund, Gislaine Parra; Fagundes, Priscila Basto; de Macedo, Douglas Dyllon Jeronimo: An Analysis of Blockchain and GDPR under the Data Lifecycle Perspective [82].

Published in: Mobile Networks and Applications (2020).

Link to original: https://doi.org/10.1007/s11036-020-01646-9.

If you want to cite the papers, please refer to the original.

For technical reasons we could not place the page where the original quote is coming from.

Abstract-Summary

"The purpose of this article is to present an analysis of the privacy principles of personal data prescribed in the General Data Protection Regulation and the treatment of data in Blockchain technology in its primary version, guided by the phases of the Data LifeCycle."

"The General Data Protection Regulation, on the other hand, presents privacy principles that contemplate the necessary treatment for data at all stages of its life cycle."

"This analysis made it possible to identify the influence that the phases of the Data Lifecycle have on the adequacy of the principles of the General Data Protection Regulation and the treatment of the data performed on the Blockchain technology associated with the phases, presenting an approach to lead the process of adapting the technology to compliance with the legislation."

Introduction

"Although there are discussions in the literature about Blockchain compliance with GDPR and the proposal of models that guide the implementation of technology to comply with the law, research that takes into account the Data Life Cycle (DLC) within this context is incipient."

"Based on the assumption that there are organizations that use the life cycle approach to guide the development of systems, the flow of information, among others, an analysis of the principles proposed by GDPR was carried out and identified in which phases of the Data Life Cycle (DLC) proposed by [1] these principles need to be observed."

"This article aims to present the results of this analysis and to encourage new studies involving the GDPR, the Data Life Cycle and the Blockchain, including research involving the proposal of models to be used in the process of adaptation to the GDPR of the organizations that use the Blockchain considering the DLC."

Background

"As it was conceived in a time before the emergence of the commercial internet and before the diffusion of business models and technologies that make use of the intense use of personal data, this directive had to undergo an update process that culminated in the current European Union General Data Protection Regulation (GDPR)."

"The regulation came into force on May 25, 2018, updating and adapting the old European Data Protection Directive to the newest forms of massive use of personal data, such as business models based on big data technologies, artificial intelligence, machine e-learning and Blockchain [2, 3]."

"According to article 4 of the GDPR, personal data is characterized by being a set of information related to a living person that can lead to their identification and that must be protected regardless of how they are stored or treated, whether in a technological or manual, provided they are organized according to pre-defined criteria [4]."

Related works

"The authors recommend 3 factors to be interpreted as guidelines for the compliance of Blockchain to GDPR considering the applied context: (1) avoid the storage of personal data on a Blockchain—they must remain stored in systems that allow them to be rectified and deleted; (2) use of pseudo-anonymization in blockchains that process personal data; (3) use of private blockchain with permission when applied in scenarios involving interorganizational workflows."

"With respect to studies involving the Data Life Cycle associated with GDPR or Blockchain technology, the research carried out in the Web of Science and IEEE Xplore Digital Library databases did not return results that could be considered as works related to this one being presented in this article, thus evidencing the originality of this study."

Methodology

"The influence of the DLC phases on the GDPR principles was identified considering that they can present different levels of performance, as shown below: Very Influential: for situations where the phase significantly influences the adequacy of the principle—represented by the symbol ●; Influential: for situations where the phase has an influence on the adequacy of the principle, but indirectly—represented by the symbol ○; Unidentified: for situations where there was no influence on the adequacy of the principle—represented by the acronym UI.; In order to identify the treatment of data carried out on Blockchain technology and associate it with the DLC phases, the functioning and functions performed by the technology were evaluated and these were related to the phases."

"With the view of this and the previous analysis that considered the influence that the phases of the DLC exert in the adequacy of the privacy principles, some treatment options for the data were presented to compose the gaps identified in each phase."

Results

"In order to adapt the principle of limitation of retention, a relevant influence was identified in the phases of collection, storage and disposal: in the phase of collection when considering that the purpose of data collection is identified, thus linking the time required to retain them to fulfill their purpose; storage, as it is at this stage that automated measures should or should not be adopted for the deletion of data that are no longer needed and that have expired the retention period and in the disposal that enables the tracking and deletion of data from all databases, whether they be the original, redundancy or backup, ensuring that the retention time is met."

"In the analysis carried out, during the collection stage, a relevant influence was identified on the adequacy of all GDPR principles: (P1) lawfulness, loyalty and transparency; (P2) limitation of purposes; (P3) minimization of data; (P4) accuracy; (P5) limitation of retention; (P6) integrity and confidentiality and (P7) responsibility."

Conclusions and future works

"The evaluation considered the data treatment performed on the Blockchain, presenting an approach for adapting a technology to the data privacy principles covered in the DLC-guided GDPR."

"It can be concluded that the DLC model proposed by [1] and the influence of its steps on the adequacy of the GDPR principles can be used as a support tool to systematize the process of adapting to the law by organizations that use or intend to adopt Blockchain technology."

"It can be observed that even in the stages that did not have a relevant influence on the adequacy of all the principles of the legislation, several data treatment actions need to be directed to adapt the technology to the legislation, since the Blockchain in its primary version, does not perform data processing data in the collection, recovery and disposal phase."

2.2.2 The Role of IS in the Conflicting Interests Regarding GDPR [83]

This is a machine-generated summary of:

Jakobi, Timo; von Grafenstein, Maximilian; Legner, Christine; Labadie, Clément; Mertens, Peter; Öksüz, Ayten; Stevens, Gunnar: The Role of IS in the Conflicting Interests Regarding GDPR [83].

Published in: Business & Information Systems Engineering (2020).

If you want to cite the papers, please refer to the original.

For technical reasons we could not place the page where the original quote is coming from.

Introduction

"With evolvement of its context, maybe also the field itself needs to adapt One sign for this simultaneous potential need and opportunity is the vivid research surrounding GDPR in the areas concerning the interdisciplinary field of BISE."

"While the aforementioned contributions stem from a certain research field or perspective, they are not only interesting, but also highly relevant for the respective other ones, because of the fact that handling the GDPR is a multi-stakeholder task by nature."

"To the ongoing discussion of the future of BISE and its relation to GDPR, we have summoned renowned experts from the fields of law, customer protection, economics, organizational and information sciences, as well as human–computer interaction to talk about how BISE research is interdependent with the GDPR in terms of contributing to an understanding of how to interpret regulation in the practice of BISE."

The EU General Data Protection Regulation Outside the Box: Competitive Advantages and Openness to Innovation

"However, this approach creates considerable legal uncertainty as neither the companies nor those affected, e.g. data subjects, can know with certainty whether or not the entrepreneur's concrete implementation of the law meets the expectations of the regulator [5]."

"Certificates and codes of conduct thus enable both the controller and the processor to reduce their legal uncertainty and to signal their GDPR conformity on the market."

"The second approach follows the classic compliance logic: A data controller or processor only fulfills the GDPR requirements to the extent that it needs proof to defend itself against a "first-time fine" and immediately implements all additional measures if the competent data protection authority demands them."

"This third strategy makes a virtue out of a necessity: A data controller or processor uses the leeway that the GDPR gives them to proactively find the best solution for their specific data processing."

Challenges with GDPR from the Enterprise Perspective—Building a Dedicated (Personal) Data Management Capability

"The difficulties with the GDPR can be explained by the changing nature of data protection regulations."

"To previous regulations that could be addressed by amending contracts and general conditions, the GDPR requires companies to fundamentally rethink the way they store and process personal data on an enterprise-wide level."

"Research in the Competence Center Corporate Data Quality (CC CDQ) reveals that the GDPR requires companies to build a dedicated data management capability [6]."

"The system capabilities require to redesign data-processing systems and are often emphasized in the GDPR debate."

"They comprise the abilities (1) to clearly identify, classify and locate personal data in system landscapes (Manage protected data scope); (2) to collect consent and ensure consent-based processing of information (Manage consent); and (3) to process data according to EU-GDPR's data rights and principles (Enable data information rights)."

Ten Critical Aspects of the European General Data Protection Regulation from the Point of View of Information Systems

"Another reason, especially for companies operating abroad, relates to so-called "escape clauses" that allow for the integration of country-specific laws and regulations in order to protect national privileges (e.g., freedom of the press)."

"Further, in many functional areas (e.g., human resources) and industries (e.g., healthcare), the GDPR conflicts with the growing number of function- and/or industry-specific rules requiring companies to keep very detailed data records."

"The manifold drawbacks of the GDPR seem to motivate German politicians to suggest exceptions, for example, related to work safety, finance/taxation, health (see also point 7 above), education (e.g., fine-grained databases to analyze reasons of early school leaving), housing and protection of tenants, homeland security, or defense."

Information Systems and the General Data Protection Regulation—A Consumer Protection Perspective

"The majority of the social media providers monitors their users' browsing activities by default and analyzes the collected data to serve personalized advertising."

"In the GDPR, users (in the GDPR called "data subject") have the right to obtain from, e.g., a service provider confirmation as to whether or not personal data concerning him or her is being processed, and, where that is the case, they have the right to access the respective personal data."

"Consumers are not able to use the downloaded data packets in order to make informed decisions regarding the transmission of their data to, for example, another social media provider."

"Only if requirements are consistently implemented by service providers or data controllers will consumers be able to exercise their rights in practice so that GDPR can achieve the desired effects."

The GDPR from a Perspective of Consumer Informatics

"IT-Security and Privacy, University of Siegen From the point of view of consumer informatics, the General Data Protection Regulation (GDPR) represents an important step towards the reorganization of data protection for a digital society."

"In times of data capitalism and the increase of AI procedures in application systems, it is important to remember that from this point of view and for a modern, liberal society the principle of informational self-determination is a great asset, which is by no means natural, but must always be defended anew."

"For this is the right of information, in which companies and authorities use a proliferation of requirements for authentication, processes, contact points and data formats that consumers have to deal with."

"Design-oriented business and consumer informatics should take up the ball and develop standardized formats for consumer data as well as reference models for the usable information process."

2.2.3 Machine Understandable Policies and GDPR Compliance Checking [84]

This is a machine-generated summary of:

Bonatti, Piero A.; Kirrane, Sabrina; Petrova, Iliana M.; Sauro, Luigi: Machine Understandable Policies and GDPR Compliance Checking [84].

Published in: KI—Künstliche Intelligenz (2020).

Link to original: https://doi.org/10.1007/s13218-020-00677-4.

If you want to cite the papers, please refer to the original.

For technical reasons we could not place the page where the original quote is coming from.

Abstract-Summary

"Towards this end, the SPECIAL H2020 project aims to provide a set of tools that can be used by data controllers and processors to automatically check if personal data processing and sharing complies with the obligations set forth in the GDPR."

Introduction

"We have seen the emergence of GDPR compliance tools (cf., [5–8]) in the form of predefined questionnaires that enable data controllers and processors to assess the compliance of services and products that process personal data."

"In order to fill this gap, SPECIAL builds upon a rich history of policy language research from the Semantic Web community (cf., [9–13]), and shows how together machine understandable policies and automated compliance checking can be used to demonstrate compliance with legal requirements set forth in the GDPR."

"We describe two different approaches to automated compliance checking used to demonstrate that: (i) data processing performed by data controllers/processors complies with consent provided by data subjects; and (ii) business processes comply with regulatory obligations set forth in the GDPR."

Requirements Analysis

"When it comes to encoding legislative requirements using machine understandable representations, such that it is possible to perform automated compliance, major considerations include: Connectedness of the various articles, paragraphs, and points, which can either explicitly refer to another piece of legislation (e.g., "scientific or historical research purposes or statistical purposes in accordance with Article 89(1)") or implicitly to knowledge about the law (e.g., "Personal data shall be: (a) processed lawfully, fairly and in a transparent manner in relation to the data subject ('lawfulness, fairness and transparency')")."

"In SPECIAL we do not try to encode the entire GDPR, but rather focus on encoding legislative obligations (relating to several articles, paragraphs, and points) such that: (i) data processing performed by data controllers/processors complies with consent provided by data subjects; and (ii) business processes comply with regulatory obligations set forth in the GDPR."

Background

"OWL provides for: (i) A rich set of relations between classes, roles and individuals (for example, owl:sameAs, owl:equivalentClass, owl:differentFrom); (ii) A number of logical operations (for example, owl:unionOf and owl:intersectionOf); (iii) Several cardinality constraints (for example, owl:someValuesFrom, owl:allValuesFrom, owl:cardinality, owl:minCardinality and owl:maxCardinality)."

"OWL 2 QL is particularly suitable for applications with lightweight ontologies and a large number of individuals, which need to be accessed via relational queries."

"OWL 2 RL provides support for applications with lightweight ontologies and a large number of individuals that make use of rule based inference and constraints mechanisms."

"We make use of owl: intersection of in order to define complex classes (e.g., a simple policy) where policy instances are composed of instances of a given list of classes (e.g., data, processing, purpose, storage, and recipients)."

Consent Compliance Checking

"Both the controller's activities and the consent of data subjects can be described by a set of simple usage policies (covering different data categories and purposes), called full (usage) policies."

"Consent can be represented with two simple policies, specified using SPECIAL's vocabularies: If HeartRate is a subclass of BiometricData and ComputeAvg is a subclass of Analytics, then the above consent allows BeFit to compute the average heart rate of the data subject in order to send her fitness recommendations."

"If the policy describes consent, then its instances represent all the operations permitted by the data subject."

"A description of (part of) the controller's activity—called business policy in SPECIAL (possibly represented as a transparency log entry)—complies with a consent policy if the former is a subclass of the latter, that is, all the possible operations described by the business policies are also permitted by the given consent."

Business Processes Compliance Checking

"And differently from the above examples, the GDPR sets obligations that are not directly related to the controller's business processes, such as the requirement that data subjects have the right to access, rectify, and delete their personal data."

"It plays a dual role, representing: preconditions authorizations specified by the business policy, e.g. if the obligation is something like getValidConsent then the derived authorizations is a rule like the specified software can read the data sources if consent has been given; obligation assertions (under human responsibility) that the data controller has set up processes for fulfilling the indicated obligations—e.g. a process to obtain consent from the data subjects—which is relevant to checking compliance with the GDPR."

"Consider the following business policy: This policy is not a subclass of the formalized GDPR (hence it does not pass the compliance check) because Religion is classified as sensitive data (it is a subclass of SensitiveData_as_per_Art9)."

The Automated Compliance Checking Algorithm

"Several implementation strategies are possible, depending on the controller's system architecture; to fix ideas, the reader may consider the following generic approach: Each of the controller's processes is labeled with a corresponding business policy that describes it, and before processing a piece of data, the business policy is compared with the data subject's consent to check whether the operation is permitted."

"Compliance checking is split into two phases: first, business policies are normalized and closed under the axioms contained in the vocabularies; in the second phase, business policies are compared with consent policies with a structural subsumption algorithm."

"This means that PLR alone can execute about 6000 compliance checks per second and more than 518 million checks per day, that is, 60 of wi-fi probing events and 40 In order to raise performance and match the frequency of events mentioned at the beginning of this section, one can re-engineer PLR using a language more performant than Java, and/or parallelize processing by means of big data architectures."

Related Work

"When it comes to automated compliance checking there are two primary bodies of work, one focusing on legal knowledge representation and reasoning, and the other on policy representation and reasoning."

"Despite superficial similarities, SPECIAL's policy framework and the many works on legal reasoning have different goals."

"Policy validation is less ambitious than legal reasoning; it is only aimed at checking whether the different properties of the policy are mutually coherent (e.g. by checking that the legal basis matches the data category), and whether all relevant parts have been included (such as the appropriate obligations in case of consent-based processing or data transfers outside the EU)."

"This difference has remarkable technical consequences: as we have already pointed out, validation in all contexts (i.e. policy comparison) is intractable in rule-based languages (that are common in the works on legal reasoning, since the seminal paper [14]), and even undecidable if rules are recursive [15]."

Conclusion and Future Work

"The overarching goal of the SPECIAL project is to develop tools and technologies that enable data controllers and processors to comply with personal data processing obligations specified in the GDPR."

"We presented the SPECIAL policy language and discussed how it can be used to encode consent, business policies, and regulatory obligations."

"We described the SPECIAL approaches to GDPR compliance checking and presented the results of our initial performance evaluation."

"Ongoing/future work includes: the optimisation of the existing compliance checking algorithm to cater for automated compliance checking for a broader set of legislative requirements; and the development of an algebra that can be used to combine multiple policies, for instance where there is a need to aggregate data from multiple data sources."

2.2.4 Enterprise Architecture Management as a Solution for Addressing General Data Protection Regulation Requirements in a Big Data Context: A Systematic Mapping Study [85]

This is a machine-generated summary of:

Georgiadis, Georgios; Poels, Geert: Enterprise architecture management as a solution for addressing general data protection regulation requirements in a big data context: a systematic mapping study [85].

Published in: Information Systems and e-Business Management (2021).

Link to original: https://doi.org/10.1007/s10257-020-00500-5.

If you want to cite the papers, please refer to the original.

For technical reasons we could not place the page where the original quote is coming from.

Abstract-Summary

"Given the wide scope of applications, the many types of processing involved, including those for purposes not yet foreseen, and the inherent privacy concerns resulting from collecting and storing personal data, the newly introduced General Data Protection Regulation (GDPR) poses specific challenges for safeguarding the security and protection of big data."

"This raises the question whether Enterprise Architecture Management (EAM), as an approach for ensuring the coherence, strategic alignment and focus on value creation of all organisational resources, offers guidance for addressing those challenges in a holistic manner, and thus provides a fruitful ground for developing an approach for complying to GDPR requirements in a Big Data context."

"This study surveys the state-of-the-art in research on security, privacy, and protection of big data."

"Further, it examines whether previous research has investigated the potential of EAM in addressing those challenges and what the main findings of those studies are."

"Our study indicates that only few researchers have explored the use of EAM practices in relation to data security and protection in a Big Data context."

"Our study does not invalidate the potential of EAM to help addressing GDPR requirements in a Big Data context."

"How EAM practices may contribute to risk management and data governance in environments where big data are being processed, is still a huge research gap, which we intend to address in our future research."

Introduction

"The concept of data protection by design as a default [18], which is key in addressing GDPR requirements, entails in practice an organisation-wide and architectural component whereby data protection is not restricted to information technology (IT) and is embedded in all business processes and their entire development cycle."

"MRQs 3 and 4 relate to our first objective and are intended to clarify how research has approached the idiosyncrasies of data security, protection and privacy in a Big Data context (MRQ3) and which specific challenges to these issues posed by the European regulatory requirements have been identified (MRQ4)."

"While MRQ 5 is instrumental for our further research, MRQ6 covers both objectives and intents to contribute to the field by outlining avenues for future research on the nexus of GDPR and Big Data, based on the understanding of the state-of-the-art that we obtain through our SMR study."

Background

"In the context of ensuring compliance with GDPR rules, EAM can prove highly beneficial for organisations as it allows to have a comprehensive overview of what is required, across the different units, departments and functions of an organization, through legal (e.g. Data Protection Impact Assessment, retention period and Record of processing activities), business (e.g. data types and flows) and technical (e.g. information security and IT applications) means and to deal methodically with all limitations [19]."

"This is particularly promising for GDPR compliance in organisations with BDA applications, as big data are frequently obtained from different sources, collected for different reasons, and processed for different purposes, and the insights gained from their processing are applied across an organisation' structure (e.g., for informing marketing campaigns, for new product/service development, for improved customer relationships, for strategic decision-making, for auditing, etc.)."

Research methodology

"As we wish to known the specific issues regarding security, privacy and protection of big data that have been addressed in the scientific literature, and enquire whether GDPR poses specific challenges regarding these issues, and further find out whether researchers have considered EAM as a possible approach to tackle these challenges, an SMR study is adequate."

"The first search string is based on the concepts of MRQ3 and MRQ4 and relates to our objective of identifying studies that address security, privacy and protection of Big Data or discuss the EU data protection regulations in relation to big data."

"As the number of studies that we can find using this string might be low, we added a third search string that looks for research on Big Data and EAM, but without explicitly requiring these studies to address issues of data security, privacy and protection."

Mapping results

"It consists of the following five dimensions: (1) data security and privacy with a focus on the two European data protection regulatory frameworks, (2) the triad of data governance, data management, and privacy risks, (3) emerging trends with regard to the data security and protection of big data, (4) the fundamental principles behind the processing of personal data in a big data context, and (5) research type."

"The integration of ethical insights into the day-to-day work and technological advancements centrally to Big Fata processing is something that has also been strongly stressed and promoted by the EU's independent data protection authority, EDPS [20]. (4) Principles related to the processing of personal data The GDPR lays down a set of six privacy principles to guide how organisations manage personal data."

"The fact that most of those papers are published from 2018 onwards indicates that the research interests in applying EAM for big data protection and privacy compliance in environments with BDA applications, is quite recent."

Discussion

"It is suggested to expand the scope of DPIA to overcome limitations on existing risk assessment models to cover the impact from social and ethical values in Big Data or BDA project contexts."

"Our future research paper that employs SLR shall therefore aim at studying data governance alongside the adequate management of data protection risks including a representative set of pertinent (DPIA and PIA) methodologies."

"Bearing in mind that Big Data problems are not solely an IT exercise but rather a strategic function, future research should therefore be enriched with papers discussing practical cases with the aim to assess the role of EAM in holistically addressing the adoption of DBA or Big Data while taking into consideration ethical and societal concerns and in the scope of conducting risk management operations with the use of a more improved version of the GDPR DPIA."

Conclusion and outlook

"The vast body of existing literature focuses mainly on solutions for technical problems like the optimisation of existing BDA algorithms or implementation of Big Data frameworks (e.g. Apache Hadoop, NoSQL, etc.) and much less on pressing organisational matters, such as risk management and data governance, which are barely addressed in the context of Big Data and BDA."

"Despite the existence of articles acknowledging the business value of EAM in a Big Data context and the positive effects of employing it during the implementation of GDPR to address several concerns or risks such as the management of the creating of RPA ensuring organisation's compliance with data subjects' rights, there are still needs for additional research to address other data protection concerns like ensuring better protection of data subjects' rights besides the ethical processing of their personal data."

2.2.5 On Enabling GDPR Compliance in Business Processes Through Data-Driven Solutions [86]

This is a machine-generated summary of:

Zaman, Rashid; Hassani, Marwan: On Enabling GDPR Compliance in Business Processes Through Data-Driven Solutions [86].

Published in: SN Computer Science (2020).

Link to original: https://doi.org/10.1007/s42979-020-00215-x.

If you want to cite the papers, please refer to the original.

For technical reasons we could not place the page where the original quote is coming from.

Abstract-Summary

"The collection and long-term retention of excessive data enables organisations to process data for insights in non-primary processes."

"General Data Protection Regulation (GDPR) abstractly defined the data processing boundaries of the personal data of European Union's citizens."

"The processing principles of GDPR, in line with the spirit of privacy by design and default, provide directions on the collection, storage, and processing of personal data."

"We are providing three solutions to enable GDPR compliance in business processes."

"We are proposing intra-process data degradation, a solution for continuous data minimisation during the course of business processes."

"We adapt process mining techniques for ascertaining compliance of business process execution to data subject rights."

"We present a scheme to utilise differential privacy technique to enable GDPR-compliant business process discovery."

Introduction

"As forward compliance, processes may need to be completely redesigned or partially overhauled to inculcate the applicable GDPR provisions, especially the data processing principles and data subject rights."

"Data minimisation is one important pillar of data processing principles and enabler of the Privacy by design and default, a requirement by Article 25 of the GDPR."

"These data elements can be degraded through precision and retained information reduction techniques such that the relevant process can still be successfully completed."

"Process mining techniques [16] provide insights on business processes relying mainly on process execution data."

"The technique has been found effective and privacy-preserving in many domains where sensitive data is processed."

"We are presenting our vision on making process discovery techniques privacy-preserving, in line with the GDPR data security requirements, by utilising differential privacy technique."

"We discuss post-execution GDPR compliance of business processes with respect to data subject rights."

Related Work

"Domain and use case specific data minimisation at the data collection stage has been addressed through different approaches in the works of [17–20]."

"The authors in [21, 22] build upon the domain generalisation graphs of [23] to propose degrading data elements by retaining only the information in the graph levels required for future processes."

"In [24], the authors demonstrate the viability of their approach by deploying data degradation for enhancing privacy in ambient intelligence."

"The approach proposed in [21–25] is time-based recurrent and inter-process, where multiple processes having distinct and mutually exclusive boundaries use the same data."

"Our proposed approach is intra-process and has three different data degradation policies."

"The life-cycle policy model of [25] utilised in [21, 22] is database-centric while our data degradation policies are more process-centric."

"The approach in [21–25] being database-centric, highlights the data degradation implementation related challenges therein."

"Our process-centric approach highlights data degradation implementation related challenges being faced from business processes perspective."

Enabling Compliance to Data Minimisation

"The event-driven data degradation policy caters for the shortcomings of the time-driven data degradation policy. (Event-driven Data Degradation Policy) In this policy, the values of sensitive data elements shall be degraded to the next possible information level as soon as the current information level is no longer required for successful execution/completion of the process, or in other words soon after the last process activity requiring the data at the current information level is executed."

"In the worst case, this type of deviations can result in performance bottlenecks where multiple process instances get synchronised which results with a spike in the data degradation loads. (Event-driven Data Degradation Policy with Margin) In this policy, the values of sensitive data elements shall be degraded to the next possible information level at suitable offsets/margins after the last process activity requiring the data at the current information level is executed."

Post-execution GDPR Compliance

"This section details on the post-execution GDPR compliance with the business process traces from two distinct angles: ascertaining compliance to GDPR's data subject rights, and practicing data security in processing of event data for mining purposes."

"Data subject rights might necessitate changes in the control-flow perspective of business processes."

"The fulfillment of data subject rights without undue delay aspect may require reworking the time perspective of business processes."

"With the event data logged for the modified business process, our GDPR compliance checking techniques can ascertain RTBF compliance in the post-execution scenario, explained in coming section."

"The event data may contain personal data of the data subjects as well as organisational resources related with the business process."

"The process discovery querying agent, residing in the non-trusted environment, can only communicate with the differential privacy engine in the trusted environment by issuing queries in order to acquire required information on the event data."

Challenges and Future Work

"We presented solutions for enabling GDPR compliance with respect to data minimisation, data subject rights in business processes, and data security while mining processes."

"The complexity of the processing environments is considered to be challenging for our intra-process data degradation approach."

"Complexity of real world business processes is another challenge for intra-process data degradation."

"Process models contain parallelism, choices and the most challenging looping constructs."

"We presented a solution for enabling GDPR compliance of business processes with respect to the data subject rights, complemented by process mining discipline."

"The complexity of the processing environments and the structural complexity of real world business processes is equally challenging in this direction."

"For the realisation of data security in mining business processes, we also see a lot of challenges."

"Effective query-synthesis is one of the major challenge in differentially-private process mining."

2.2.6 Protecting Citizens' Personal Data and Privacy: Joint Effort from GDPR EU Cluster Research Projects [87]

This is a machine-generated summary of:

de Carvalho, Renata M.; Del Prete, Camillo; Martin, Yod Samuel; Araujo Rivero, Rosa M.; Önen, Melek; Schiavo, Francesco Paolo; Rumín, Ángel Cuevas; Mouratidis, Haralambos; Yelmo, Juan C.; Koukovini, Maria N.: Protecting Citizens' Personal Data and Privacy: Joint Effort from GDPR EU Cluster Research Projects [87].

Published in: SN Computer Science (2020).

Link to original: https://doi.org/10.1007/s42979-020-00218-8.

If you want to cite the papers, please refer to the original.

For technical reasons we could not place the page where the original quote is coming from.

Abstract-Summary

"This paper provides information on ongoing and recent developments in this area in the European Union (EU) space."

"We start by providing an overview of EU's General Data Protection Regulation (GDPR) and proceed by identifying challenges concerning GDPR implementation, either technical or organizational."

"We consider the work currently being done by a set of EU projects on the H2020 DS-08-2017 topic, namely BPR4GDPR, DEFeND, SMOOTH, PDP4E, PAPAYA and PoSeID-on, which address and aim at providing specific, operational solutions for the identified challenges."

Introduction

"The European Union, first through the Data Protection Directive [26], then through the General Data Protection Regulation [27], has recognized the importance for citizens of the value of data privacy and the necessity for the privacy-enabled management of personal data."

"Our research focuses on the following projects: Business Process Re-engineering and functional toolkit for GDPR compliance (BPR4GDPR) Data Governance for Supporting GDPR (DEFeND) GDPR Compliance Cloud Platform for Micro Enterprises (SMOOTH) Methods and tools for GDPR compliance through Privacy and Data Protection Engineering (PDP4E) PlAtform for PrivAcY preserving data Analytics (PAPAYA) Protection and control of Secured Information by means of a privacy enhanced Dashboard (PoSeID-on) These projects aim at finding solutions for "both technological as well as organizational challenges for organizations which have to implement novelties such as the right to data portability, the right to be forgotten, data protection impact assessments and the various implementations of the principle of accountability" [28]."

General Data Protection Regulation (GDPR)

"The assigned control of personal data to individuals in the EU with the addition of new rights for EU data subjects is impacting the manner in which organizations are dealing with personal information."

"GDPR has altered the ways for collecting and managing personal information, including the definition of new roles in organizations handling with data subjects, such as: Data controller "is a natural or legal person, public authority, or body that alone or jointly with others determines the purposed and means of the processing of personal data."

"Data protection officer is the role designated by controller and processor whose responsibility is overseeing an organization's data protection strategy and implementation, for ensuring that it is complying with the GDPR's requirements."

"Information technology plays a key role in data governance, systems strategies and management, to accomplish personal data requirements, enhancing information security and developing breach-awareness capabilities aligned with those of the organization."

Challenges Concerning GDPR

"The challenges, either technical or organizational, include, among others: interpretation of GDPR requirements; operational adaptation towards privacy-aware and compliant business practices; holistic data views and processing actions inventory; enforcement of security means; management of the relations with third parties and the data subjects, and enforcement of rights thereof; last but not least, significant resources are required and, whereas big companies may have money and resources to invest, both human and monetary, this does not necessarily apply to SMEs."

"It is therefore important, as indicated by the EU call "Secure societies—Protecting freedom and security of Europe and its citizens" (topic DS-08-2017: Cybersecurity PPP: Privacy, Data Protection and Digital Identities), "to develop tools and methods to assist organizations to implement GDPR..."."

"The goal of the PoSeID-on Project ("Protection and control of Secured Information by means of a privacy enhanced Dashboard") is to develop a transparent ecosystem for personal data protection, in line with GDPR with respect to digital security."

Solutions Proposed by Each Research Project

"The main aim of the DEFeND project is to deliver an innovative data privacy governance platform, which will facilitate scoping and processing of data and data breach management and will support organizations towards GDPR compliance."

"The results of the different tools are also related: the risk management and the requirements engineering tools provide complementary views (risk-oriented and goal-driven, respectively) of the attributes that shall be met and validated by the design; the assurance tool captures artefacts produced by other tools as evidences for compliance, etc. The main objective of PAPAYA is to design and develop a platform of privacy preserving analytics modules that allows the outsourcing of analytics operations into untrusted cloud servers while protecting the privacy of the data."

Conclusion

"Although not final, the solutions proposed and discussed in this paper will already provide means for: (1) allowing citizens to track their personal data, manage permissions and view the risk of data exposure; (2) dealing with encryption of data that enables analytics operations to be performed without exposure of the data; (3) providing methods to assist organizations in implementing security and legal aspects; (4) enabling privacy and data protection to be considered in the development process from different perspectives; (5) caring about the impacts (specially financially) on SMEs and MEnts in adopting GDPR; and (6) supporting SMEs to become GDPR-compliant by automatically re-engineering their processes to enforce GDPR."

"Despite the work being carried out, several other aspects of GDPR remain open and were not discussed in this paper, such as purpose limitation, data minimization and storage limitation."

"As these are also important to protect citizens' privacy, they will be addressed by future work/projects."

2.2.7 Modeling Data Protection and Privacy: Application and Experience with GDPR [88]

This is a machine-generated summary of:

Torre, Damiano; Alferez, Mauricio; Soltana, Ghanem; Sabetzadeh, Mehrdad; Briand, Lionel: Modeling data protection and privacy: application and experience with GDPR [88].

Published in: Software and Systems Modeling (2021).

Link to original: https://doi.org/10.1007/s10270-021-00935-5.

Copyright of the summarized publication:

If you want to cite the papers, please refer to the original.

For technical reasons we could not place the page where the original quote is coming from.

Abstract-Summary

"In Europe and indeed worldwide, the General Data Protection Regulation (GDPR) provides protection to individuals regarding their personal data in the face of new technological developments."

"Although the GDPR is highly beneficial to individuals, it presents significant challenges for organizations monitoring or storing personal information."

"We present a complete GDPR UML model as a first step toward designing automated methods for checking GDPR compliance."

"We provide (1) the GDPR conceptual model we developed with complete traceability from its classes to the GDPR, (2) a glossary to help understand the model, (3) the plain-English description of 35 compliance rules derived from GDPR along with their encoding in OCL and (4) the set of 20 variations points derived from GDPR to specialize the generic model."

Introduction

"In another recent work [29], starting from the work we have done in this paper, (i.e., considering the identification of the interdependencies between legal basis and data subject rights), we developed an automated AI-based method for checking whether a given privacy policy complies with the provisions of GDPR."

"The level of abstraction of such a representation has to be suitable for ensuring a consistent implementation and interpretation of the regulation, national laws and case law In this paper, we draw on model-driven engineering (MDE) [30] for building a machine-analyzable representation of the GDPR as a first step toward the development of future automated methods for assessing GDPR compliance."

"In order to address this gap, in a previous conference paper [31], we tackled the following three research questions (RQ): RQ1: How can we develop a generic and adaptable model-based representation of the GDPR to support automated compliance checking?"

GDPR overview

"If a business is subject to the GDPR, it has to identify itself as either a data controller or data processor."

"The responsibilities of a given business under the GDPR vary depending on whether it is a processor or a controller and depending on the kind of data processed."

"Processors notably have to: (i) implement adequate technical and organizational measures to keep personal data safe and secure, and, in cases of data breaches, notify the controllers; (ii) appoint a statutory data protection officer and conduct a formal impact assessment for certain types of high-risk processing; (iii) keep records about their data processing; and (iv) comply to the GDPR restrictions when transferring personal data outside the EU."

Toward a model-based approach for automated GDPR compliance checking

"We develop, through a qualitative study, the following: (i) a generic model of the GDPR's main concepts and relationships, (ii) generic OCL constraints that verify GDPR compliance, (iii) a glossary to facilitate the understanding of the GDPR and (iv) the variation points describing specific situations where the generic representation needs to be adapted to a given domain or organizational context."

"This is an intermediate step to create representations of legal and technical documents in the form of an instance of the specialized model."

"Stated otherwise, step 3 generates a model instance providing a structured representation of the legal and technical documents that have a bearing on GDPR compliance."

"In step 4, the model instance generated from step 3 is checked against the specialized OCL constraints obtained from step 2."

Modeling the GDPR

"This modeling activity addresses RQ1 and yields: (1) a UML class model (CM) that captures the GDPR's key concepts and their relationships (see Sect. 1 of Appendix A); (2) a set of 35 OCL constraints over the CM reflecting the GDPR's obligations (see Sect. 3 of Appendix A for the plain-English version and Appendix B for their OCL version)."

"Each session started with the authors in attendance presenting to the experts, using slides or printed documentation, the new parts of the GDPR that had been modeled in the form of diagrams and/or natural language descriptions of OCL constraints."

"Additional effort, including by legal experts, will be required for defining OCL constraints over the specialized model, noting that these constraints necessarily refer to legal material (e.g., EMS laws, GDPR case law and domain adaptations) that is more complex and fragmented than the GDPR."

Lessons learned

"We observed that modeling the GDPR, whether at a generic or specialized level, necessitates substantial legal knowledge and expertise that may go beyond the GDPR itself, e.g., knowledge of the Article 29 Working Party."

"OCL, which we use to formally express the GDPR rules, was challenging and intimidating to legal experts, despite our attempts to explain the meaning of the constraints."

"Both previous tables facilitated the validation of the OCL constraints by legal experts."

"Another observation from our GDPR modeling experience is that both analysts and legal experts often needed to consult specific articles to refresh their memory."

"Legal experts appreciated the intuitive and GDPR-traceable approach of our model artifacts that allows them to maintain the integrity of the artifacts by reducing arduous tasks such as going back to read a specific article of the GDPR related to a class (without traceability)."

Future directions

"The DSRL should be, on the one hand, expressive enough to be useful for the precise specification of GDPR compliance checking rules and, on the other hand, understandable enough to be readily used by legal experts."

"To support automated compliance checking, the rules specified in the DSRL should be automatically translatable into OCL so that the rules can be checked directly over instantiations of a specialized GDPR model."

"A main criterion to fulfill regarding goal decomposition would be to ensure that the decomposition process makes progress toward a set of concrete claims for which meaningful evidence about satisfaction (in term of model specialization) could be collected."

"The metadata identified will be the basis for the model-based representation of the legal and technical documents to be checked."

"An automatic instantiation process will convert the metadata extracted with NLP and ML for a given document into a model-based representation, i.e., the instance of a specialized model."

Limitations and threats to validity

"Consider the following details that were included in another recent work [29] where we automated the activity of checking GDPR compliance of privacy policies: (a) the concept of adequacy decisions between the EU and a territory (e.g., Andorra, the Bailiwick of Jersey, etc.), specific sectors (e.g., the commercial organizations from Canada, Argentina, etc.) and a country (e.g., Japan and New Zealand), (b) the legal basis contract and more specifically whether the provision of personal data for this legal basis is a statutory requirement, a contractual requirement or a requirement necessary to enter into a contract, (c) the appropriate safeguards, i.e., binding corporate rules or EU model clauses that would allow the controller to share the collected personal data to recipients outside Europe, and (d) the specific derogation in terms of unambiguous consent to allow the controller to share the collected personal data outside Europe."

Related work

"We distinguish two categories of studies related to the work presented in this paper: proposals that report on (1) Modeling the GDPR and (2) Checking Compliance."

"The third column "GDPR Coverage" shows the degree of coverage of the GDPR in each paper: (a) Complete, when the authors discuss a solution that involve the totality of the GDPR, (b) Partial, when the authors deal with a specific issue addressed by the GDPR, i.e., checking privacy policy compliance, and (c) Not Applicable (NA), when the GDPR is not involved in the paper."

"Burmeister and others [32] present an approach based on enterprise architecture (EA) models to help with checking GDPR compliance."

"The authors describe an approach for using their proposed model as a tool to develop an organizational privacy policy along with an illustration of compliance-rule extraction."

Conclusion

"The key motivation behind this research is to pave the way for the creation of automated, model-based GDPR compliance analysis solutions."

"Our research resulted in the development of a generic GDPR model alongside a detailed and well-defined strategy for specializing this model according to different contexts and for satisfying the requirements of different types of GDPR-related analysis."

"We presented several artifacts: (1) a GDPR conceptual model with full traceability to the GDPR, (2) a glossary to help explain the conceptual model, (3) 35 compliance rules extracted from GDPR and defined both in plain English and in OCL and (4) a set of 20 GDPR variation points to specialize the generic model."

"We also suggested potential strategies for solving the challenges we found in our work and promoting longer-term research focus on the model-based analysis of GDPR compliance."

2.2.8 Data Governance Through a Multi-DLT Architecture in View of the GDPR [89]

This is a machine-generated summary of:

Zichichi, Mirko; Ferretti, Stefano; D'Angelo, Gabriele; Rodríguez-Doncel, Víctor: Data governance through a multi-DLT architecture in view of the GDPR [89].

Published in: Cluster Computing (2022).

Link to original: https://doi.org/10.1007/s10586-022-03691-3.

If you want to cite the papers, please refer to the original.

For technical reasons we could not place the page where the original quote is coming from.

Abstract-Summary

"The centralization of control over the processing of personal data threatens the privacy of individuals due to the lack of transparency and the obstruction of easy access to their data."

"Having direct control over the flow of their personal data would not only favor their privacy but also a "data altruism", as supported by the new European proposal for a Data Governance Act."

"After an in-depth analysis of the tensions between the GDPR and DLTs, we propose the following components: (1) a personal data storage based on a (possibly decentralized) file storage (DFS) to guarantee data sovereignty to individuals, confidentiality and data portability; (2) a DLT-based authorization system to control access to data through two distributed mechanisms, i.e. secret sharing (SS) and threshold proxy re-encryption (TPRE); (3) an audit system based on a second DLT."

Introduction

"The control, direct or indirect, that individuals currently exercise over their personal data is conditioned by the centralized platform-based personal information management techniques, which are then concentrated in a few internet service providers (ISPs) for the purpose of exploring, filtering and obtaining data of interest [33]."

"Dedicated technologies can help either companies to comply with GDPR (and similar) and individuals to exercise their rights, with particular regard to address two main issues: the lack of transparency in the management of personal information and the inability to access and make interoperable personal data."

"We stem from this vision and propose a decentralized approach, based on distributed ledger technologies (DLTs) and decentralized file storage (DFS), to manage data sharing and access."

"The main benefit of such a decentralized architecture is that, by bringing together regulations and technologies, it provides individuals with the ability to record their data in some interoperable personal data spaces (PDS) [34], guarantees data sovereignty and enables users to control what personal data they want to share [35–38]."

Background and related work

"In the broader scope of DLT-based data sharing and access control systems, we may find a subset of personal data management solutions that: (1) require an additional effort to comply with regulations such as GDPR; (2) include architectural components designed following the logic of "Privacy by Design" [39]."

"The main purpose of a DLT-based system, again, is to provide transparency in the process of accessing personal data, but simultaneously enable users to control their own data [34, 38, 40, 41]."

"Truong and others [42] provide a DLT-based GDPR-compliant personal data management solution where consent is handled through a token and data is stored and served through a Database Management System."

"Sovrin [43] is an open source identity network built on a layered architecture involving a public permissioned DLT that only stores identity transactions, without personal data, and where only trusted institutions, i.e. banks, universities, governments, can be DLT nodes."

Multi-DLT architecture for a personal information management system

"We provide an overview of the architecture of a multi-DLT-based PIMS, with regards to personal data storage and sharing."

"It may be difficult to identify which participants in a DLT are the data controllers, processors or subjects, because often the information does not necessarily flow directly from one point to the other, i.e., users to providers [43]."

"Even if the encrypted personal data are only stored in a (D)FS, this actor is a data controller because such data can be considered as pseudonymous (and not completely anonymous), and thus, in principle, subject to the GDPR [44, 45]."

"All the servers within an authorization system act as joint controllers in a shared responsibility approach [46, 47] and the legal basis for the processing of personal data is guaranteed by mutual agreements."

Architecture components design

"In our model, authorization servers (i.e. data controllers) are protected from this threat because they maintain the ACL and its history thanks to the (semi-)private DLT, and thus can demonstrate the data owner's malicious behavior. (v) The same holds if the data owner modifies the smart contract data with the aim of tampering the ledger, i.e. the history of the modification of the smart contract is kept by all the servers through the DLT. (vi) Finally, the data owner can try to repudiate some of the actions performed, however all the owner interactions with the systems involve the use of a public key or an address in the form of a digital signature, thus not being repudiated. (i) The data consumer can attempt to illegitimately access to personal data and thus threat the subject privacy."

Performance evaluation

"Since we are interested in performance concerning messages uploading from the user's point of view, and since users' devices generally send requests to only one node, we focused on the DFS node response latency only, instead of testing the response of the whole network."

"To assess the time needed to distribute data and encrypt/decrypt them, we specifically focused on two aspects: DFS service providers response we conducted our performance evaluation of the three different types of DFS nodes with the purpose of conducting a stress test."

"Even in these tests, we can see the effect of SS keys generation and distribution, since a larger number of nodes means an increase in the time needed to reach all of them."

"An adequate deployment of DFS nodes (e.g. maintained by the same nodes as the authorization system) based on the number of users requesting access to the service can properly support a PIMS."

Conclusions

"The access to the data is controlled through two distributed mechanisms, i.e. secret sharing (SS) and threshold proxy re-encryption (TPRE); an audit DLT, that consists of a permissionless DLT that provides proof of a correct audit for the authorization system."

"Results from our performance evaluation show that: (i) up to a certain overload, a proprietary service, where a dedicated node is in charge of running an IPFS node, appears to provide stronger assurances for responsiveness and reliability; (ii) TPRE is faster when increasing the size of data to encrypt/decrypt and it is also more scalable, as better behaves as the number of nodes and the threshold value increases while executing the protocol; (iii) however, TPRE has the drawback of requiring the generation of re-encryption keys for each new data consumer, while SS does not. (iv) We also analyzed the transaction throughput of a private Ethereum blockchain using Proof-of-Authority, confronting smart contracts operations in terms of gas."

2.2.9 Modeling Ecosystems of Reference Frameworks for Assurance: A Case on Privacy Impact Assessment Regulation and Guidelines [90]

This is a machine-generated summary of:

Ruiz, Alejandra; Martin, Yod-Samuel; Martinez, Jabier; Quintans, Jacobo; Mockly, Guillaume; Gyrard, Amelie; Crepax, Tommaso: Modeling ecosystems of reference frameworks for assurance: a case on privacy impact assessment regulation and guidelines [90].

Published in: Software and Systems Modeling (2022).

Link to original: https://doi.org/10.1007/s10270-022-01061-6.

If you want to cite the papers, please refer to the original.

For technical reasons we could not place the page where the original quote is coming from.

Abstract-Summary

"To assure certain critical quality properties (e.g., safety, security, or privacy), supervisory authorities and industrial associations provide reference frameworks such as standards or guidelines that in some cases are enforced (e.g., regulations)."

"As several frameworks might apply for same systems, certain overlaps appear (e.g., regulations for different countries where the system will operate, or generic standards in conjunction with more concrete standards for a given industrial sector or system type)."

"We propose the use of modelling for alleviating the complexity of these reference frameworks ecosystems, and we provide a tool-supported method to create them for the benefit of different stakeholders."

"The European GDPR regulates the movement and processing of personal data, and, contrary to available software engineering privacy guidelines, articles in legal texts are usually difficult to translate to the underlying processes, artefacts and roles that they refer to."

"We modelled the legal requirements of the GDPR on data protection impact assessments, and then, we selected the ISO/IEC 29134, a mainstream engineering guideline for privacy impact assessment, and, taking a concrete sector as example, the EU Smart Grid Data Protection Impact Assessment template."

"We provide a qualitative evaluation from legal experts and privacy engineering practitioners to report on the benefits and limitations of this approach."

Introduction

"We claim that reference framework models can make legal texts like the GDPR more "operational" from an engineering perspective, by operationalizing the legal text into engineering process activities and bridging them to those specified by technical standards."

"To support our claim, we take as case study the GDPR Article 35 on DPIA, and confront it with ISO/IEC 29134, the EU Smart Grid DPIA template, and the state-of-the-art CACM metamodel (Common Assurance and Certification Metamodel) [48] for modeling reference frameworks, we aim to answer the following research questions: RQ1: To what extent can information privacy law and privacy protection standards for conducting DPIAs be modelled from a development process perspective?"

"RQ2: To what extent do the reference framework models of GDPR DPIA match to the ISO/IEC 29134, and to the EU Smart Grid DPIA template?"

Background

"Eclipse OpenCert [49] tool can be used to support the assurance activities allowing the modelling of reference frameworks, equivalence mapping models, and many other functionalities for assurance project management that are out of the scope of this paper such as the creation of assurance cases and evidence models, each of them with their own metamodel besides CACM."

"Apart from reference framework models, mappings between one another aim to reduce assurance efforts thanks to the identification of overlapping assurance requirements and activities derived from the compliance targets (law/standard/policy in the selected ecosystem)."

Related work

"The representation of GDPR as a workflow proposed by PrOnto is clearly aligned with the result of our approach, aiming to represent the key concepts and legal axioms of GDPR in a model that can later be used for compliance checking and risk analysis."

"The more significant difference between this and our approach is given by the fact that the described UML class model is specifically tailored for the GDPR, whereas the model proposed in this paper is open and can be used to develop new reference frameworks and new mappings, derived from different pieces of legal text."

"With this approach, the GDPR is modeled as a set of rules, and said model does not result, again, in a workflow that can be followed to provide assurance."

Method

"A team of two researchers, experts on privacy (including the GDPR and ISO/IEC 29134), was the main responsible for creating the models."

"For the equivalence map modelling, the reference frameworks of ISO/IEC 29134 and Smart Grid DPIA template were systematically analysed (iterating through all its elements) and manually identifying the full and partial mappings to the GDPR DPIA model."

"We used this approach to capture the elements of the process model underlying the execution of a DPIA, from its description in the GDPR."

"The relation between the GDPR clauses and the process stages is not linear, some elements are not described explicitly, but implied by the text, etc. The different GDPR DPIA paragraphs were modelled and then consolidated in a complete model."

"Reference framework modelling starts by annotating the text of the regulation to be modelled with elements from reference framework meta-model, and then, they were manually interpreted, with expert judgement, to create the reference framework model."

Modelling results

"Thanks to the creation of the GDPR DPIA and the ISO/IEC 29134 reference frameworks using the CACM metamodel, we were able to tackle the creation of the equivalence mapping model."

"In our GDPR model, from the data controller's perspective, three groups of activities can be detected: those that are carried out before the DPIA proper (decide whether DPIA is needed, consult with data subjects, experts, etc.), the production of the impact analysis itself (including risk elicitation, analysis, control measures, etc.), and the activities that use such impact analysis as an input (publication, review, consultation with supervisory authorities if needed, etc.)."

"In this case we do not refer to the GDPR clause itself but to the elements from our reference framework (as the correspondence is not so direct), and (c) the type of mapping: Whether the mapping is full or partial and discusses what is missing from the source (ISO/IEC 29134) in the target (GDPR)."

Evaluation and Discussion

"To evaluate these aspects, the following questions will be considered: (1) whether the CACM metamodel can be easily translated to an engineering process and (2) whether the mapping proposed between ISO/IEC 29134 and GDPR DPIA is representative of both documents, more specifically in terms of coverage and ordering of activities."

"One relevant aspect that was considered by both experts (engineering and legal) is that this approach of modelling the privacy-related ecosystem (reference frameworks and equivalence maps) is promising in improving the communication among both points of view."

"The validity of our results is inherently limited by (1) the choice of a specific metamodel (CACM) to define reference frameworks, and (2) its application to a single information privacy law and two guidelines (GDPR, ISO/IEC 29134 and EU Smart Grid DPIA template)."

Conclusion and future work

"Application to further regulations Another potential avenue for future work consists in extending the reference models herein presented with those derived from regulations recently issued by other jurisdictions (Nevada Privacy Law, California Consumer Privacy Act, Brazil's General Data Protection Law, etc.), each having their own similarities and differences with regard to GDPR; and so might we consider other EU laws which deal with information privacy (e.g., national level privacy regulations, ePrivacy Directive, NIS Directive)."

"The GDPR itself allows for other accompanying legal instruments that can be modelled as extensions to our reference framework, ranging from opening clauses which allow for national derogations in some specific aspects and adequacy decisions that recognize laws from third countries, to quasi-regulations that supplement GDPR with hermeneutics helping in its interpretation (rulings from the Court of Justice of the European Union and guidance from the European Data Protection Board), to co-regulations that refine further stipulations e.g. addressing concrete techniques (codes of conduct, codes of practice, voluntary certification schemes) and self-regulations regarding international transfers (binding corporate policies and standard contractual clauses)."

2.2.10 Against Explainability Requirements for Ethical Artificial Intelligence in Health Care [91]

This is a machine-generated summary of:

Kawamleh, Suzanne: Against explainability requirements for ethical artificial intelligence in health care [91].

Published in: AI and Ethics (2022).

Link to original: https://doi.org/10.1007/s43681-022-00212-1.

If you want to cite the papers, please refer to the original.

For technical reasons we could not place the page where the original quote is coming from.

Abstract-Summary

"I challenge this Explainability Imperative (EI) by considering the following question: does the use of epistemically opaque medical AI systems violate existing legal standards for informed consent?"

"If not, then based on at least one metric of ethical medical practice (informed consent), explainability is not required for the ethical use of AI in healthcare."

"I show that the use of epistemically opaque AI applications is compatible with meeting accepted legal criteria for informed consent."

"I conclude that the use of black-box AI systems does not violate patients' rights to informed consent, and thus, with respect to informed consent, explainability is not required for medical AI."

Introduction

"The use of epistemically opaque or "black box" AI systems in healthcare is taken to be unethical because it violates foundational principles of medical ethics, such as transparency, and the legal requirements for informed consent."

"My argument against the Explainability Imperative for AI systems is based on the U.S. legal requirements for informed consent for human agents."

"My aim is to define ethical standards that are informative and useful to practitioners and patients, I will be arguing against the ethical demand for explainability through the lens of existing legal requirements in the U.S. I take the legal requirements for informed consent as defining a minimal criteria for ethical healthcare and providing practical guidance in health care contexts."

"If the use of epistemically opaque AI systems prevented one from meeting legal requirements for informed consent, then it would clearly be unethical to use such systems in clinical settings."

Informed consent

"As an instrument, informed consent is intended to enable patients to conduct their own evaluation of the risks and benefits associated with a medical intervention and to make informed decisions about their own bodies, health, and future."

"While informed consent involves other legal requirements (e.g., having the capacity to make decisions and being free to do so from other's manipulation and control), I will be focusing on the information requirement for informed consent: the patient should be adequately informed of the risks and benefits of a given medical intervention."

"Legally, a breach of duty to secure informed consent requires a demonstration of: "(1) failure to disclose a specific risk in violation of the governing standard; (2) materialization of that risk; (3) "causation"—that is, if the risk had been disclosed, the patient, or a prudent person in the patient's position, would not have proceeded as she did; and (4) that no exception, like emergency, excuses the failure to disclose" Wilson [50] In short, the information required concerns the risks and benefits of: treatment recommended, alternative treatments possible, and no treatment."

Explainable AI for informed consent

"Grote and Berens [51] claim that epistemic opacity effectively reduces the evaluation of medical AI systems by medical experts to a guessing game: "in the clinical setting, if an algorithm provides the clinician with a prognostic/diagnostic decision, the rationale of that very decision remains elusive…without knowing why a decision has been made, its evaluation turns into a guessing game for the clinician." (Grote 2021, 5) But exactly what sort of explanation or information is ethically required but unavailable for epistemically opaque AI systems?"

"If physicians do not fully understand (yet) how to explain an AI system's predictions or errors, how could this knowledge deficit impact the quality of an informed consent process and medical care more generally?" (Schiff and Borenstein [52], 140)."

"Wadden [53] argues that the ethical use of AI in healthcare, especially with respect to informed consent rights, requires both explainability and understandability."

"For, Wadden, AI in healthcare must be explainable and understandable to both medical experts and patients in order to uphold informed consent standards in healthcare (Wadden [53], 4)."

[Section 4]

"A concrete way to explore this question is to compare the type of information that the epistemically opaque AI system can provide to justify its diagnosis with (1) the type of information that a human expert provides in the assessment portion of the medical documentation record and with (2) the type of information legally required for informed consent."

"In the assessment portion of their note, they justify their diagnosis in two ways: (1) they refer to the presence of macular drusen in the medical images provided and (2) the absence of indicators for alternative eye diseases like macular edema, hemorrhage, etc. This assessment is a form of post-hoc reasoning used to support the treatment plan by highlighting the salient evidence for the diagnosis in the medical image—the presence of drusen in a retinal image—and ruling out alternative diagnoses."

Reconsidering the EI

"While a patient needs to know the risks and benefits of an eye surgery or treatment (like retinal injections) which a medical expert recommends based on a diagnosis made using an AI system, the patient does not need to know the risks and benefits pertaining to the agent, human or machine, making the diagnosis."

"Just as there are doubts that post-hoc explanations or explainable models of epistemically opaque AI systems may not be factive, but merely approximate, similar doubts arise with respect to the reasons experts give for their own decisions."

"A patient or peer expert bases their second-person knowledge of human diagnostic reasoning on a medical expert's first-person knowledge or self-report of how they reason from a clinical indicator to a diagnosis."

"It is the presumed reliability of human experts' self-report of reasons for a diagnosis that grounds most demands to explainable AI."

Concluding remarks

"The concern is that epistemically opaque AI systems violate patient's right to certain types of information needed for informed consent."

"There are two key assumptions at play in this argument: (1) there is a distinct type of epistemic opacity associated with AI systems that calls for strict(er) explainability requirements and, conversely, (2) patients ordinarily need and do have access to detailed explanations of diagnoses for informed consent."

"Human medical experts and their diagnostic reasoning is just as epistemically opaque as medical AI systems."

"The explanations provided by epistemically opaque AI systems are on par with human explanations of medical diagnoses."

"If human experts' own clinical reasoning was readily transparent to themselves and easily and frequently communicated to patients, then explainable AI would be a straightforward research project: simply unpack the explanations which human experts provide to patients and implement them in an algorithm."

2.2.11 *Using Artificial Intelligence to Support Compliance with the General Data Protection Regulation [92]*

This is a machine-generated summary of:

Kingston, John: Using artificial intelligence to support compliance with the general data protection regulation [92].

Published in: Artificial Intelligence and Law (2017).

Link to original: https://doi.org/10.1007/s10506-017-9206-9.

If you want to cite the papers, please refer to the original.

For technical reasons we could not place the page where the original quote is coming from.

Abstract-Summary

"The General Data Protection Regulation (GDPR) is a European Union regulation that will replace the existing Data Protection Directive on 25 May 2018."

"A last-minute rush to become compliant is therefore expected, and numerous companies are starting to offer advice, checklists and consultancy on how to comply with GDPR."

"In such an environment, artificial intelligence technologies ought to be able to assist by providing best advice; asking all and only the relevant questions; monitoring activities; and carrying out assessments."

"The paper considers four areas of GDPR compliance where rule based technologies and/or machine learning techniques may be relevant: Following compliance checklists and codes of conduct; Supporting risk assessments; Complying with the new regulations regarding technologies that perform automatic profiling; Complying with the new regulations concerning recognising and reporting breaches of security."

"The requirements that GDPR (or organisations that need to comply with GDPR) state for explanation and justification of reasoning imply that rule-based approaches are likely to be more helpful than machine learning approaches."

Introduction

"The General Data Protection Regulation (GDPR) is a European Union regulation that will replace the existing Data Protection Directive 95/46/EC on 25 May 2018."

"The GDPR is directly applicable in each Member State and will also apply to organisations outside the European Union that process data about EU citizens."

"The GDPR does relax a couple of requirements that were in the previous Data Protection Directive: Controllers will no longer be required to register their processing activities with a Data Protection Authority (DPA) in each member state."

"GDPR will become law some time before the earliest date when the UK might leave the European Union; some of its provisions apply to data processors in non-EU states; and the UK Information Commissioner's Office has stated that it will introduce a law similar to GDPR post-Brexit."

Artificial intelligence technologies: rule-based and machine learning

"Many of them can be loosely classified into two categories: rule-based technologies and machine learning technologies."

"The AI system can easily give explanations of its reasoning if required, by describing the rules that were triggered; the information that triggered them; and their conclusions."

"The X-MATE system [60] applies rule-based technology to mortgage application assessment."

"Individual rules identify combinations of information that increase the risk of mortgage default (or rather, increase one of three risks: that the applicant is poor at budgeting; that the applicant's disposable income may drop significantly due to changing circumstances; or deliberate fraud is being attempted)."

"More recent examples of the commercial application of rule based technologies include identifying procurement fraud [61] and monitoring large numbers of gas turbines, steam turbines, and generators [62]."

"An example of machine learning is an analysis conducted of decisions made by the European Court of Human Rights which attempted to predict the outcome of cases by applying machine learning techniques to the textual content and the components of cases [63]."

Preparing for GDPR: an AI-supported checklist

"The checklist says: Organisations should implement a training programme covering data protection generally and the areas that are specifically relevant to their organisations, and implement a policy for determining when training should take place and when refresher training should be carried out and a process for recording when training has been completed The relevant section of the GDPR is Article 39: Data Protection Officers are under a specific obligation to implement appropriate training."

"An AI program that supported decisions about complying with the training clause would start by asking questions to determine whether organisations needed to comply with the requirement to designate a DPO."

"c. Warrant: The organisation is in breach of the GDPR if their DPO does not offer such training."

"b. The case cited is an exception c. Conclusion: Therefore the organisation is exempted from the Article 39 requirement for the DPO to offer training."

"The Warrant from the Legal Claim asserts that the organisation is in breach of the GDPR if their DPO does not offer suitable training."

Risk assessment

"Article 23, which deals with the requirement to implement privacy protection at the design stage, states that the chosen privacy measures should reflect the risk and context of the controller's processing activities, as well as the available technology and cost of implementation."

"They must either select appropriate measures to mitigate those risks, or they must identify a suitable code of conduct/certification and adapt their processing activities to comply with it (Articles 40 and 42)."

"The linking between levels of risk and appropriate mitigation measures should reflect "the available technology and cost of implementation" as well as the processing activities."

"If precedents are available stating that a particular mitigation measure is suitable for a particular level of risk, and the company is happy to expend the money and time required to implement that measure, then that would be a good legal justification for applying that measure."

Giving adequate replies to requests for information about automatic profiling

"If their personal data is being used for "automatic decision making including profiling" then they are also entitled to "meaningful information about the logic involved, as well as the significance and the envisaged consequences of such processing for the data subject."

"Article 22 spells out the purpose of these requirements: that "the data subject shall have the right not to be subject to a decision based solely on automated processing, including profiling, which produces legal effects concerning him or her or similarly significantly affects him or her."

"Does GDPR require that data subjects are given a full explanation of how their profile score was calculated in order to take decisions about them?"

"Rule based AI systems can meet this requirement far more easily by providing an explanation of the rules that have been triggered and the consequences of each for the data subject's profiling score."

Identifying and assessing a potential or actual data breach

"Another area where AI technology might be of use to organisations during their data processing activities is in the identification and assessment of potential or actual breaches."

"Since the exception clause in Article 33 focusses on the likely effect of the breach on data subjects, rather than on the company's own policies, the AI system will need minimal customisation to a company's particular circumstances; so it is likely that an AI software firm might develop a package that multiple organisations can purchase and use."

"Since GDPR appears to be largely based on strict liability, and there seems no reason to consider an AI risk assessor as an 'innocent agent' that has been incited to illegal activity unknowingly (i.e. used in a way that it was not designed for), then there is a significant likelihood that an AI system that makes an incorrect recommendation would indeed be found liable for the breach of GDPR."

Conclusion

"The GDPR raises significant business risks for organisations that carry out data processing."

"The GDPR regularly requires explanations of reasoning, or data processing organisations may require explanations of an AI system's reasoning to convince them that the legal arguments it is making should stand up to scrutiny."

"In the one aspect of GDPR that actually requires explanations of logic to be given to data subjects who request it, it may be undesirable to give full explanations for reasons of business confidentiality."

"Some of the decisions that the GDPR requires are one-time decisions for a data processing organisation, which means that the only organisations likely to make a business case for an AI system to support those decisions are consultancies who advise multiple organisations on GDPR compliance."

2.2.12 An AI Framework to Support Decisions on GDPR Compliance [93]

This is a machine-generated summary of:

Lorè, Filippo; Basile, Pierpaolo; Appice, Annalisa; de Gemmis, Marco; Malerba, Donato; Semeraro, Giovanni: An AI framework to support decisions on GDPR compliance [93].

Published in: Journal of Intelligent Information Systems (2023).

Link to original: https://doi.org/10.1007/s10844-023-00782-4.

If you want to cite the papers, please refer to the original.

For technical reasons we could not place the page where the original quote is coming from.

Abstract-Summary

"The Italian Public Administration (PA) relies on costly manual analyses to ensure the GDPR compliance of public documents and secure personal data."

"The main aim of this work is to design a framework that can be effectively adopted to check whether PA documents written in Italian meet the GDPR requirements."

"The main outcome of our interdisciplinary research is INTREPID (artI ficial iNT elligence for gdpR compliancE of P ublic admI nistration D ocuments), an AI-based framework that can help the Italian PA to ensure GDPR compliance of public documents."

"We set the foundations for a text classification methodology to recognise the public documents published by the Italian PA, which perform data breaches."

"We show the effectiveness of the framework over a text corpus of public documents that were published online by the Italian PA."

Introduction

"This paper takes up the challenge and investigates how AI techniques can actually help to automate (or semi-automate) the data protection workflows of the PA, in order to reduce risks of breaches of security in public documents."

"We focus on the problem of detecting breaches of security related to unlawful disclosure of health information in public documents, according to the GDPR standards, which is one of the main causes of administrative fines."

"The main outcome of our investigation is an AI framework—INTREPID (artI ficial iNT elligence for gdpR compliancE of P ublic admI nistration D ocuments)—that is designed to process the public documents produced by the Italian PA and classify them as compliant or non-compliant with the GDPR standards."

"The corpus contains both 45 GDPR-compliant documents that correctly protect personal data and 45 non GDPR-compliant documents that unlawfully disclose health information."

Related work

"Inspired by these studies, AI techniques are explored for the GDPR compliance in the automation of the legal evaluation of privacy policies, which are the primary channel through which companies inform users about their data collection and sharing practices."

"Recent developments in NLP have led to an increased interest in the exploration of NLP approaches for text processing in various data protection problems, e.g. text anonymization, in contexts where text data (e.g. police reports, healthcare dossiers) are either sensitive by nature or protected by privacy laws (e.g. GDPR)."

"We study how the recognised named entities can be injected into a text data engineering step, in order to train a classification model able to analyse the GDPR-compliance of Italian PA documents."

Corpus preparation

"The corpus was annotated by two independent annotators, experienced GDPR lawyers, who read the text thoroughly, located the target entities, highlighted them on the annotation platform, and chose categories from a predetermined list: administration, health, location, omissis, organisation and person."

"To the standard categories (person, location, organisation), we introduced some domain-specific categories, to recognise roles in public administration and disease names, which may be relevant for identifying GDPR non-compliant documents."

"The size of the partial match is the number of characters that are different between two overlapped annotations with the same category."

"The number of annotations that match between the annotators achieves an elbow value with the size of the partial match set equal to 5."

"2 of GDPR declares that data ceases to be personal when they are made anonymous, and an individual is no longer identifiable, we performed a process of anonymization that replaced any identified or identifiable information with artificial identifiers."

The INTREPID framework

"We calculate frequency features from named entity categories (administration, health, location, omissis, organisation and person) by counting the number of times a named entity category appears in the document."

"That, in the current version of the proposed framework, the BoNE model allows us to extract six features, one for each named entity category, from each document."

"The bi-gram X−Y occurs in a document if and only if there exists an entity x named with X and an entity y named with Y that have been identified in the text by the NER so that the distance between x and y is max_d at most, measured as the number of tokens between x and y. The distance is computed by ignoring stop-words."

"That the BoNNEG model allows us to extract thirty-six features from each document, one for each bi-gram of named entity categories."

"The BoWNE model extracts six features from each document, that is, one feature for each named entity category."

Experimental evaluation

"We analysed the effect of lemmatization, stopword removal and lowercase transformation on the accuracy performance of classifiers trained with BoW-based features by varying the classification algorithm."

"We note that the highest accuracy performance with the BoW-based features is achieved by training the XGBoost classifier with the SVM classifier as a runner-up."

"In NER+SVM, the highest accuracy performance is achieved in the configuration that trains the SVM classifier by accounting for the bag of named entities of the BoNE feature group."

"In NER+RF, the highest accuracy performance is achieved in the configuration that trains the RF classifier with the word information combined with the named entities (BoWNe feature group)."

"The highest accuracy performance with NER-based features is achieved training the SVM classifier having the RF classifier as runner-up."

Discussion

"With the goal to support the Italian PA ensuring the GDPR compliance of public documents and securing personal data, we formulated INTREPID, a framework based on AI to automate the detection of breaches of security in PA documents."

"We prepared a corpus of Italian PA documents for both training and evaluation by using an appropriate pipeline to balance the need of replacing any identified or identifiable information with artificial identifiers and the fact that the GDPR check does not apply to anonymized information."

"Locating data breach positions within documents."

"It allows us to recognise a PA document that may be non-compliant with the GDPR standards, but this is done without locating the position data breaches within documents."

"The classification model of the proposed framework has been trained with data breaches related to unlawful disclosure of health information."

Conclusions

"We describe an AI framework to learn a text classification model from this labelled text corpus, so that the learned model can be used to predict a new public document as compliant or non-compliant with the GDPR standards."

"We leveraged these NER features to enrich the traditional BoW representation of the text documents and train a classifier to label documents as compliant or non-compliant with the GDPR standards."

"We assessed the effectiveness of the proposed framework in terms of agreement of the annotation predictions of the NERs with the annotations by domain experts, as well as sensitivity of the accuracy of the text classification model to the groups of extracted features."

"The next stage of this research will be to extend the evaluation of the effectiveness proposed framework by including new documents possibly referred to different categories of data breaches in the text corpus and improve on the performance of NER models using our annotated corpus."

2.2.13 Future of Data Analytics in the Era of the General Data Protection Regulation in Europe [94]

This is a machine-generated summary of:

Kolasa, Katarzyna; Ken Redekop, W.; Berler, Alexander; Zah, Vladimir; Asche, Carl V.: Future of Data Analytics in the Era of the General Data Protection Regulation in Europe [94].

Published in: PharmacoEconomics (2020).

Link to original: https://doi.org/10.1007/s40273-020-00927-1.

If you want to cite the papers, please refer to the original.

For technical reasons we could not place the page where the original quote is coming from.

Abstract-Summary

"In the European Union, the value assessment based on real-world data has become challenging as individuals have recently been given the right to have their personal data erased in the case of consent withdrawal or when the data are regarded as being no longer necessary."

"This act may limit the usefulness of data in the future as it may introduce information bias."

"Among healthcare stakeholders, this has become an important topic of discussion because it relates to the importance of data on one side and to the need for personal data protection on the other side, especially when it comes to "personal data related to the physical or mental health of a natural person, including the provision of health care services, which reveals information about his or her health status"."

"At the forefront of these discussions are data protection issues as well as the population's trust in digital services."

"It has become increasingly important to ensure that individuals are properly introduced into complex environments with multiple sources of Big Data for clinical and behavioral purposes to provide an optimal balance between societal and individual benefits."

"A number of issues are considered and discussed, based upon the authors' experience, with the aim of helping the reader better understand the implications of the use of Big Data and the importance of data protection in the coming years."

Introduction

"The amount of available data is growing in the healthcare sector as well."

"Many examples are already available that illustrate how much Big Data are reshaping decision-making processes in healthcare."

"Beyond the clinical records of an individual's medical history, diagnoses, laboratory results, prescriptions, and healthcare service consumption, there is a growing amount of behavioral data to contend with."

"There is an ever-growing number of significant examples of how behavioral data can contribute to decision-making processes in healthcare."

"With the growing amount of available data, the concept of "data-driven innovation" has been launched."

"To our knowledge, there has been limited discussion regarding its impact on the growing role of Big Data in decision making in healthcare."

"Both issues are especially important for evidence generation as a basis for health technology evaluation in the new era of Big Data."

How to Ensure General Data Protection Regulation (GDPR) Compliant Research in the Era of 'Patient-Centric' Data Ownership?

"While the initial proposal of the Commission provided the option to carry out scientific research on a legal basis other than consent, the European Parliament Committee on Civil Liberties introduced a revision requiring that "consent should always form the correct basis for the processing of personal health data in a research context unless such research serves a purpose of 'exceptionally high public interest'"."

"The consent provided for the primary processing of personal data may be withdrawn at any time, and this introduces a factor of uncertainty in a research project."

"Some provisions in the GDPR provide the possibility to the EU Member States to further legislate on some specific points, such as the possible derogations to data subject rights in the frame of scientific research or processing of data concerning health, genetic, and biometric data (Article 89.2 of the GDPR) [54]."

How Can we Ensure Free Data Sharing Across Multiple and Diverse Health Data Sources?

"Sharing personal data is a processing activity that is subject to the rules of the GDPR unless the data have been adequately anonymized."

"As a solution to create a widely used and accepted data format, the integration profile process has been proposed as a way to enable end-to-end interoperability by sharing structured (and unstructured) data between the point-of-care systems [55]."

"With automated data collection from connected medical and wellness devices, the patient, and especially the empowered patient, can now contribute actively and create part of the Big Data needed for research and public health [56]."

"The blockchain would allow for secure sharing of electronic medical information such as genomic data, clinical trial data, hospital, outpatient clinic, visiting nurse and immunization records, imaging and laboratory results, as well as pharmacy records, health data from mobile devices, wearables, and the "Internet of Things" at the click of a button."

Conclusions

"Big Data provide a multitude of opportunities to further develop the concept of value in the healthcare sector."

"Only when giving sufficient care to data security and other data protection principles, will we succeed to ensure that Big Data works to benefit all of the healthcare system stakeholders and, more importantly, the patients."

"The pseudonymization will inevitably help the Big Data industry to develop methods to grant access to the personal health data for third parties."

"In due course, it will hopefully enable Big Data-driven health innovation and advance interoperability frameworks while taking into account ethical and security risks in a new digital era."

"It has also increased the trend of renovation of laws and regulations beyond the EU, which will hopefully allow safer re-use of patient-driven data for research and public health globally."

2.2.14 General Data Protection Regulation in Health Clinics [95]

This is a machine-generated summary of:

Lopes, Isabel Maria; Guarda, Teresa; Oliveira, Pedro: General Data Protection Regulation in Health Clinics [95].

Published in: Journal of Medical Systems (2020).

Link to original: https://doi.org/10.1007/s10916-020-1521-0.

If you want to cite the papers, please refer to the original.

For technical reasons we could not place the page where the original quote is coming from.

Abstract-Summary

"The focus on personal data has merited the EU concerns and attention, resulting in the legislative change regarding privacy and the protection of personal data."

"The General Data Protection Regulation (GDPR) aims to reform existing measures on the protection of personal data of European Union citizens, with a strong impact on the rights and freedoms of individuals in establishing rules for the processing of personal data."

"This work presents the evolution of the applicability of the Regulation (EU) 2016/679 six months after its application in Portuguese health clinics."

Introduction

"According to another author [57], the main innovations of the General Data Protection Regulation are: 1. New rights for citizens: the right to be forgotten and the right to a user's data portability from one electronic system to another. 2. The creation of the post of Data Protection Officer (DPO). 3. Obligation to carry out Risk Analyses and Impact Assessments to determine compliance with the regulation. 4. Obligation of the Data Controller and Data Processor to document the processing operations. 5. New notifications to the Supervisory Authority: security breaches and prior authorisation for certain kinds of processing. 6. New obligations to inform the data subject by means of a system of icons that are harmonised across all the countries of the EU. 7. An increase in the size of sanctions. 8. Application of the concept 'One-stop-shop' so that data subjects can carry out procedures even though this affects authorities in other member states. 9. Establishment of obligations for new special categories of data. 10. New principles in the obligations over data: transparency and minimisation of data."

Research methodology

"The choice of an appropriate data collection to characterise the implementation of the GDPR in medical clinics fell on the survey technique, since it enables a clear, direct and objective answer to the questions asked to the respondents."

"The aim of the survey was to characterise the current state of health clinics with regard to the implementation of the GDPR, in other words, determine their level of knowledge and preparation regarding the issue of personal data protection and privacy, as well as their evolution since the conduction of the last survey."

"The two following groups contained questions concerning the GDPR applicability, preceded by the paramount core question: 'has the clinic implemented the measures imposed by the GDPR yet?'"

"After responding to this central question and when the answer was negative, respondents were asked whether they intended to implement such measures, since they were not in compliance with the regulation, and if so, whether the implementation process was already in motion."

Results

"After the conclusion of these implementation stages, a compliance assessment must be conducted periodically since the data is not immutable and even the company business and activity may undergo changes which may make the measures initially implemented inadequate to the new circumstances."

"The time taken to implement the GDPR will always depend on the complexity of the company's business activity, its organisational maturity, the volume and variety of the personal data used, the adequacy and flexibility of its information systems and on all its stakeholders' availability and willingness."

"Following on from this you will need to appoint your GDPR program team, identify and assess relevant Personal Data Processing activities, prioritise a set of remediation actions, establish a centralised Personal Data register, educate Personal Data Handlers and Data Processors and update your Data Protection policies and Privacy Notices."

Conclusion

"The implementation of GDPR can be a challenge for Health Clinics."

"Most of the information which was previously shared in paper is currently shared digitally, thus posing new digital challenges and threats concerning security and privacy, namely regarding personal data protection in a society increasingly digital [58]."

"There are currently 28 laws of data protection based on the 1995 EU Data Protection Directive, which was implemented over 20 years ago and is gradually being replaced by the new GDPR [59]."

"Digital revolution brings along inevitable concerns regarding users' data security, privacy and protection, especially as far as health and clinical information is concerned [58, 60, 61]."

"It might be relevant to carry out further research to determine to what extent this GDPR, although targeted at data protection, might not be as well a booster for the digital transformation of health clinics."

2.2.15 The GDPR and the Research Exemption: Considerations on the Necessary Safeguards for Research Biobanks [96]

This is a machine-generated summary of:

Staunton, Ciara; Slokenberga, Santa; Mascalzoni, Deborah: The GDPR and the research exemption: considerations on the necessary safeguards for research biobanks [96].

Published in: European Journal of Human Genetics (2019).

Link to original: https://doi.org/10.1038/s41431-019-0386-5.

If you want to cite the papers, please refer to the original.

For technical reasons we could not place the page where the original quote is coming from.

Abstract-Summary

"Along with the set of carefully outlined data subjects' rights, the GDPR provides for a two-level framework to enable derogations from these rights when scientific research is concerned."

"By directly invoking provisions of the GDPR on a condition that safeguards that must include 'technical and organisational measures' are in place and second, through the Member State law Although these derogations are allowed in the name of scientific research, they can simultaneously be challenging in light of the ethical requirements and well-established standards in biobanking that have been set forth in various research-related soft legal tools, international treaties and other legal instruments."

"We review such soft legal tools, international treaties and other legal instruments that regulate the use of health research data."

"These instruments were also reviewed to provide guidance on possible safeguards that should be followed when implementing any derogations."

Introduction

"Along with the set of carefully outlined data subjects' rights, the GDPR also provides for a two-level framework to enable derogations from these rights when scientific research is concerned."

"Although these derogations are allowed in the name of scientific research, they can simultaneously be seen as challenging in light of the ethical requirements and long-standing protection standards for participants in biobanking that have been established in various research-related soft legal tools, international treaties and other legal instruments."

"The recognition of ethical standards is here related to broad consent and there is no explicit requirement to consider derogations in line with these 'recognised ethical standards for scientific research'."

"Research that could be deemed legal under the GDPR or Member State laws that permits further derogations, might not necessarily be in line with ethical standards that are currently required by research ethics committees (RECs)."

The ethical concerns related to biobanking

"Concerns about research aims may be unrelated to 're-identifiability' but rather related to possible uses in research, potentially against one person's ethical beliefs (use of health and genetic data to profile families and new generations, gender profiling, race profiling, communities discrimination, biological weapons based on genetic specificities etc.)."

"Concerns about possible actors involved in the use of data (insurances, private companies etc.) are also major issues as an individual's willingness to participate in research is often based on trust in specific institutions."

"Recent instruments show that consent (previously the main legal basis in which to lawfully conduct research) has been in fact paired with strong governance measures and third party oversight mechanisms to be adopted by institutions to regulate sharing, accessing and use of the data."

The ethical rip-off: GDPR perspectives on data subject and biobanking

"For example, a Member State may attribute particular value to biobank research; they may limit a data subjects right to control the use of their data in research by removing the consent requirement for the processing of genetic data in biobanking, provided the national law respects the principle of proportionality, the essence of the right of data protection, and provides for suitable and specific measures to safeguard the rights and interests of data subjects."

"Article 89(1) GDPR enables processing of genetic data for scientific research purposes, if there are appropriate safeguards for the rights and freedoms of the data subject."

"These derogations relate to the data subject's access rights (Art 15), right to rectification (Art 16), right to restriction of processing (Art 18), as well as right to object (Art 21), and as specified in Article 89(2) GDPR, they can be applied if the derogations are necessary for research purposes."

Methods

"Soft legal tools, international treaties and other legal instruments (hereinafter collectively called 'instruments') that are influential on the regulation of health research were identified."

"Our review was limited to instruments that may be relevant to some or all Member States of the EU, which are either directly applicable to the Member States or to certain professions within the Member States."

"Our review did not include international consortia guidelines."

"All instruments were imported into Nvivo 11 and coded using thematic content analysis."

Results

"The WMA Taipei Declaration seeks to regulate health databases and biobanks and provides more details on the requirements for consent: participants have to be informed about the purpose, the risks and burdens, storage and use of data and material, the nature of the data or material to be collected, the procedures for return of results including incidental findings, the rules of access to the health database or biobank, the protection of privacy, the governance arrangements, procedures to inform participants about the impact of anonymisation of data, their fundamental rights and safeguards as established in the Declaration, and when applicable, commercial use and benefit sharing, intellectual property issues and the transfer of data or material to other institutions or third countries (Article 12)."

Discussion

"What emerges from the instruments reviewed is that there are broadly three levels to a governance structure that Member States should follow: a national legal and ethical framework; independent and interdisciplinary review and oversight of the research; and local policies on data sharing and protection."

"While each research project that seeks to collect and use personal data are currently subject to independent ethics review, the subsequent use and access of this data may not, with no body ensuring that the rights of data subjects are protected."

"The law is a minimal standard to which researchers must comply with and the Article 29 Working Party Guidelines on consent state that (where consent is the legal basis for processing) 'consent for the use of personal data should be distinguished from other consent requirements that serve as an ethical standard or procedural obligation' [62]."

Conclusion

"This review makes it clear that a full implementation of the derogations as provided for under the GDPR may render the research unethical and not in line with individuals interests."

"These instruments also suggest that clear governance procedures and policies on the use and re-use of personal data can go some way towards ensuring that there are necessary safeguards in place to ensure the protection of personal data."

"By following these necessary safeguards, biobanks can ensure that they may continue to conduct research, which ensuring the protection of personal data."

2.2.16 General Data Protection Regulations (2018) and Clinical Research: Perspectives of Patients and Doctors in an Irish University Teaching Hospital [97]

This is a machine-generated summary of:

Davey, Matthew G.; O'Donnell, John P.M.; Maher, Elizabeth; McMenamin, Cliona; McAnena, Peter F.; Kerin, Michael J.; Miller, Nicola; Lowery, Aoife J.: General data protection regulations (2018) and clinical research: perspectives of patients and doctors in an Irish university teaching hospital [97].

Published in: Irish Journal of Medical Science (2021).

Link to original: https://doi.org/10.1007/s11845-02102789-8.

If you want to cite the papers, please refer to the original.

For technical reasons we could not place the page where the original quote is coming from.

Abstract-Summary

"To evaluate the knowledge and expectations of patients and doctors regarding GDPR and implications for future clinical research."

"Three hundred fifty were patients (68.8%) and 159 were doctors (31.2%)."

"Three hundred thirty-four participants were aware of GDPR (65.7%): 116 doctors (73.0%) and 218 patients (62.3%, P = 0.018)."

"71.1% of doctors were willing to allow their personal data to be processed anonymously as part of a clinical research project compared to 43.4% of patients (P < 0.001)."

"80.2% of patients believed explicit consent is needed before using personal data in clinical research in comparison to 60.4% of doctors (P < 0.001)."

"Level of education impacted awareness of GDPR (P < 0.001); a higher level of education among patients increased GDPR familiarity (P < 0.001), however failed to impact doctor familiarity (P = 0.117)."

"GDPR has introduced complexity to the processing and sharing of personal data among researchers."

"This study has identified differences in the perception of GDPR and willingness to consent to data being used in clinical research between doctors and patients."

"Measures to adequately inform prospective research participants on data processing and the evolving landscape of data protection regulation should be prioritised."

Introduction

"The strict interpretation of some of these new legislative changes (in particular, the addition of the Health Research Regulations (HRR) in the Republic of Ireland (ROI)) has the potential to impede clinical research by creating new barriers to overcome in terms of patient consent as well as a lack of clarity in relation to the capacity of international collaborators to share research data [63, 64]."

"The efforts of the HRR coincide with GDPR in relation to health-related research, outlining suitable and specific safeguards required when processing personal data for health research in the ROI [64]."

"The aim of this study was to evaluate the views and perspectives of patients and doctors as to how personal data is processed following the introduction of GDPR and their preferences in this regards as well as the potential implications of its introduction for future clinical research."

Methods

"Each participant was asked to participate in the study having been addressed by one of the four independent data processors (EM, CM, MGD, and JPOD) during October 2019–September 2020."

"Doctors working in GUH were invited to partake in the study while at work in the hospital."

"Patients attending medical and surgical outpatient clinics were also invited to partake in the study at the time of their hospital visit."

"No participants were recruited to undertake the study through electronic surveys, and no incentive was offered to any participants for their participation in the survey."

Results

"Of these, doctors were more likely to have agreed for their data to be used in clinical research (51.6% vs. 22.0%, P < 0.001 †)."

"Doctors were significantly more likely to believe explicit consent should be obtained to gather information from a patient's medical notes and used in clinical research projects (80.2% vs. 60.4%, P < 0.001 †)."

"Doctors with higher levels of education were willing to provide their anonymised data in clinical research (P = 0.014 χ^2), however level of education did not alter patient willingness to do so (P = 0.219, χ^2)."

"Higher level of education among patients failed to increase their willingness have their data shared in an international research collaboration (P = 0.358, χ^2), however did impact doctors willingness (P < 0.001, χ^2)."

Discussion

"In spite of this, results from the present study suggest only 1/6th of doctors believe they are 'very familiar' with GDPR regulations; this highlights the potential for significant data breaches in the setting of clinical research, and the fundamental requirement for an in-depth understanding of GDPR for any stakeholder in the clinical research [64]."

"This study has identified differences in the perception of GDPR and willingness to consent to data being used in clinical research between doctors and patients."

"Measures to adequately inform prospective research participants on data processing and the evolving landscape of data protection regulation should be prioritised, with emphasis upon the conscious processing of patient data in a transparent manner, with the formal implementation of GDPR education strategies to envisage how clinical research may be carried out in a safe and GDPR compliant manner in the future."

2.2.17 GDPR: An Impediment to Research? [98]

This is a machine-generated summary of:

Clarke, Niamh; Vale, Gillian; Reeves, Emer P.; Kirwan, Mary; Smith, David; Farrell, Michael; Hurl, Gerard; McElvaney, Noel G.: GDPR: an impediment to research? [98].

Published in: Irish Journal of Medical Science (2019).

Link to original: https://doi.org/10.1007/s11845-019-01980-2.

If you want to cite the papers, please refer to the original.

For technical reasons we could not place the page where the original quote is coming from.

Abstract-Summary

"The recent introduction of the General Data Protection Regulation and Health Research Regulations has been an area of significant concern for those engaged in clinical research."

"These European regulations, following subsequent interpretation by Ireland's Department of Health, now place Ireland in a unique position which differs substantially from other European countries and may prove a significant impediment to Irish clinical research, depriving Irish patients of timely access to potentially life-saving treatments and making Ireland less attractive to pharmaceutical companies engaged in this area."

"At the very least, the regulations, as applied in Ireland, will place a significant extra burden of work on Ireland's clinical researchers and at their worst will force individuals and institutions out of the clinical research field, which will result in significant loss to the Irish knowledge economy and lead to the detriment of patient care."

"We explore what exactly is proposed by Europe's GDPR and by Ireland's Health Research Regulations."

Introduction to GDPR

"Many of the principles that form GDPR reflect the core principles of the Directive and the definition of personal data, as outlined in Article 4(1) of GDPR, includes "any information relating to an identified or identifiable natural person ('data subject')"."

"The definitions of data processors and data controllers under Article 2 of the Directive are virtually identical to the definitions now contained in Article 4 of the GDPR."

"A data controller is an individual or legal person(s) such as a company, department, or organisation, which under Article 4 of GDPR "determines the purposes and means of the processing of personal data"."

"The formal definition (GDPR Article 4) states that a processor means a natural or legal person, public authority, agency, or other body that processes personal data on behalf of the controller."

"The principles of GDPR Article 5, regarding personal data processing, apply to data processors just as much as they apply to data controllers."

Roles and responsibilities of the data controller

"The data controller must adhere to what is stipulated under Article 5 GDPR, which states that personal data must be processed lawfully, fairly, and in a transparent manner ("lawfulness, fairness and transparency")."

"With respect to data subjects, it is important that transparent information is provided to the intended subjects by the data controller on the methods by which their data will be processed."

"Article 35 GDPR requires that data controllers carry out an impact assessment (DPIAs) for "high-risk processing" and implement measures to mitigate a risk."

"Data processors are required to inform data controllers of any data breach, which must be reported to the office of the Data Protection Commissioner (DPC) within 72 h where there is a risk to the rights of the data subject."

The data subject

"GDPR has made significant advances on the rights of the data subject and includes right to rectification of inaccurate data in a timely manner (Article 16), right to be forgotten or right to erasure of personnel data (Article 17), and right to object to processing of personal data (Article 21)."

"The rights of data subjects under GDPR extend to accessing their own personal data (Article 15)."

"Within 1 month of receipt of such a request, the data controller must respond and upon verification of the identity of a data subject, should provide, at no cost, a copy of the requested personal data in a concise, transparent, and easily accessible form."

"The response time may be extended by 2 months if a request is complex, and the controller may charge a reasonable fee for further copies of personnel data that is undergoing processing."

GDPR and Health Research Regulations

"Inclusion of several miscellaneous provisions (Regulations 14–16) The HRR were signed into law by Ireland's Minister of Health on August 8, 2018 and relate to processing of personal data for health research."

"The HRR lists the suitable and specific safeguards required when processing personal data for health research in Ireland."

"This is a unique Irish addition to GDPR, because the European regulation, which took effect on May 25, 2018, did not require explicit consent, and allowed for data processing without consent subject to safeguards."

"The subsequent Irish legislation enshrines explicit consent for data processing; and in cases where this is not possible, the researcher must apply to the National Consent Declaration Committee (CDC) (https://hrcdc.ie/)."

"The introduction of an Irish requirement for explicit consent is unique among member states in the EU and will inevitably lead to restrictions on health research in Ireland."

GDPR and HRR: the ongoing challenges for research in Ireland

"Retrospective chart reviews: the latest update from the Department of Health regarding this important type of research was published on 26th November 2018 on the Health Research Board website and states: "As regards Retrospective Chart Reviews carried out for research purposes, and having consulted with the Data Protection Commission, it has been determined that the requirement for explicit consent will commence on 1st May 2019."

"Biobanks/archival material: In this area, re-consenting is a major issue, and not using the millions of valuable and carefully documented tissues archived in Irish Pathology Departments and in the many designated disease-specific biobanks for research poses the most serious threat to health research progression and subsequent future treatment for Irish patients affected by a wide variety of health conditions such as epilepsy, cancer, heart disease, and potentially fatal childhood skin disorders, to name but a few."

Ongoing unanswered questions

"It should be also noted and taken into account by the CDC that the majority of research was in fact in compliance with Article 9(2)(j) and the safeguards outlined in Article 89 of GDPR, namely pseudonymisation, minimisation, and anonymisation implemented where appropriate."

"Apart from the original and obvious role of the Research Ethics Committees (REC) to review ethical aspects of research applications, there seems to be some misunderstanding as to who should ensure compliance."

"The data controller and the institution which employs the researcher has overall responsibility, but accountability is inadvertently diverting to RECs to provide guidance, advice and information, a role for which they are not trained and are poorly resourced."

"It is the role of the DPO to advise the data controller and/or institution of their legal requirements to ensure they are GDPR and HRR compliant."

Recommendations

"Permit chart reviews for medical research by healthcare professionals involved in the provision of clinical care for the cohort of patients under review."

"Seek broad consent from patients about to undergo surgical and medical resections to give broad consent for biobanking of their tissues which are surplus to the diagnostic pathology procedures with the understanding that research conducted on these tissues will have been placed before and approved by a research ethics committee."

"Triage and allocate work of the CDC to sub-committees with mandates to provide guidelines in each of these 3 areas: (a) Biobank/Archival Material (Pathology archives—material collected for clinical purposes in which obtaining consent for individual projects is no longer feasible, for one of the reasons set out above) (b)

Retrospective chart reviews (low-risk huge public benefit) and provision of guidelines on the use of the electronic patient record in research. (c) Studies for which consent adhered to the previous legislation."

Conclusion

"GDPR and HRR have the potential to ensure better patient protection in our health system but the application of these processes to Ireland as set out by the Department of Health is problematic, challenging, resource-intensive, and costly."

"The Department of Health has taken a unique and arguably restrictive approach to data protection in Ireland which is quite at variance from our European colleagues and which if followed through as outlined will impact negatively on patient care and clinical research in Ireland."

"We have outlined the potential benefits and challenges of GDPR and HRR and have suggested solutions in the Irish context which we feel would safeguard patients' rights while at the same time protecting their access to newer treatments and diagnostics."

2.2.18 Hypocrisy Around Medical Patient Data: Issues of Access for Biomedical Research, Data Quality, Usefulness for the Purpose and Omics Data as Game Changer [99]

This is a machine-generated summary of:

Tantoso, Erwin; Wong, Wing-Cheong; Tay, Wei Hong; Lee, Joanne; Sinha, Swati; Eisenhaber, Birgit; Eisenhaber, Frank: Hypocrisy Around Medical Patient Data: Issues of Access for Biomedical Research, Data Quality, Usefulness for the Purpose and Omics Data as Game Changer [99].

Published in: Asian Bioethics Review (2019).

Link to original: https://doi.org/10.1007/s41649-01900085-3.

If you want to cite the papers, please refer to the original.

For technical reasons we could not place the page where the original quote is coming from.

Abstract-Summary

"Whether due to simplicity or hypocrisy, the question of access to patient data for biomedical research is widely seen in the public discourse only from the angle of patient privacy."

"This goal can only be achieved by extracting research insight from patient data in addition to working on model organisms, something that is well understood by many patients."

"Most biomedical researchers working outside of clinics and hospitals are denied access to patient records when, at the same time, clinicians who guard the patient data are not optimally prepared for the data's analysis."

"Clinicians and bioinformaticians, each group with their own interests, have to join forces with the goal to generate medical data sets both from clinical trials and from routinely collected electronic health records that are, as much as possible, free from errors and obvious inconsistencies."

"This opens up opportunities for re-analyses and cross-trial studies at future time points with more sophisticated analyses of the same data, the collection of which is very expensive."

Introduction

"When a non-clinical researcher wants to test a hypothesis or study trends with medical data for the first time, the access to the data is typically barred due to arguments of patient data privacy."

"Bringing the privacy argument into the forefront of public discussion is also hypocritical as is allows hiding complexities of extracting benefits from medical records from the public eye."

"The access to medical patient data is in the center of a conflict between the vested interests of various groups: (i) the patients who do not want their data being misused but who are generally happy to trade in privacy for the prospects for better biomedical science and improved therapy, (ii) the clinicians who collect and guard the medical data and who might fear that others consider the data or its analysis messy, (iii) the bioinformaticians who cannot move forward scientifically without the data."

The Issue of Patients' Data Access in the Era of Omics Technologies Entering Healthcare

"The reasons are manifold and include aspects of evidence-based medical therapy, increasing amount of clinical, laboratory, imaging, and especially omics data collected per patient over the time of observation, the need of their analysis, the complexity of treatment options, legalities and documentation issues, etc. Yet, the process is primarily driven by the need to provide services of increasing sophistication to an ever larger cohort of patients (that ultimately encompasses all population living on a given territory) for affordable prices."

"As a byproduct, large electronic databases of medical patient data are emerging (Electronic Health Records, or EHRs) that offer new, unprecedented opportunities for biomedical research."

"To individual gains, society will benefit due to the extension of health span and of duration of work life, decreasing costs for care and cure, etc. Medical institutions might become the main source of data for studying genome-phenotype relationships that will outclass traditional biological laboratories in the future."

Access to Medical Patient Data in Biomedical Research: the Patients' View in the Matter of Data Privacy Versus Medical Innovation

"Ethical considerations implemented in professional standards dating back to the oath of Hippocrates and enhanced by PDPA-type law and punishment mechanisms prohibit the use of medical data in ways that might result in damaging the patients' individual and group interests."

"The right for patient data privacy clashes with the more basic need of patients to live and to prevent death and disability, to receive efficient medical treatment for their condition when the costs for state-of-the-art treatment are exorbitant or, especially, when the current level of medical science cannot deliver effective care even at the symptomatic level due to the limit of knowledge and insight into pathogenetic mechanisms."

"Patients' data privacy and the need for medical innovation conflict."

"Without access to patient data for research, progress in the medical sphere is impossible."

Access to Medical Patient Data in Biomedical Research: the Patients' Views with Respect to Consent

"The current practices in requesting and documenting the consent of patients are not very suitable for supporting large-scale medical research on patient data."

"Leaving the full burden and risk of the patients' consent on the shoulder of researchers and their institutions who have very limited resources is not an arrangement that will widen the scale of medical data research, something that is not in the society's best larger interest."

"Some societal support is needed to make clinical data research possible and a standardization of consent documentation and procedures would help."

"An alternative would be to request much broader consent that allows storing the collected data in anonymized form in clinical databases for all kinds of medical research including future efforts [77, 78, 79, 80, 81]."

Access to Medical Patient Data in Biomedical Research: the Patients' Views with Respect to Data Anonymization

"Various pieces of anamnesis data taken together might suffice to discover the patient's identity, especially if linked with other databases, for example diverse public data repositories, telephone books, social media such as Facebook, etc. The introduction of omics data finally compromises the concept of anonymization as the genome sequence and all derived omics data are ultimately individual and are always sufficient to find the patient in a cohort."

"It might be problematic to put a ledger to every elementary piece of data as the computational overhead can become prohibitive but, for example, applying them only to larger blocks such as the patient entry as a whole could be a compromise solution. (ii) There are sensitivities of organizations and nation states that do not allow their patient data to leave the limits of their territory."

"The authors of this article think that block-chain ledgers, federated databases, and encrypted computation will have their role in medical patient data studies when simply no other access to these data is possible."

Access to Medical Patient Data in Biomedical Research: the Physicians' Views

"The public discussion usually forgets that collecting medical observation data for studies is extremely expensive and one of the most costly modes of life science research [82, 83]."

"It is desirable to provide public incentives to private investors to supply their clinical trial data to research depositories once they have exhausted their own analysis effort."

"In an extreme case, the clinical party might assume that their factual ownership of the data implies ownership of all medically relevant conclusions and that the non-clinical researchers provide only a technical service to them without creative contributions."

"It becomes critical to achieve the rapport of the clinical researchers in the medical data analysis process, to dutifully acknowledge their contribution in data collection and its primary quality control and data analysis."

Access to Medical Patient Data in Biomedical Research: the Bioinformaticians' Views

"Whereas clinical data analysis might have only involved cursory biostatistical expertise in addition to the clinical contribution a few decades ago, the large amount, the heterogeneity, and complexity of medical data, especially in cases with omics and certain imaging components, require additional input from professionals in the area of information technology and computational biology/bioinformatics."

"Of transforming medical data into actionable medical insight, the bioinformaticians' group contributes two important functions, namely (i) the creation of electronically analyzable datasets and (ii) the analysis and interpretation of this data (in collaboration with the clinical partners)."

"The creation of error-free, long-living databases for medical patient data are necessary for reproducible biomedical research and reliable scientific conclusions; thus, it is an ethical imperative on its own (besides its economic importance) as erroneous data will eventually support useless, wrong or even detrimental clinical conclusions."

Access to Medical Patient Data in Biomedical Research: Experiences at the Bioinformatics Institute Singapore (BII)

"Data collation from various consortium partners in a multi-centric effort is not a trivial process, and it involves stages of alignment, data cleansing, and dataset versioning."

"With the processing pipelines implemented as part of our own in-house developed database software TIMS (Translational Information Management System), the records uploaded by the clinical partners will be aligned along patient codes so that different data from one and the same patient can be jointly analyzed."

"Depending on the type of data, we check completeness of records, data error margins, the likelihood of the data points being really collected from the same patient (for example, via plausible levels of sequence identity in various datasets involving DNA/RNA sequencing, mismatches in medical images, etc.), consistency of various groups of data items (e.g., gender- and age-specific trends in laboratory data, physiological limits), etc. If inconsistencies cannot be resolved automatically, request lists are generated for our clinical partners to review."

Usefulness of Patient Datasets for Biomedical Research: What Can Go Wrong in Patients' Records Does Indeed Go Wrong Sometimes and This Has Consequences for the Conclusions

"When developing our software pipelines for clinical data processing, we had brainstormed about possible inconsistency checks in patients' records, implemented these ideas into programs and we were surprised to see that, with enough incoming data, instances for every such error were indeed found and, in some cases, their frequency would have led astray data analysis using the raw, uncorrected data."

"As the individual genome sequence leaves a signature in all types of omics data, it can be established very reliably whether all omics datasets with the same patient label are actually from the same patient (or from another patient in the same database or even from yet another person)."

"Patients with high discordance in their genomic information across different platforms will be permuted across all patients' omics data and identify the potential matching."

Usefulness of Patient Datasets for Biomedical Research: Clinical Trials, Electronic Health Records, Abbreviated Medical Record Sets and the Importance of Billing Information

"The clinical partners invest enormous efforts to collect good patient cohorts that lead to long time series of data recorded in a variety of modalities."

"We think that research work on EHRs in the foreseeable future cannot compete with the outcome achieved from well-curated clinical trial data analysis due to data quality issues."

"The groups of computational biologists, bioinformaticians, and biostatisticians that work with clinicians over clinical trial data belong to a very privileged but small group of scientists as they have access to clinical data of their patient cohort in their entirety."

"The whole breadth of scientific work principles such as application of non-standard analysis approaches, the gazing over outlier data and the comparison of thought outcomes from studying various clinical diagnosis modalities and time series is possible."

"Almost all of the scientific studies with patient data work with records where billing data is stripped off."

Discussion and Conclusions

"The question of access to patient data for biomedical research is widely seen only from the angle of patient privacy."

"The outcome, namely that the majority of biomedical researchers has no or very abbreviated access to clinical data, is not good for the patients as the extraction of valuable insight with impact for diagnosis and treatment is delayed or hindered at all."

"It appears that currently, as exemplified by the 2018 cyberattack on SingHealth, a major health care provider in Singapore, hackers have easier access to patient data than researchers [84] when the hackers have no intentions to improve biomedical insight or healthcare."

"It is generally not the patients who create the major obstacle for the patients' data reaching biomedical researchers and it is hypocritical and simplistic to represent them as the core of the problem."

2.2.19 The European Parliament Proposal for the New EU General Data Protection Regulation May Severely Restrict European Epidemiological Research [100]

This is a machine-generated summary of:

Nyrén, Olof; Stenbeck, Magnus; Grönberg, Henrik: The European Parliament proposal for the new EU General Data Protection Regulation may severely restrict European epidemiological research [100].

Published in: European Journal of Epidemiology (2014).

Link to original: https://doi.org/10.1007/s10654-014-9909-0.

If you want to cite the papers, please refer to the original.

For technical reasons we could not place the page where the original quote is coming from.

Articles 6.2 and 83: shaky pillars forming the legal foundation

"Article 83, however, is remarkably meagre; all it says is that processing of personal data for the purposes of historical, statistical or scientific research is allowed only if these purposes cannot be otherwise fulfilled using anonymous data and that "data enabling the attribution of information to an identified or identifiable data subject is kept separately from the other information under the highest technical standards, and all necessary measures are taken to prevent unwarranted re-identification of the data subjects" (i.e., pseudonymisation is mandatory)."

"Our interpretation is that 6.2 overrides 6.1 and that Article 6 approves processing of personal data for scientific research purposes, even in the absence of consent."

"A general problem with the Article 6.2–83 axis is that while it implies that the relevant conditions in the two Articles will fully determine the scope for the processing of personal data for historical, statistical or scientific research purposes, derogations for research appears in Articles 5e, 9.2i, 17.3c, and 81.2."

An amendment to Article 81 is a serious threat to large-scale epidemiological research

"Article 9.2i lays down that processing of sensitive personal data, including data concerning health, is allowed when necessary for historical, statistical and scientific research purposes subject to the conditions referred to in Article 83."

"A very unfortunate amendment by LIBE to Article 81, dealing with processing of data concerning health, notably for the due operation of health care services, has materially disturbed the original apparent symmetry between Articles 6, 9, 81, and 83."

"The revised Article 81.2 says that "processing of personal data concerning health which is necessary for historical, statistical or scientific research purposes shall be permitted only with the consent of the data subject, and shall be subject to the conditions and safeguards referred to in Article 83"."

"An additional drawback of the reliance on Member State law for exemption from the obligatory consent will be that the intended uniformity of research-related legislation throughout Europe will not be attained, maintaining existing obstacles for free movement of research data across European borders."

Uncertainties about the future of health registers

"The status of the high-quality health registers—epidemiological crown jewels for public health statistics and public health policy in several European countries and essentially indispensable resources in health research by virtue of their completeness and virtual absence of bias—might become a cliffhanger."

"The registers will first stumble on the previously mentioned LIBE amendment requiring consent (Article 81.2), and then remain at the mercy of national legislation, which may or may not waive the obligatory consent but cannot remove the obligatory pseudonymisation or the right for the data subjects to object."

"Even if national laws will support the collection of personal data on health without consent of the data subjects, the release of health register data for research conducted by other researchers may require consent."

Mandatory pseudonymisation: not a trivial issue

"Obligatory pseudonymisation (data enabling identification of specific data subjects being kept separately from the other information) might be seen as a small and reasonable concession, but if strictly interpreted the consequences for epidemiological research may be detrimental."

"If indirect identification is to be counted as "data enabling attribution of information to a data subject", then research databases must be stripped of considerable amounts of information in order to adhere to the requirement of pseudonymisation, possibly rendering many—if not most of them—useless for epidemiological research."

"There are no explicit provisions regarding the lawfulness of, or procedures for, warranted re-identification for the purpose of e.g. record linkages, quality control of data, or verification of conducted research."

"It would be helpful if Article 83 would explicitly state that the pseudonymisation requirement can be lawfully waived during checking or matching operations and also acknowledge that processing of identifiable personal data is sometimes necessary for sustaining the highest quality in epidemiological research."

What next?

"The LIBE amendment represents a hardline stand with only few concessions specifically for research, adapted to Member States with the least favourable conditions for large-scale epidemiology, but admittedly with some option for Member States to relax the provisions."

"As pointed out in amendment proposals from the European Parliament's Committee on Industry, Research and Energy and Committee on Legal Affairs, an exemption from the purpose limitation in Article 5(b), corresponding to the existing exemption in the current Directive 95/46/EC, should be reintroduced."

"Article 6 ought to be revised so that it becomes clear that 6.2 (establishing the lawfulness of processing of personal data for the purposes of historical, statistical or scientific research) overrules 6.1."

"Other provisions relevant to scientific research (exemption from the "storage minimization principle" in 5e, exemption from the prohibition against processing of sensitive data in 9.2i, exemption from the "right to erasure" in 17.3c, and the hopefully reintroduced exemption from "purpose limitation") should be moved to Article 83."

[Section 6]

"The GDPR is planned to replace the 1995 Directive 95/46/EC, which constitutes the present European legal framework for processing of personal data."

"This new binding Regulation will lay the legal foundation for future European epidemiology based on personal data, including register-based research."

"The proposed text threatened to restrict currently existing possibilities to produce scientific evidence based on European data analysis and, in turn, to impede efforts to improve public health and welfare in the union and elsewhere."

"In Directive 95/46/EC there was an exemption for research, namely that further processing of data for historical, statistical or scientific purposes is not to be considered as incompatible with the original purpose as long as Member States provide appropriate safeguards."

"This exemption was omitted in LIBE's amendments, dramatically reducing the scope for data sharing between research groups and severely restraining the use of retrospective (historic) cohort study designs."

2.2.20 Clinical Trial Data Transparency in the EU: Is the New Clinical Trials Regulation a Game-Changer? [101]

This is a machine-generated summary of:

Zemła-Pacud, Żaneta; Lenarczyk, Gabriela: Clinical Trial Data Transparency in the EU: Is the New Clinical Trials Regulation a Game-Changer? [101].

Published in: IIC—International Review of Intellectual Property and Competition Law (2023).

Link to original: https://doi.org/10.1007/s40319-023-01329-4.

If you want to cite the papers, please refer to the original.

For technical reasons we could not place the page where the original quote is coming from.

Abstract-Summary

"The benefits of access to clinical trial data are related to their inestimable value from the perspective of clinical trial participants, society as a whole, public health systems and scientific progress."

"In light of the development of innovative data analysis technologies, access to raw clinical trial data opens up an ever-widening array of possibilities: it can profoundly facilitate machine data analysis for, inter alia, hypothesis generation, risk modelling, counterfactual simulation and—finally—drug repurposing and development."

"536/2014 (CTR) and introduction of the Clinical Trials Information System (CTIS) were heralded as ensuring a level of transparency in clinical trials that is sufficient to contribute to protecting public health and fostering the innovation capacity of European medical research, while recognizing the legitimate economic interests of sponsors."

"This paper presents the hitherto binding rules for the disclosure of clinical trial data and, against this background, their new framework, introduced by the CTR."

Introduction

"Promises of unparalleled transparency levels raised concerns on the part of research-based pharmaceutical companies, which are uncertain of the depth of changes to existing rules on protection for their confidential clinical trial data and a potentially negative impact of such changes on their competitive position."

"This paper presents the hitherto binding rules for the disclosure of clinical trial data and, against this background, their new framework, introduced by the CTR."

"By carrying out a detailed analysis of the new rules for publishing clinical trial data in the CTIS, particularly their scope, timing of publication, and relevance for various stakeholders in the pharmaceutical sector, this paper aims to assess whether they fulfil the CTR's objectives."

"As the new rules on clinical trial data disclosure only came into force at the end of January 2022, and thus many questions remain unanswered, this paper maps out areas for further research."

Importance of Clinical Trials Results and the EU Policy in this Area

"Within the first stage, the issue of availability of data relates mainly to trial registration, as well as reporting and publication of trial results upon completion of their respective phases, whereas access within the second stage concerns clinical study reports (CSRs) and individual patient data (IPD)."

"There are also policies and guidelines issued by the European Medicines Agency (EMA): specifically, Policy 0043 for request-based access to documents related to medicinal products and Policy 0070 for publication of clinical trial data, as well as the guidance document on the identification of commercially confidential information (CCI) and personal data within the structure of the marketing authorization application, issued by the Heads of Medical Agencies (HMA) together with the EMA."

"The pharmaceutical industry's concerns regarding the potential misuse of disclosed information by competitors in support of third-party medicinal product authorization applications without the consent of the sponsor/MA holder have resurfaced following the introduction of the CTR, particularly in connection with the introduction of requirements for earlier publication of clinical trial data."

The Principle of Disclosure in the CTR

"Although, as indicated above, the CTR is not the first EU policy initiative aimed at increasing public access to clinical trials data, it substantially shifts the transparency requirements to earlier phases of clinical trials, establishing transparency needs throughout the whole process of clinical development, whilst previous efforts were focused on access to dossiers of already authorized medicinal products after this process had been completed."

"Confidentiality can be justified on the following grounds: (a) protecting personal data in accordance with Regulation (EU) 2018/1725; (b) protecting commercially confidential information, in particular by taking into account the status of the marketing authorization for the medicinal product, unless there is an overriding public interest in disclosure; (c) protecting confidential communication between Member States in relation to the preparation of the assessment report; (d) ensuring effective supervision of the conduct of a clinical trial by Member States."

Rules of Protection for Commercially Confidential Information in the CTR

"The EMA implementing guidelines provide that CCI is to be understood as (a) any information contained in the data or documents submitted to the CTIS database, (b) that is not in the public domain or publicly available, and (c) where disclosure may undermine the legitimate economic interest or competitive position of the clinical trial sponsors or marketing authorization applicants/holders."

"The EMA's "Draft Guidance on how to approach the protection of personal data and commercially confidential information in documents uploaded and published in the Clinical Trial Information System (CTIS)" is the first official document that aims to ensure a common understanding of what may be considered CCI under the new Regulation and, specifically, in reference to documents submitted to the CTIS, and to increase consistency in the CCI identified across various types of information."

The Deferral Mechanism

"With regard to the above, however, it must be pointed out that some information is made publicly available regardless of the trial category and even in case of deferral, namely: EU clinical trial number, sponsor name and address, nature of clinical trial (e.g. bioequivalence in 24 healthy volunteers), decision outcome on the trial application and date of decision, date of start of the trial, dates of start and end of recruitment, date of end of the trial in the Member State(s) in the EEA, and globally (including early termination of the trial), principal investigator curriculum vitae, and suitability of the facilities."

"Additional rules for transparency are established in Art 17(2) for medicinal products of relevance to the public health emergency which have received marketing authorization—in such a case, the Agency publishes, inter alia, the clinical trial data submitted to the EMA in support of the application, where possible within two months of the MA."

Enforcement of the New CTR Rules on Disclosure and Deferrals

"Even taking into account the EMA's future guidance documents, it seems that national authorities will have the final say in the decision on deferral of publication."

"The interpretation of policies and guidelines may thus differ in particular countries, which may in turn leave room for diverging decisions on publication of certain clinical trial data or its deferral."

"The EMA's rules on deferral allow clinical trial sponsors to defer certain trial information for up to 10 years after the end of a trial by establishing a complex classification system without providing clear criteria on the prerequisites for derogation from the transparency principle."

"Sponsors are "encouraged to submit trial documents with relevant redactions, instead of requesting long deferrals for their publication", out of concern that "extensive deferrals could significantly reduce the utility of clinical trial data in CTIS"."

Conclusions—the True Impact of CTIS on Transparency of Clinical Trial Data

"The new legislation should not be understood as a direct revision of existing measures, as the scope of its application is different from the EMA's policies: whereas the policies covered centrally authorized products only, the CTR's scope includes investigational medicinal products regardless of whether they have marketing authorization—this means that any rules on transparency, disclosure, deferrals and redactions in terms of clinical trial data are now to be uniformly applied by Member States as the process of assessment and supervision of clinical trials is harmonized by the CTR throughout the EU."

"When assessing the newly established rules on publication of documents in the CTIS from the point of view of transparency of clinical trial data for the sake of the public health interest, it is vital to stress that requiring sponsors to make decisions on deferrals and redactions prospectively at the time of submission of an initial application, without establishing a corrective mechanism within which these decisions could be altered depending on various circumstances, contradicts the assumed flexibility of the introduced mechanisms for CCI protection."

2.2.21 Enablers and Barriers to the Secondary Use of Health Data in Europe: General Data Protection Regulation Perspective [102]

This is a machine-generated summary of:

Vukovic, Jakov; Ivankovic, Damir; Habl, Claudia; Dimnjakovic, Jelena: Enablers and barriers to the secondary use of health data in Europe: general data protection regulation perspective [102].

Published in: Archives of Public Health (2022).

Link to original: https://doi.org/10.1186/s13690-022-00866-7.

If you want to cite the papers, please refer to the original.

For technical reasons we could not place the page where the original quote is coming from.

Abstract-Summary

"We aimed to provide an overview of the General Data Protection Regulation (GDPR) enablers and barriers to the secondary use of health data in Europe from the research we conducted in the Joint Action InfAct (Information for Action!) WP10 Assessing and piloting interoperability for public health policy, as well as to provide an example of a national-level case study on experiences with secondary use of health data and GDPR on an example of the Austrian COVID-19 data platform."

"GDPR was seen as an enabler to the secondary use of health data in Europe when it comes to user rights over their data, pre-existing laws regarding data privacy and data sharing, sharing anonymized statistics, developing new data analysis approaches, patients' trust towards dealing with their health data and transparency."

"GDPR was seen as a barrier to the secondary use of health data in Europe when it comes to identifiable and individual-level data, data sharing, time needed to complete the process, workload increase, differences with local legal legislations, different (and stricter) interpretations and access to data."

"The results of our analysis show that GDPR acts as both an enabler and a barrier for the secondary use of health data in Europe."

Background

"We aimed to provide an overview of the GDPR enablers and barriers to the secondary use of health data in Europe from the research we conducted in the Joint Action InfAct (Information for Action!) on Health Information co-funded by the Third EU Health Programme (2014–2020) which includes 40 partners in 28 countries with the goal of strengthening national and EU health information systems by establishing a sustainable EU health information system infrastructure."

"The main objective of this work is to provide insights into the relation between a regulation—GDPR—and data practice—secondary use of health data."

"To illustrate this, the article presents a country-specific case study from Austria on experiences with secondary use of health data and GDPR along the lines of the national COVID-19 data platform that provides data to national and international researchers."

Methods

"We have conducted semi-structured in-depth interviews with participants from European initiatives, projects and organizations that have dealt with cross-border health data sharing, linkage and management to find out what are the legal, GDPR-related enablers and barriers to cross-border health data exchange in Europe and how are European countries dealing with secondary use of health data since the implementation of the GDPR."

"During the European Joint Action Infact we identified a number of European initiatives, projects and organizations that have dealt with cross-border health data sharing, linkage and management, leading to 17 personal, structured interviews with leading experts in population and digital health in Europe."

"The interview instrument was developed with a specific aim of being used for semi-structured interviewing techniques involving InfAct researchers, as interviewers, and representatives of cross-country health data exchange projects in scope, as interviewees."

Results

"GDPR implementation was easier in countries with pre-existing legislation on data privacy and data sharing regulation."

"This made complying with the new GDPR regulation easier, while the countries which did not have pre-existing laws regarding data privacy and data sharing had to implement and enforce the new regulation."

"There are much more concerns about data protection, partly due to implementation of the GDPR, which makes it more difficult to share data for scientific purposes."

"Implementing GDPR represents a major work burden in projects which deal with cross-border health data exchange as they have limited budgets from research funding and limited personnel and the legal issues take much more time and work than it is available which restricts carrying out the project simultaneously while and there is a lack of funding to set up data and information exchange systems which would be compliant with the GDPR."

Discussions

"We conducted our interviews to find out what are the legal, GDPR-related enablers and barriers to cross-border health data exchange in Europe and how are European countries dealing with secondary use of health data since the implementation of the GDPR."

"In our interviews we included participants from all levels of project hierarchy to gain a complete insight into possible GDPR-related enablers and barriers to cross-border health data exchange in Europe."

"GDPR is seen as a barrier when it comes to dealing with identifiable and individual-level data and different GDPR interpretations and implementations in different countries, as well as an unclear understanding of it are seen as a barrier for cross-border health data exchange in Europe."

Conclusions

"Both the implementation and the interpretation of the GDPR varied a lot between EU member states (and within different institutions within the member states) which represents a clear barrier."

"Since the implementation of the GDPR there are much more concerns about data protection which makes it more difficult to share data for scientific purposes."

"While GDPR in itself does not prohibit data sharing for scientific purposes, different interpretations and unclear understanding of it may cause difficulties."

"When it comes to dealing with identifiable and individual-level data, GDPR represents an issue with health data which is narrowly defined as it could be a way of identifying individuals and it makes health research on rare diseases more difficult as it is a field where the data is potentially identifiable."

2.2.22 *The Use of Human Tissue in Epidemiological Research; Ethical and Legal Considerations in Two Biobanks in Belgium [103]*

This is a machine-generated summary of:

Truyers, Carla; Kellen, Eliane; Arbyn, Marc; Trommelmans, Leen; Nys, Herman; Hensen, Karen; Aertgeerts, Bert; Bartholomeeusen, Stefaan; Hansson, Mats; Buntinx, Frank: The use of human tissue in epidemiological research; ethical and legal considerations in two biobanks in Belgium [103].

Published in: Medicine, Health Care and Philosophy (2009).

Link to original: https://doi.org/10.1007/s11019-009-9230-y.

If you want to cite the papers, please refer to the original.

For technical reasons we could not place the page where the original quote is coming from.

Abstract-Summary

"This paper discusses the legal implications of setting up two new biobanks in Belgium."

"Belgium has no specific legislation that regulates storage of human tissue and related databases."

"Several issues concerning the protection of individuals with regard to the processing of personal medical data are discussed from the existing privacy legislation."

Introduction

"Interest has risen for an additional step in the process of medical data collection; biobanking is the storage of human tissue material for research or quality control of health care."

"Biobanking raises several ethical issues, such as the tension between the rights of the individuals versus progress in research, the need to provide adequate information before giving consent, and raising awareness of future unforeseen studies for which currently no specific consent can be asked for [85]."

"We want to discuss the legal and ethical implications on setting up two new biobanks in Belgium: one will obtain biopsies from bladder cancer and haematological tissues in relation to cancer registration and the second will deal with the collection of human tissue (blood samples) in an existing general practice based morbidity registration network."

Setting

"The purpose of the project is to obtain a large database providing data on incidence and prevalence of diseases in Belgium, as well as on used diagnostic tests and treatments."

"Intego has already received approval for the collection of medical data discussed above (but not yet for tissue collection) from the Commission for the Protection of Privacy and also from the ethical review board of the medical school of the Katholieke Universiteit Leuven [86]."

"To the existing morbidity registration database the possibility is investigated to collect tissue samples for research purposes."

"LIKAR collects cancer data of all inhabitants of Limburg, a province in the north–east of Belgium (a population of 798,583)."

"Data collection started in 1996 and is provided by all the pathological, cytological and haematological laboratories located in the province and some outside the province that regularly examine samples from Limburg inhabitants."

Definitions

"Personal data is referred to when any information relating to an identified or identifiable person is used."

"There exist essentially three kinds of personal data: anonymous, encoded or identifiable and identified data."

"Anonymous data cannot be connected to an identified or identifiable person."

"Encoded or identifiable data can only be connected to an identified or identifiable person by means of a code."

"The last category is identified data, which refers to personal data [87]."

Regulations and guidelines

"Art 2 of the Belgian Privacy Law expresses the basic idea: every natural person is entitled to the protection of his fundamental rights and freedoms and in particular their right to privacy with respect to the processing of personal data."

"We believe, however, that the guidelines are generally consistent with practices using broad and future consents for biobank research on the conditions that (1) personal information related to the research is safely handled, (2) donors of biological samples are granted the right to withdraw consent and, (3) new research studies or changes regarding the legal or ethical authority of a biobank are subject to approval by an ethical review board [88]."

"The notification should at least contain information about why biological samples are being collected, used and stored, indicate possible risks, describe that personal data are encoded and protected in accordance with appropriate legislation and inform that there is a possibility to withdraw consent at any time."

2.2.23 Strengthen Electronic Health Records System (EHR-S) Access-Control to Cope with GDPR Explicit Consent [104]

This is a machine-generated summary of:

de Carvalho Junior, Marcelo Antonio; Bandiera-Paiva, Paulo: Strengthen Electronic Health Records System (EHR-S) Access-Control to Cope with GDPR Explicit Consent [104].

Published in: Journal of Medical Systems (2020).

Link to original: https://doi.org/10.1007/s10916-020-01631-5.

If you want to cite the papers, please refer to the original.

For technical reasons we could not place the page where the original quote is coming from.

Abstract-Summary

"To cope with this need, in this article, an adaptation of existent Role-Based Access Control (RBAC), including patient-centric control, is described."

"The revisited feature of existing administrative and supporting RBAC functions allows exclusive control orchestrated by the patient as sole information owner, including the ability to encrypt their data for confidentiality purposes."

"The additions mimic a Discretionary Access Control (DAC) capability using existing user group membership to vet access over symmetric keys bind to patient's data via the associated PERMS matrix."

Introduction

"[65–67] Since EHR-S is a specific type of system within the realm of the Health Information System (HIS) ecosystem, devoted to more intensive interaction and use by health-care professional personnel during servicing, it requires information security measures to protect and regulate access to patient data."

"As health-care services and EHR-S functionalities have evolved, different researchers have pointed out gaps or failures in RBAC-protected systems that are designed to cope with the needs of contemporary access."

"In this particular paper, we address the need for additional access authorization to be granted by data owners (patient consent) to health-care professionals prior to using role-authorized operations for handling private EHR data."

"An important extra challenge to be considered at any adaptation endeavor is that using RBAC based solution as an access-control to vet users limited interaction over EHR data, the number of roles added to achieve more granular restriction builds up system decision processing in terms of complexity."

Objective and methods

"The main goal is to provide for access-control decision flow that embraces RBAC core functions with minimal changes."

"For that, we propose rewriting functions in a way that the authorization process can include owner discretion and management while keeping the remaining RBAC characteristics and functionalities."

"A simple Proof of Concept (PoC) is developed to test and demonstrate the newly designed functions behavior and the resulting authorization orchestration."

"In addressing owner discretion capabilities on RBAC, we present our system changes and RBAC remodeling, which consisting of two newly designed core functions for RA and UA operations within security model."

Addressing owner discretion capabilities on RBAC

"Patient ID, an especial role and access-controlled objects (folders and cryptographic key files) are named using this unique keyword so that the regular RBAC function based on centralized management can reach the whole set of objects, but discretionary control to the patient is promoted restricting interactions that match the same owner_id."

"The access requestor ID that was priorly sent to the information owner during consent collection is used to bind this user with the only role within EHR-S scope that is capable of interacting with a patient's personal folder contents."

"Role and owner_id are used to check if the inner objects within PERMS match before performing the special UA, formally: During PHR related operations, the sole objective is granting system user's with the necessary permissions to retrieve the patient key."

"In our particular PoC, both key-material and key are stored into a patient personal folder, which restricts access to the special PHR role only and related users."

Conclusion

"We have demonstrated a simple PoC that illustrates that DAC basic capabilities required for GDPR and other privacy-related compliances can be achieved using the existing RBAC functions with minor changes in the access-control."

"PHR data secrecy is maintained and controlled by the information owner as the only actor capable of allowing others to access private data that is protected using cryptographic functions."

"The suggested RBAC function additions mimic a Discretionary Access Control (DAC) capability using existing user group membership process to vet access over symmetric keys bind to patient's data via the associated PERMS matrix."

2.2.24 Analysis of Solutions for a Blockchain Compliance with GDPR [105]

This is a machine-generated summary of:

Godyn, Mateusz; Kedziora, Michal; Ren, Yingying; Liu, Yongxin; Song, Houbing Herbert: Analysis of solutions for a blockchain compliance with GDPR [105].

Published in: Scientific Reports (2022).

Link to original: https://doi.org/10.1038/s41598-02219341-y.

If you want to cite the papers, please refer to the original.

For technical reasons we could not place the page where the original quote is coming from.

Abstract-Summary

"The aim of this paper was to perform an analysis of the state-of-the-art solutions of the permissioned blockchain compliance with the General Data Protection Regulation (GDPR), including the implementation of one of the analyzed methods and the own solution."

"This paper covers the subject of GDPR and its impact on already existing blockchain databases to determine the domain of the problem, including the necessity to introduce mutability in the data structure to comply with the "right to be forgotten"."

"In the experimental part, attempts were made to research and implement the Reference-based Tree Structure (RBTS), including the performance tests."

Introduction

"To make the blockchain comply with GDPR and not be susceptible to deletion, there should be a way to manipulate the data in it, which is a difficult problem because the core of the blockchain is immutability: without it, chains of blocks cannot be validated and therefore stored as true in a distributed network [68]."

"Many of their goals are based on GDPR, but it is difficult to see the situation when after losing the key, other participants would propose modification on the transaction sender's behalf leaving the personal data in the blockchain without any way to delete it."

"In that way, the root of the tree takes the role of the genesis block and allows linking to it up to 16 subchains depending on the business operations like transferring currency to organization A by person A, transferring between organizations A and B, persons A and B, etc. Kuperberg's idea is quite easy to comprehend and to work with if one omits problems like how to handle data integrity, authorized voting, etc. that are not strictly related to the blockchain itself."

Experiment results

"Another transaction will be formed as an agreement between A and B referencing the previously created transaction and block and put into a new block mined and placed on subchain A, B. Finally, the state of the blockchain will be shown, the participant B will leave the blockchain, and to verify the state, the blockchain will be presented again in the console."

"Time to delete a subchain does not exceed 160 μs for lengths up to 2000 blocks."

"Inspecting context chains with the highest deletion times has shown no correlation between the time to delete and the number of participants in the chain."

"The proposed implementation considers this and makes time to delete any of the context chains not bound to total number of subchains in blockchain."

"This implementation does not impact performance for write operations to the blockchain, and deleting particular chains on demand is quick."

Conclusion and future work

"The aim of this paper was to analyze existing methods for enabling compliance between blockchain and the General Data Protection Regulation introduced by the European Parliament, then propose and implement proof-of-concept for one of the covered methods."

"To achieve GDPR compliance, it is clear that some form of mutability in the blockchain is necessary to be able to fulfill potential deletion requests using the individual's right to be forgotten."

"The key advantages of the proposed solution is a unique design that allows deleting data from the blockchain 'silently' without interaction from other blockchain nodes and no impact on the performance of writing data into the database."

"The solution has some downsides, though, like the use of a very simple blockchain without interfaces for distributed testing, or the only consensus in form of a Proof of Work algorithm."

2.2.25 Survey of NCHDs in Ireland to Assess Their Views and Opinions in Relation to Participation in Health Research and the Impact of New Irish Data Protection Regulations [106]

This is a machine-generated summary of:

Wallace, Rachel; Greene, Elaine: Survey of NCHDs in Ireland to assess their views and opinions in relation to participation in health research and the impact of new Irish data protection regulations [106].

Published in: Irish Journal of Medical Science (2020).

Link to original: https://doi.org/10.1007/s11845-020-02185-8.

If you want to cite the papers, please refer to the original.

For technical reasons we could not place the page where the original quote is coming from.

Abstract-Summary

"Since August 2018, Irish health research is regulated by the European General Data Protection Act (GDPR) 2016 with the (Irish) Data Protection Act and Health Research Regulations (HRR) 2018."

"This project aimed to collect the views of NCHDs in Ireland in relation to health research and the impact of GDPR and HRR."

"Internationally barriers to engagement with health research have been identified; however, to our knowledge, no previous studies have been conducted in Ireland."

"Ninety-nine percent had previously participated in research, and 87% acknowledged the importance for doctors to be involved in research."

"Forty-four percent of NCHDs stated they would consider travelling abroad to complete research."

"Our results confirm that NCHDs in Ireland are enthusiastic about engaging in research."

"Engaging in research abroad could be further explored as a factor impacting emigration."

"Our findings show that NCHDs in Ireland are concerned about the impact of GDPR and HRR and are seeking further training."

Introduction

"Protection of patient confidentiality is at the core of health research and is heavily governed in Ireland under data protection law, international human rights law, common law, constitutional law, professional regulation and ethics in the conduct of research [69, 70]."

"General Data Protection Regulations (GDPR) commencement and practice in the European Union (EU) began on May 28, 2018, and August 2018 saw the commencement of the Health Research Regulations 2018 in Ireland."

"Ireland then interpreted these regulations and published the Data Protection Act 2018 with the Health Research Regulations 2018 (HRR)."

"The new regulations have resulted in significant changes for current practice in health research, discussing these in detail is beyond the scope of this article."

"The aim of this study was to collect and report the views and attitudes of NCHDs in Ireland in relation to health research."

Methods

"All Irish colleges and faculties were emailed, the study was explained, and we requested that they distribute the survey to their NCHD members."

"The "Doctors Res" and "Enough is Enough-Irish NCHDs Oppose Current Working Conditions in the HSE" public Facebook groups were used to distribute the survey in attempts to reach NCHDs whose faculties did not circulate the survey."

"All NCHDs currently working in Ireland were eligible to complete the survey."

"As colleges and faculties responded and distributed the survey at staggered intervals, it remained open over a three-month period."

Results

"When asked if they would like to be involved in research, 89% (n = 170) responded yes."

"Of the respondents, 178 (93%) felt that the new regulations would have implications for research in Ireland."

"Envisioning more generalised impacts following the implementation of the regulations, 98% of the respondents speculated that health research will be more challenging to conduct."

"82% of respondents anticipated that there will be new barriers and challenges to conducting research and more specifically sharing of data with our international colleagues as a result of these regulations."

"Eighty-six percent of respondents forecast that patients who do not have the capacity to consent will be excluded from participating in research."

"Pertaining to the conducting of research, 80% of participants agreed that patient recruitment will become more challenging as a result of the regulations."

Discussion

"While there have been recent national attempts to improve structured support and funding for research in Ireland, for example the Health Research Board (HRB) Irish Clinical Academic Training (ICAT) programme, incorporating suggested strategies more consistently into training schemes may make research a more attractive option for a wider group of NCHDs [71]."

"We have already identified that this small sample of NCHDs are enthusiastic and involved in research, doctors who would take an opportunity to receive training if available."

"A lack of training and education may result in a lack of confidence in trainees, as they are fearful of conducting research incorrectly and potentially breaching regulations."

"To explore NCHDs awareness and attitudes towards the potential impact of the new regulations on health research in the future, a number of statements were given that encompass views of the research community."

Conclusion

"These positive attitudes however will need to be met at training and service level with further support and resources that enable NCHDs to meaningfully invest themselves in research."

"Specific training and guidance in relation to GDPR and HRR is also needed nationally to allow NCHDs to confidently continue conducting research."

"Further research is needed in relation to Irish NCHD's adaptation to these new regulations and resultant implications going forward."

2.2.26 Research Under the GDPR—A Level Playing Field for Public and Private Sector Research? [107]

This is a machine-generated summary of:

Quinn, Paul: Research under the GDPR—a level playing field for public and private sector research? [107].

Published in: Life Sciences, Society and Policy (2021).

Link to original: https://doi.org/10.1186/s40504-021-00111-z.

Copyright of the summarized publication:

If you want to cite the papers, please refer to the original.

For technical reasons we could not place the page where the original quote is coming from.

Abstract-Summary

"Such research is not the domain of a single type of organization but can be conducted by a range of different entities in both the public and private sectors."

"The use of personal data may be indispensable for many forms of research, the data protection framework will play an important role in determining not only what types of research may occur but also which types of actors may carry it out."

"This article looks at the role the EU's General Data Regulation plays in determining which types of actors can conduct research with personal data."

1. **Introduction**

"This article aims to demonstrate which legal bases may be of most use to various types of research actor and aims in particular to contrast the position of researchers based in public sector institutions like universities with others working in the private sector e.g. for large commercial entities."

"Whilst as sections 6 and 7 discuss, certain legal bases (e.g. allowing for research that is in the 'public interest' or for 'scientific research') may be in theory open to both public and private actors alike, the de facto reality surrounding their use means that private sector entities may find it more difficult to use them."

"A further important possibility is the seemingly very broad option to further process any personal data a data controller may be in possession of for purposes of scientific research so long as it had a valid legal base for the possession of the data in the first place."

2. **The use of personal data is becoming more common and complex in the age of big data**

"The use of such large forms of secondary data may have considerable advantages, including the analytic power that such large datasets often bring and also potentially the ability to avoid problems surrounding the primary collection of data, e.g. administrational, practical, ethical etc. Whilst this world of readily available big data might facilitate research in many ways it also brings more research within the purview of data protection."

"The ability of researchers to access such data may vary depending on the type of entity they are and the context of the research in question (further discussed in section 6). (ii) Many forms of big data may contain personal data (even where not immediately obvious)."

"In the big data world it may not always be intuitively obvious whether large data sets contain personal data within them."

3. **The implications of data protection for research**

"In terms of its general application, it does not apply to specific contexts, but rather to most instances where personal data is being processed (including but thus not limited to research)."

"This can be used to allow a broader form of consent for the use of personal data in scientific research than may be possible for other forms of processing. (ii) Data subject rights Data controllers must facilitate a number of important data subject rights, including in research contexts."

"One important consideration for researchers is that the legal base selected for the processing of personal data in research may determine the extent to which data subject rights apply given the ability the GDPR allows Member States to limit them in instances of scientific research (discussed further in section 4b). (iii) Administrative requirements The GDPR foresees a number of important administrative requirements that are likely to apply to researchers, particularly when they use sensitive data."

4. **The GDPR provides a wealth of legal bases for researchers**

"Another important feature of this legal base that may often make it attractive to researchers is that Member States can, through legislation, limit a number of data subject rights normally incumbent upon instances of processing of personal data (discussed further below in (b))."

"Having to contact data subjects even in instances where consent is not used as the legal basis for processing would often present serious problems for researchers that opt for other bases such as public interest or legitimate interests (where such bases are indeed available)."

"Below however the second potential option of a valid legal base existing simply for 'further processing for scientific research' is more puzzling and raises serious questions about the potential limitation of data subject rights in a wide range of contexts."

5. **Potential legal bases for research with sensitive data**

"Whilst these are important requirements that should not be underestimated, they will not be explored further in this article given they apply to most forms of research with sensitive data irrespective of the identity of the data controller or the legal base used."

"As with non-sensitive data, the remaining focus will be on the legal bases available to researchers given that they may in reality be differently available to different types of actors engaged in research."

"The most important grounds for research are: As for non-sensitive personal data, consent is usually considered as one of the most important bases for the use of sensitive data [96]."

"As with the use of the public interest legal base for non sensitive data (see section 4), there is however important conditionality attached [97, 98]."

6. **An unlevel playing field?**

"The availability of the a legal base that does not require consent is thus important for many forms of research. (ii) The importance of the scientific research exception Legislation exits in all EU Member States that permits universities to process both non sensitive and sensitive personal data for aims of scientific research [99]."

"As section 5 discussed, some ethics bodies may be more hostile than others concerning the use of the scientific research exception (or more specifically the processing of personal data in the absence of consent) [100]."

"Whilst large and important public sector entities such as universities, university hospitals and large research institutions may have been considered important enough to merit legislation facilitating the use of a personal data for research, this may not be the case for commercial entities [101]."

7. **Are private sector interests favoured by the GDPR?**

"The sheer amount of data that many large commercial actors may possess, taken together with the existence of a legal grounds such as 'legitimate interests' or 'further processing for scientific research' and the relative weakness of their forms of ethical review may often provide an advantage in terms of the ability to conduct research, particularly with big data."

"This factor renders the available advantage to private sector researchers of these options less significant than it might seem, especially given the importance of sensitive data to many forms of research."

"A reduced ability to process sensitive research data (without resorting to consent as a legal base) is therefore an important de facto impediment to research in the private sector that should not be underestimated."

"This places a severe limitation on the ability of private sector researchers to be able to rely on 'legitimate interests' as a base in the future, especially in areas of highly valued research (that often require sensitive data)."

Conclusion

"This factor, taken together with the availability of the 'legitimate interest' ground for private entities (which is not available for public research institutions) and the comparative superior access that some commercial organisations may have to large pools of customer data mean that the de facto advantage public institutions possess over private sector competitors to conduct research is less than it would appear on a simple reading of the GDPR."

"Not being able to use the legal bases it can for research with non-sensitive data to conduct research with sensitive data means that private entities will often be forced to rely on explicit consent as a base for processing where universities may be able to use one of the research exceptions outlined in the GDPR."

"This problem will arguably become increasingly important for research in the future and mean that researchers (in both the public and private realms) will have to seek a legal base for processing sensitive data."

2.2.27 A Framework for GDPR Compliance for Small- and Medium-Sized Enterprises [108]

This is a machine-generated summary of:

Brodin, Martin: A Framework for GDPR Compliance for Small- and Medium-Sized Enterprises [108].

Published in: European Journal for Security Research (2019).

Link to original: https://doi.org/10.1007/s41125-019-00042-z.

If you want to cite the papers, please refer to the original.

For technical reasons we could not place the page where the original quote is coming from.

Abstract-Summary

"The EU's General Data Protection (GDPR) is an EU regulation that affects everyone in the EU and all organisations outside the EU that wants to do business with the EU."

"This work uses design science to develop a framework for SMEs to adapt to GDPR."

"The framework was empirically evaluated in three different types of organisations, resulting of GDPR compliance according to their Data Protection Officers."

"The paper also highlights some of the most important changes in GDPR compared to its predecessor, Directive 95/46/EC of the European Parliament and of the Council of 24 October 1995 on the protection of individuals with regard to the processing of personal data and on the free movement of such data (DIR95)."

Introduction

"The paper is structured as follows: section one introduces GDPR and presents what is new in GDPR compared to its predecessor—directive 95/46/EC of the European Parliament and of the Council of 24 (October 1995) on the protection of individuals with regard to the processing of personal data and on the free movement of such data (DIR95)."

"DIR95 states that personal data shall be processed lawfully and fairly, whereas GDPR adds in a transparent manner in relation to the data subject."

"GDPR also introduce a special section for processing of children's personal data."

"A new right in GDPR is the Right to data portability which allows the data subject to obtain and reuse their personal data for their own purposes across different services if the processing is carried out by automated means and the legal ground is consent or contract."

Method

"Design science deals with two challenges; to address a practical problem in a specific organisational setting and to construct and evaluate an artefact (in this case a low-cost method, with accompanying tools) that addresses the class of problems identified (complying with a new data protection law)."

"As the name design science implies, design is the central part of the research process and the development of the artefact take place in a design search or development process [102–108]."

"When the artefact is designed it must be evaluated [103–108]."

"The artefact was developed during an iterative process with evaluations in practice followed by adjustments of the framework."

"It is important to communicate the final result [103, 105] (Hevner and Chatterjee 103; Peffers and others 105), and the final version of the framework was communicated to interested organisations and to the research community through this article."

Framework

"Managing information on mobile devices is close to the work that is needed for GDPR, since both are about achieving control of the organisation's data and protecting it at all times."

"The first analysis is information analysis, which is conducted in one or more workshops, to get an overview of all personal data processing in the organisation."

"When the organisation knows what kind of personal data they process and where it is stored it is time for information classification followed by an information security analysis."

"Information classification—requiring the classification scheme to be updated to include personal data according to GDPR, and some information needs to be re-classified."

"Policies that manage processing of personal data needs to be updated, for instance the personal data policy; which clarify what kind of information that is collected, how it is collected and how it is used and stored."

Empirical Evaluation

"A register of all personal data processing was created with the help of previous work and an information analysis."

"Out was to update work routines where personal data are processed to make sure that all personal data are processed in accordance with GDPR."

"The next task was to create or updated processes, routines, and policies are that are affected by GDPR—for instance, processes in the information management system regarding how to start a new data processing that contains personal data, and procurement of IT systems."

"Communication was intranet-based (where a news item was published and templates uploaded), and through the information management system (where the process regarding personal data processing was updated according to decisions in the design phase)."

"During the implementation, all employees received an GDPR education to get a basic understanding of the regulation and how the organisation has chosen to work with personal data processing from now on."

Theoretical Evaluation

"Specifying data needs and usage is the first implication, which is addressed in the first analysis, information analysis."

"The challenge for SMEs is to get a picture of all personal data processing, its legal basis and that the processing is in accordance to GDPR."

"Information analysis results in a list of all personal data processing and the answer to the implication for each process, it also shows if there is any data processing in international context (the second implication)."

"An important aspect of demonstrating compliance with GDPR requirements is to ensure security by technical and organisational measures."

"Building privacy through data protection by design and default is about implementing technical and organisational measures and procedures to ensure protection of the rights of all data subjects and compliance with the GDPR."

"It is also important for organisations to provide information to data subjects about their rights, for instance by a personal data policy."

Conclusion

"The framework was developed using design science, and the compliance results was reviewed and approved in each of the organisations by both experts and senior management."

"The result was also evaluated against the twelve practical implications for organisations in their road to GDPR compliance identified in scientific literature [109]."

"The result suggests that the framework can be of great help to SMEs and provide a secure, low-cost process for GDPR compliance."

"Although the framework has only been evaluated in three organisations, the result indicates that it is effective."

"GDPR applies throughout the EU/EEA but this framework is only tested in Sweden."

2.2.28 Deontic Modality in the GDPR Based Finnish Privacy Notices in the Light of the Transparency Principle [109]

This is a machine-generated summary of:

Rydzewska-Siemiątkowska, Joanna: Deontic Modality in the GDPR Based Finnish Privacy Notices in the Light of the Transparency Principle [109].

Published in: International Journal for the Semiotics of Law—Revue internationale de Sémiotique juridique (2022).

Link to original: https://doi.org/10.1007/s11196-022-09939-w.

If you want to cite the papers, please refer to the original.

For technical reasons we could not place the page where the original quote is coming from.

Abstract-Summary

"As the Regulation requires that all information about data subjects' rights be conveyed in a clear and plain language (transparency principle), this study focuses on the linguistic means of expressing rights and obligations in the GDPR based privacy notices."

"The research corpus consists of fourteen privacy notices of Finnish design companies which were analysed in order to identify deontic expressions."

"The qualitative analysis of the expressions with the help of Easy Language Meter criteria for Finnish revealed that privacy notices show simplifying tendencies like choice of modal expressions, the use of certain subjects and the way of addressing the reader."

"These comply with the Plain Language principles and thus the transparency principle highlighted in the GDPR."

Introduction: Rights and Obligations Under the GDPR

"It aims at strengthening the right of individuals to the protection of personal data and harmonises privacy laws across Europe."

"The right to the protection of personal data is considered a fundamental right as expressed in the Charter of Fundamental Rights of the European Union (Article 8(1)) [72]."

"For the sake of terminological clarity, when referring to the GDPR the term 'data protection' will be used according to the meaning constituted in the European law, as opposed to the concept of 'information privacy', used in the American tradition [2]."

"The Finnish version of the General Data Protection Regulation was selected as a research material in this study due to the salient role of the GDPR in setting data protection law worldwide [73]."

"The point of departure is the analysis of deontic modality means in the Articles 15–21 of the GDPR and privacy notices, respectively."

Privacy Notice as a Source of Rights and Obligations

"The GDPR introduces a catalogue of such rights of the individuals and obligations of certain entities, and privacy notices summarise them."

"Issuing such a notice means adhering to the 'right to be informed' or in other terms to the controller's obligation to inform data subjects (e.g. customers of an online shop) about the details of processing of their data."

"The difference is that a 'privacy notice' is a public document issued for the data subjects, whereas the 'privacy policy' can be described as a rather internal document for the entity's employees which holistically describes the company's mission [74]."

"It has to be stressed that no such terms as 'privacy notice' or 'privacy policy' are mentioned in the GDPR."

"The meaning of the term will refer to any public document issued by the company which fulfils the informative obligation of the entity towards its customers (data subjects)."

Deontic Modality

"The usage and frequency of the aforementioned expressions differ depending on the genre of the text (e.g. the Criminal Code of Finland 'Rikoslaki' uses a stronger necessive construction on tehtävä not the verb tulee to impose obligation) and the language register."

"Rydzewska-Siemiątkowska [75] examined the synonymy of the aforementioned obligation verb tulee and the necessive construction on tehtävä basing on a survey among Finnish natives and showed that despite the same normative character and near-synonymous meaning of these forms, the expression on tehtävä was perceived more obligatory by the informants."

"Laine has shown that the aforementioned necessive construction on tehtävä is widely used in the simplified texts whereas the obligation verb tulee is a prevalent modal form in the original version of the official document."

Plain Language Background and Principles

"The Finnish name for Plain Language, selkeä kieli, is close to selkokieli, which is a compound word denoting an easier variation of the plain language, namely the so called Easy Language."

"While there has been no legislation on the status of Easy Finnish, the significance of the two language varieties for the Finnish society is also clearly emphasised for the first time in the Finnish government resolution from 2021."

"It states that: "The aim is to increase the use of Easy Finnish and Easy Swedish in the activities of ministries and administrations, in communication and event planning, as well as it is to increase the competences concerning the Easy Language and Plain Language" [76]."

"Plain Language is close to the Advanced Easy Finnish which is the most demanding version of Easy Language [77]."

Principle of Transparency in the GDPR

"A detailed way of compliance with the transparency rule can be found in Recital 39 which states that [78]: "[t]he principle of transparency requires that any information and communication relating to the processing of those personal data be easily accessible and easy to understand, and that clear and plain language be used. (…)" [the highlight is author's]."

"Giving consent to data processing is also subject to the transparency principle."

"All things considered, complying with the principle of transparency is often a matter of constant bargaining which an apt remark illustrates: [t]here is an inherent tension in the GDPR between the requirements on the one hand to provide the comprehensive information to data subjects which is required under the GDPR, and on the other hand do so in a form that is concise, transparent, intelligible and easily accessible [73, 79]."

Data and Methodology

"The research material for this study consists of a corpus of fourteen design companies' privacy notices."

"The main line of business of the companies included into the analysis is manufacturing of ceramic household and decorational articles ("keraamisten talous- ja koriste-esineiden valmistus" in Finnish) and bears the Standard Industrial Classification number 23,410."

"The list of companies for the analysis excluded businesses that either did not have their own webpage or did not include any privacy notice on their webpage at all."

"The corpus of Finnish companies' privacy notices was scrutinised for deontic modality."

"The modal expressions in the privacy notices were scrutinised and analysed against chosen criteria of the Easy Language Metre developed for Finnish which are connected with the deontic expressions."

"Privacy notices mentioned exactly those rights, hence they were included into the analysis."

Linguistic Analysis of Rights and Obligations in the GDPR

"FI: Rekisteröidyllä on oikeus saada rekisterinpitäjä poistamaan rekisteröityä koskevat henkilötiedot ilman aiheetonta viivytystä, ja rekisterinpitäjällä on velvollisuus poistaa henkilötiedot ilman aiheetonta viivytystä, edellyttäen että jokin seuraavista perusteista täyttyy: (...) EN: The data subject shall have the right to obtain from the controller the erasure of personal data concerning him or her without undue delay and the controller shall have the obligation to erase personal data without undue delay where one of the following grounds applies: The controller's obligation which is formulated in a clear explicit way in the GDPR with the phrase rekisterinpitäjällä on velvollisuus 'controller shall have the obligation' in the privacy notice can be modified into a passive voice without any modal: FI: (...) epätarkat, virheelliset tai vanhentuneet henkilötiedot poistetaan tai oikaistaan viipymättä."

Discussion

"The aim of the article was to examine the most common indicators of deontic modality in the GDPR text and in the Finnish privacy notices."

"Although privacy notices use similar deontic structures, on the contrary, they revealed features which are untypical for the language of law It can be concluded that modality is often expressed in a more direct way than in the legislative text."

"The lack of the obligation verb tulee in the GDPR on the one hand, and its many occurrences in privacy notices, on the other, constitute a distinctive feature between privacy notices and a legislative text in which there is a tendency to avoid the use of explicit modal expressions to impose obligation [80]."

"A privacy notice may be perceived as a text belonging to a 'juristic language', i.e. especially in Polish legal tradition understood as a language that is a metalanguage towards the legislative language [81]."

Final Remarks

"This study provides useful insight into the deontic modality in the new type of an informative text which is the privacy notice and shows how the language of law can be made more accessible."

"The analysis drew upon identifying and comparing several patterns of deontic modality with regard to two different text genres and language registers and additionally referring to some Plain Language principles."

"The first genre is the legislative text of the GDPR, and the second one, a privacy notice."

"Although this study provided only a preliminary linguistic analysis on modality, it has revealed certain modal patterns in privacy notices."

"All things considered, a privacy notice is a new and an interesting text genre which is yet to be explored, with its modal means and its potential to become more accessible in the spirit of Plain Language principles."

References

1. Sant'Ana RCG (2016) Data life cycle: A perspective from the Information Science. Informação & Informação 21(2):116–142. https://doi.org/10.5433/1981-8920.2016v21n2p116
2. Hoofnagle CJ, Van der Sloot B, Borgesius FZ (2019) The European Union general data protection regulation: what it is and what it means. Information & Communications Technology Law 28(1):65–98. https://doi.org/10.1080/13600834.2019.1573501
3. Mercer S (2020) The limitations of European data protection as a model for global privacy regulation. AJIL Unbound 114:20–25. https://doi.org/10.1017/aju.2019.83
4. GDPR (General Data Protection Regulation) (2016) Regulation (EU) 2016/679 of the European Parliament and of the Council. https://eur-lex.europa.eu/eli/reg/2016/679/oj. Accessed 24 April 2020
5. Eifert M, Hoffmann-Riem W, Schmidt-Aßmann E, Voßkuhle A (2012) Regulierungsstrategien. In: Grundlagen des Verwaltungsrechts. Bd. I "Methoden – Maßstäbe – Aufgaben – Organisation", 2nd edn. Beck, München
6. Labadie C, Legner C (2019) Understanding data protection regulations from a data management perspective: a capability-based approach to EU-GDPR, In: Internationale Tagung Wirtschaftsinformatik 2019. https://wi2019.de/wp-content/uploads/Tagungsband_WI2019_reduziert.pdf. Accessed 10 Feb 2020
7. Agarwal S, Steyskal S, Antunovic F, Kirrane S (2018) Legislative compliance assessment: framework, model and gdpr instantiation. In: Annual privacy forum. Springer, Cham, pp 131–149
8. Information Commissioner's Office (ICO) UK (2017) Getting ready for the GDPR. https://ico.org.uk/for-organisations/resources-and-support/data-protection-self-assessment/getting-ready-for-the-gdpr/
9. Microsoft Trust Center (2017) Detailed GDPR Assessment. http://aka.ms/gdprdetailedassessment
10. Nymity. GDPR Compliance Toolkit. https://www.nymity.com/gdpr-toolkit.aspx
11. Bonatti PA, Coi JLD, Olmedilla D, Sauro L (2010) A rule-based trust negotiation system. IEEE Trans Knowl Data Eng 22(11):1507–1520. https://doi.org/10.1109/TKDE.2010.83
12. Jajodia S, Samarati P, Sapino ML, Subrahmanian VS (2001) Flexible support for multiple access control policies. ACM Trans Database Syst (TODS) 26(2):214–260

13. Kagal L, Finin T, Joshi A (2003) A policy language for a pervasive computing environment. In: Proceedings POLICY 2003. IEEE 4th international workshop on policies for distributed systems and networks. IEEE, pp 63–74
14. Uszok A, Bradshaw JM, Jeffers R, Suri N, Hayes PJ, Breedy MR, Bunch L, Johnson M, Kulkarni S, Lott J (2003) KAoS policy and domain services: toward a description-logic approach to policy representation, deconfliction, and enforcement. In: Proceedings POLICY 2003. IEEE 4th international workshop on policies for distributed systems and networks. IEEE, pp 93–96
15. Woo TYC, Lam SS (1993) Authorizations in distributed systems: a new approach. J Comput Secur 2(2–3):107–136. https://doi.org/10.3233/JCS-1993-22-304
16. Sergot MJ, Sadri F, Kowalski RA, Kriwaczek F, Hammond P, Cory HT (1986) The British nationality act as a logic program. Commun ACM. https://doi.org/10.1145/5689.5920
17. Bonatti PA (2010) Datalog for security, privacy and trust. In: Datalog Reloaded—First International Workshop, Datalog 2010. https://doi.org/10.1007/978-3-642-24206-9_2
18. Rubinstein I, Good N (2013) Privacy by design: a counterfactual analysis of Google and Facebook privacy incidents. Berkeley Technol Law J 28:1333–1413. https://doi.org/10.2139/ssrn.2128146
19. Huth D, Burmeister F, Matthes F, Schirmer I (2020) Empirical results on the collaboration between enterprise architecture and data protection management during the implementation of the GDPR. In: 2020 Hawaii international conference on system sciences (HICSS), 2020. https://doi.org/10.24251/HICSS.2020.715
20. EDPS (2018) Ethics: European Data Protection Supervisor. https://edps.europa.eu/data-protection/our-work/ethics_en. Accessed 10 March 2018
21. van der Aalst WMP. Process mining: data science in action. 2nd ed. Berlin: Springer; 2016. https://doi.org/10.1007/978-3-662-49851-4
22. Antignac T, Le Métayer D. Trust driven strategies for privacy by design. In: IFIP international conference on trust management, IFIPTM 2015. Cham: Springer; 2015, pp. 60–75
23. Anciaux N, Nguyen B, Vazirgiannis M. Miminum exposure in classification scenarios; 2011
24. Anciaux N, Boutara D, Nguyen B, Vazirgiannis M. Limiting data exposure in multi-label classification processes. Fundamenta Informaticae. 2015;137(2):219–36
25. Anciaux N, Nguyen B, Vazirgiannis M. Limiting data collection in application forms: a real-case application of a founding privacy principle. In: 2012 tenth annual international conference on privacy, security and trust, PST 2012. IEEE; 2012, pp. 59–66
26. Anciaux N, Bouganim L, Van Heerde H, Pucheral P, Apers PMG. Data degradation: making private data less sensitive over time. In: Proceedings of the 17th ACM conference on information and knowledge management, CIKM 2008; 2008, pp. 1401–1402
27. Anciaux N, Bouganim L, Van Heerde H, Pucheral P, Apers PMG. Instantdb: enforcing timely degradation of sensitive data. In: 2008 IEEE 24th international conference on data engineering, ICDE 2008. IEEE; 2008, pp 1373–1375
28. Hilderman RJ, Hamilton HJ, Cercone N. Data mining in large databases using domain generalization graphs. J Intell Inf Syst. 1999,13(3):195–234
29. van Heerde HJW, Anciaux N. Data degradation to enhance privacy for the Ambient Intelligence. CTIT Technical Report Series 11/06-74; 2006
30. Anciaux N, Bouganim L, Van Heerde H, Pucheral P, Apers P. The life-cycle policy model; 2008
31. European Union. Directive 95/46/EC of the European parliament and of the council on the protection of individuals with regard to the processing of personal data and on the free movement of such data. https://eur-lex.europa.eu/legal-content/en/TXT/?uri=CELEX%3A31995L0046. Accessed 17 Dec 2019
32. European Union. Regulation (EU) 2016/679 of the european parliament and of the council of 27 April 2016 on the protection of natural persons with regard to the processing of personal data and on the free movement of such data, and repealing directive 95/46/ec (general data protection regulation). https://eur-lex.europa.eu/eli/reg/2016/679/oj. Accessed 17 Dec 2019

33. European Union. Horizon 2020 work programme 2016–2017 14. secure societies—protecting freedom and security of Europe and its citizens (part 14—page 74). https://ec.europa.eu/research/participants/data/ref/h2020/wp/2016_2017/main/h2020-wp1617-security_en.pdf. Accessed 17 Dec 2019
34. European Union: The GDPR: New opportunities, new obligations. Justice and Consumers (2018)
35. Torre, D., Abualhaija, S., Sabetzadeh, M., Briand, L.C., Baetens, K., Goes, P., Forastie, S.: An AI-assisted approach for checking the completeness of privacy policies against GDPR. In: Proceedings of 28th IEEE International Conference on Requirements Engineering (RE'20) (2020)
36. Brambilla, M., Cabot, J., Wimmer, M.: Model-Driven Software Engineering in Practice, 2nd edn. Morgan & Claypool Publishers (2016)
37. Torre, D., Soltana, G., Sabetzadeh, M., Briand, L.C., Auffinger, Y., Goes, P.: Using models to enable compliance checking against the GDPR: an experience report. In: 22nd ACM/IEEE International Conference on Model Driven Engineering Languages and Systems, MODELS 2019, Munich, Germany, September 15–20, 2019, pp. 1–11 (2019)
38. Burmeister, F., Drews, P., Schirmer, I.: A privacy-driven enterprise architecture meta-model for supporting compliance with the general data protection regulation. In: Bui, T. (ed.) 52nd Hawaii International Conference on System Sciences, HICSS 2019, Grand Wailea, Maui, Hawaii, USA, January 8–11, 2019, pp. 1–10. ScholarSpace (2019)
39. Pariser, E.: The filter bubble: what the internet is hiding from you. Penguin UK, London (2011)
40. European Commission: A European strategy for data (2020)
41. European Parliament: European Parliament resolution of 3 October 2018 on distributed ledger technologies and blockchains: building trust with disintermediation (2017). https://www.europarl.europa.eu/doceo/document/TA-8-2018-0373_EN.html. Accessed 3 July 2022
42. Giannopoulou, A.: Data protection compliance challenges for self-sovereign identity. In: Prieto, J., Pinto, A., Das, A.K., Ferretti, S. (eds.) Blockchain and applications, pp. 91–100. Springer, Cham (2020)
43. Zichichi, M., Ferretti, S., D'Angelo, G., Rodríguez-Doncel, V.: Personal data access control through distributed authorization. In: 2020 IEEE 19th International symposium on network computing and applications (NCA), pp. 1–4. IEEE (2020)
44. Zichichi, M., Ferretti, S., D'Angelo, G.: On the efficiency of decentralized file storage for personal information management systems. In: Proc. of the 2nd International Workshop on Social (Media) Sensing, Co-located with 25th IEEE Symposium on Computers and Communications 2020 (ISCC2020), pp. 1–6. IEEE (2020)
45. Cavoukian, A.: Privacy by design. Take the challenge. Information and privacy commissioner of Ontario, Canada (2009)
46. Zichichi, M., Ferretti, S., D'Angelo, G.: A distributed ledger based infrastructure for smart transportation system and social good. In: 2020 IEEE 17th Annual Consumer Communications & Networking Conference (CCNC), pp. 1–6. IEEE (2020)
47. Aiello, M., Cambiaso, E., Canonico, R., Maccari, L., Mellia, M., Pescapè, A., Vaccari, I.: Ippo: A privacy-aware architecture for decentralized data-sharing. arXiv preprint arXiv:2001.06420 (2020)
48. Truong, N.B., Sun, K., Lee, G.M., Guo, Y.: Gdpr-compliant personal data management: a blockchain-based solution. IEEE Trans. Inf. Forensics Secur. 15, 1746–1761 (2020). https://doi.org/10.1109/TIFS.2019.2948287
49. Foundation, T.S.: Innovation meets compliance: data privacy regulation and distributed ledger technology. Technical report, The Sovrin Foundation (2020)
50. Finck, M., Pallas, F.: They who must not be identified-distinguishing personal from non-personal data under the GDPR. Int. Data Privacy Law 10(1), 11–36 (2020). https://doi.org/10.1093/idpl/ipz026
51. Article 29 Working Party: Opinion 05/2014 on Anonymisation Techniques (2014). https://ec.europa.eu/justice/article-29/documentation/opinion-recommendation/index_en.htm. Accessed 3 July 2022

52. CNIL—Commission nationale de l'informatique et des libertés: Solutions for a responsible use of the blockchain in the context of personal data (2018). https://www.cnil.fr/sites/default/files/atoms/files/blockchain_en.pdf. Accessed 3 July 2022
53. Rieger A, Guggenmos F, Locki J, Fridgen G, Urbach N (2019) Building a blockchain application that complies with the eu general data protection regulation. MIS Quarterly Executive 18(4):263–279. https://doi.org/10.17705/2msqe.00020
54. Ruiz, A., Lopez, A., Juez, G., Martinez, C., de la Vara, J., Álvarez, J., Parra, E., Alonso, L., Lopez, B., Puri, S., et al.: AMASS platform validation, D2.9, 2019, https://www.amass-ecsel.eu/sites/amass.drupal.pulsartecnalia.com/files/D2.9_AMASS-platform-validation_AMASS_Final.pdf
55. Eclipse, Opencert platform, https://www.eclipse.org/opencert (2021) (Accessed April, 2021)
56. Wilson, Robin Fretwell. 2016. The Promise of Informed Consent. Edited by I. Glenn Cohen, Allison K. Hoffman, and William M. Sage. Vol. 1. Oxford University Press. https://doi.org/10.1093/oxfordhb/9780199366521.013.53
57. Grote, T., Berens, P.: On the ethics of algorithmic decision-making in healthcare. J. Med. Ethics 46(3), 205–211 (2020). https://doi.org/10.1136/medethics-2019-105586
58. Schiff, D., Borenstein, J.: How should clinicians communicate with patients about the roles of artificially intelligent team members? AMA J Ethics 21(2), E138–E145 (2019). https://doi.org/10.1001/amajethics.2019.138
59. Wadden, J.J.: Defining the undefinable: the black box problem in healthcare artificial intelligence. J. Med. Ethics. (2021). https://doi.org/10.1136/medethics-2021-107529
60. Kingston J (1991) X-MATE: Creating an interpretation model for credit risk assessment. In: Research and development in expert systems VIII: proceedings of 11th annual technical conference of the BCS specialist group. Cambridge University Press, pp 165–174
61. Dhurandhar A, Ravi R, Graves B, Maniachari G, Ettl M (2015) Robust system for identifying procurement fraud. In Proceedings of the twenty seventh conference on innovative applications in artificial intelligence (IAAI-15), Austin, Texas, Jan 25–29 2015, pp 3896–3903
62. Thompson ED, Frolich E, Bellows JC, Bassford BE, Skiko EJ, Fox MS (2015) Process Diagnosis System (PDS)—A 30 Year History. In: Proceedings of the twenty seventh conference on innovative applications in artificial intelligence (IAAI-15), Austin, Texas, Jan 25–29 2015, pp 3928–3934
63. Aletras N, Tsarapatsanis D, Preotiuc-Pietro D, Lampos V (2016) Predicting judicial decisions of the European Court of Human Rights: a natural language processing perspective. Peer J Comput Sci. https://doi.org/10.7717/peerj-cs.93
64. Attardi, G., Basile, V., Bosco, C., Caselli, T., Dell'Orletta, F., Montemagni, S., Patti, V., Simi, M., & Sprugnoli, R. (2015). State of the art language technologies for italian: the EVALITA 2014 perspective. Intelligenza Artificiale, 9(1), 43–61. https://doi.org/10.3233/IA-150076
65. De Felice, I., Dell'Orletta, F., Venturi, G., Lenci, A., & Montemagni, S. (2018). Italian in the trenches: linguistic annotation and analysis of texts of the great war. In E. Cabrio, A. Mazzei, & F. Tamburini (Eds.) Proceedings of the 5th Italian conference on computational linguistics, CLiC-it 2018, CEUR Workshop Proceedings, (vol. 2253 pp. 1–5)
66. Regulation (EU) 2016/679 of the European Parliament and of the Council of 27 April 2016 on the protection of natural persons with regard to the processing of personal data and on the free movement of such data, and repealing Directive 95/46/EC (General Data Protection Regulation). Available from: https://eur-lex.europa.eu/eli/reg/2016/679/oj
67. Hoerbst A, Ammenwerth E. Quality and certification of electronic health records: an overview of current approaches from the US and Europe. Appl Clin Inform. 2010;1(2):149–64
68. Continua Design. Guidelines. 2017. Available from: https://www.pchalliance.org/continua-design-guidelines. Accessed 29 Apr 2019
69. Díaz, E., The new European Union General Regulation on Data Protection and the legal consequences for institutions. Church, Communication and Culture 1:206–239, 2016
70. SPMS—Serviços Partilhados do Ministério da Saúde, Privacidade da informação no setor da saúde, 2017.

71. Goubau, T.: How GDPR Will Change Personal Data Control and Personal Data Control an Affect Everyone in Construction. https://www.aproplan.com/blog/construction-news/gdpr-changes-personal-data-control-construction, last Accessed 2018/07/20
72. Martins, J., Gonçalves, R., Branco, F., Pereira, J., Peixoto, C., Rocha, T. How Ill Is online health care? an overview on the Iberia peninsula health care institutions websites accessibility levels. New Advances In Information Systems And Technologies, 445:391–400. Springer, 2016
73. Martins, J., Gonçalves, R., Oliveira, T., Cota, M., and Branco, F., Understanding the determinants of social network sites adoption at firm level: A mixed methodology approach. Electron. Commer. Res. Appl. 18:10–26, 2016
74. Article 29 Data Protection Working Party. Guidelines on consent under Regulation 2016/679. WP259 rev.01.
75. Chico V (2018) The impact of the general data protection regulation on health research. Br Med Bull 128(1):109–118. https://doi.org/10.1093/bmb/ldy038
76. Clarke N, Vale G, Reeves EP et al (2019) GDPR: an impediment to research? Ir J Med Sci 88(4):1129–1135
77. Coughlin, Steven S. 2011. Broad consent and biorepositories for molecular epidemiology and genomics research. International Journal of Molecular Epidemiology and Genetics 2 (4): 401–402
78. Faden, Ruth R., Nancy E. Kass, Steven N. Goodman, Peter Pronovost, Sean Tunis, and Tom L. Beauchamp. 2013. An ethics framework for a learning health care system: A departure from traditional research ethics and clinical ethics. The Hastings Center Report 43 (S1): S16–S27. https://doi.org/10.1002/hast.134
79. Karlsen, Jan Reinert, Jan Helge Solbakk, and Søren Holm. 2011. Ethical endgames: Broad consent for narrow interests; open consent for closed minds. Cambridge Quarterly of Healthcare Ethics 20 (4): 572–583. https://doi.org/10.1017/S0963180111000314
80. Sheehan, Mark. 2011a. Broad consent is informed consent. BMJ 343: d6900. https://doi.org/10.1136/bmj.d6900
81. Sheehan, Mark. 2011b. Can broad consent be informed consent? Public Health Ethics 4 (3): 226–235. https://doi.org/10.1093/phe/phr020
82. Collier, Roger. 2009. Rapidly rising clinical trial costs worry researchers. CMAJ. 180 (3): 277–278. https://doi.org/10.1503/cmaj.082041
83. Martin, Linda, Melissa Hutchens, Conrad Hawkins, and Alaina Radnov. 2017. How much do clinical trials cost? Nature Reviews. Drug Discovery 16 (6): 381–382. https://doi.org/10.1038/nrd.2017.70
84. Tham, Irene, Rachel Au-Yong, Tin May Linn, and Rodolfo Pazos. 2018. SingHealth cyber attack: How it unfolded. Straits Times, 20 July 2018. Available from https://graphics.straitstimes.com/STI/STIMEDIA/Interactives/2018/07/sg-cyber-breach/index.html
85. Bartholomeeusen, S., et al. 2005. The denominator in general practice, a new approach from the Intego database. Family Practice 22: 442–447
86. Cancer control using population-based registries and biobanks. 2007. http://www.cancerbiobank.org/
87. Hansson, M.O. 1998. Balancing the quality of consent. Journal of Medical Ethics 24: 182–187
88. Wendler, D. 2006. One-time general consent for research on biological samples. BMJ 332: 544–547
89. Kloss LL, Brodnik MS, Rinehart-Thompson LA. Access and disclosure of personal health information: A challenging privacy landscape in 2016–2018. Yearb Med Inform 2018; 60–66
90. Kloss L. AHIMA Press; 2015. Implementing information governance, lessons from the Field; pp. 111–113.
91. HealthIT.gov . Computable Privacy. 2016. Available from: https://www.healthit.gov/policy-researchers-implementers/computable-privacy
92. Farshid, S., Reitz, A. & Roßbach, P. Design of a forgetting blockchain: A possible way to accomplish gdpr compatibility. In Proceedings of the 52nd Hawaii International Conference on System Sciences (2019)

93. Medical Council (2019) Guide to Professional Conduct and Ethics for Registered Medical Practitioners (Amended) 8th Edition 2019. https://www.medicalcouncil.ie/News-and-Publications/Reports/Guide-to-Professional-Conduct-and-Ethics-for-Registered-Medical-Practitioners-Amended-.pdf. Accessed 12 September 2019
94. World Medical Association (2006) Declaration of Geneva (Amended) May 2006. https://www.wma.net/wp-content/uploads/2018/07/Decl-of-Geneva-v2006-1.pdf. Accessed 20 September 2019
95. Health Research Board Irish Clinical Academic Training (ICAT) Programme. https://icatprogramme.org/. Accessed 27 September 2019
96. Corrigan, O. 2003. Empty ethics: The problem with informed consent. Sociology of Health & Illness 25: p768–p792
97. Quinn, P. 2017. The Anonymisation of research data—A pyric victory for privacy that should not be pushed too hard by the EU data protection framework? European Journal of Health Law 24 (4): 1–21
98. Hallinan, D., and M. Friedewald. 2015. Open consent, biobanking and data protection law: Can open consent be 'informed' under the forthcoming data protection regulation? Life Sciences, Society and Policy 11 (1). https://doi.org/10.1186/s40504-014-0020-9
99. Molnár-Gábor, F. 2018. Germany: A fair balance between scientific freedom and data subjects' rights? Human Genetics 137: 619–626
100. Mostert, M., A. Bredenoord, M. Biesaart, and J. Van Delden. 2016. Big data in medical research and EU data protection law: Challenges to the consent or anonymise approach. European Journal of Human Genetics 24: 956–960
101. Taylor M,·Wallace S, Prictor P "United Kingdom: Transfers of genomic data to third countries", Human Genetics, (2018), 137, 637–645 In the UK for example, national legislation is seen as providing a broad discretion for public sector organizations to process genetic data without consent.
102. Gregor S, Jones D (2007) The anatomy of a design theory. J Assoc Inf Syst 8(5):312–335
103. Hevner A, Chatterjee S (2010) Design research in information systems, vol 22. Springer, New York
104. Nunamaker J, Chen M, Purdin T (1991) Systems development in information systems research. J Manag Inf Syst 7:89–106
105. Peffers K, Tuunanen T, Rothenberger MA, Chatterjee S (2007) A design science research methodology for information systems research. J Manag Inf Syst 24(3):45–78
106. Rossi M, Sein MK (2003) Design research workshop: a proactive research approach. In: 26th Information systems research seminar in Scandinavia (IRIS), Haikko, Finland
107. Vaishnavi VK, Kuechler W (2007) Design science research methods and patterns: innovating information and communication technology. Taylor & Francis Group, New York
108. Walls JG, Widmeyer GR, El Sawy OA (1992) Building an information system design theory for vigilant EIS. Inf Syst Res 3(1):36–59
109. Tikkinen-Piri C, Rohunen A, Markkula J (2017) EU general data protection regulation: changes and implications for personal data collecting companies. Comput Law Secur Rev 34(1):134–153
110. Charter of Fundamental Rights of the European Union OJ 2012 C 326/391
111. IT Governance Privacy Team. 2020. EU General Data Protection Regulation (GDPR)—An Implementation and Compliance Guide, Fourth Edition: Vol. Fourth edition. ITGP
112. Irwin, Luke. 2021. Is your organisation's privacy notice GDPR compliant? https://www.itgovernance.eu/blog/en/is-your-organisations-privacy-notice-gdpr-compliant. Accessed 28 February 2022
113. Rydzewska-Siemiątkowska, Joanna. 2016. Explaining synonymy between obligation expressions in Finnish legal language–results of a survey. Comparative Legilinguistics. https://doi.org/10.14746/cl.2016.28.4
114. Nationalspråkstrategin, Kansalliskielistrategia: Statsrådets principbeslut, Valtioneuvoston periaatepäätös 2021. Helsinki: Valtioneuvosto

115. Leskelä, Leealaura. 2021. Easy Language in Finland. In Handbook of Easy Languages in Europe, eds. Camilla Lindholm and Ulla Vanhatalo, 149–190. Frank & Timme
116. Regulation, (EU) 2016/679 of the European Parliament and of the Council of 27 April 2016 on the protection of natural persons with regard to the processing of personal data and on the free movement of such data, and repealing Directive 95/46/EC (General Data Protection Regulation), OJ 2016 L 119/1
117. Guidelines on Transparency under Regulation 2016/679. 2018. https://ec.europa.eu/newsroom/article29/items/622227/en. Accessed 9 February 2022
118. Kanner, Antti. 2011. Kansallisen lainsäädännön ja EU-säädösten kieli nesessiivisten rakenteiden valossa. Master's thesis. University of Helsinki
119. Mattila, Heikki, E. S. 2017. Vertaileva oikeuslingvistiikka. Helsinki: Alma Talent
120. Freund, Gislaine Parra; Fagundes, Priscila Basto; de Macedo, Douglas Dyllon Jeronimo An Analysis of Blockchain and GDPR under the Data Lifecycle Perspective. Mobile Networks and Applications (2020). https://doi.org/10.1007/s11036-020-01646-9
121. Jakobi, Timo; von Grafenstein, Maximilian; Legner, Christine; Labadie, Clément; Mertens, Peter; Öksüz, Ayten; Stevens, Gunnar The Role of IS in the Conflicting Interests Regarding GDPR. Business & Information Systems Engineering (2020). https://doi.org/10.1007/s12599-020-00633-4
122. Bonatti, Piero A.; Kirrane, Sabrina; Petrova, Iliana M.; Sauro, Luigi Machine Understandable Policies and GDPR Compliance Checking. KI—Künstliche Intelligenz (2020). https://doi.org/10.1007/s13218-020-00677-4
123. Georgiadis, Georgios; Poels, Geert Enterprise architecture management as a solution for addressing general data protection regulation requirements in a big data context: a systematic mapping study. Information Systems and e-Business Management (2021). https://doi.org/10.1007/s10257-020-00500-5
124. Zaman, Rashid; Hassani, Marwan On Enabling GDPR Compliance in Business Processes Through Data-Driven Solutions. SN Computer Science (2020). https://doi.org/10.1007/s42979-020-00215-x
125. de Carvalho, Renata M.; Del Prete, Camillo; Martin, Yod Samuel; Araujo Rivero, Rosa M.; Önen, Melek; Schiavo, Francesco Paolo; Rumín, Ángel Cuevas; Mouratidis, Haralambos; Yelmo, Juan C.; Koukovini, Maria N. Protecting Citizens' Personal Data and Privacy: Joint Effort from GDPR EU Cluster Research Projects. SN Computer Science (2020). https://doi.org/10.1007/s42979-020-00218-8
126. Torre, Damiano; Alferez, Mauricio; Soltana, Ghanem; Sabetzadeh, Mehrdad; Briand, Lionel Modeling data protection and privacy: application and experience with GDPR. Software and Systems Modeling (2021). https://doi.org/10.1007/s10270-021-00935-5
127. Zichichi, Mirko; Ferretti, Stefano; D'Angelo, Gabriele; Rodríguez-Doncel, Víctor Data governance through a multi-DLT architecture in view of the GDPR. Cluster Computing (2022). https://doi.org/10.1007/s10586-022-03691-3
128. Ruiz, Alejandra; Martin, Yod-Samuel; Martinez, Jabier; Quintans, Jacobo; Mockly, Guillaume; Gyrard, Amelie; Crepax, Tommaso Modeling ecosystems of reference frameworks for assurance: a case on privacy impact assessment regulation and guidelines. Software and Systems Modeling (2022). https://doi.org/10.1007/s10270-022-01061-6
129. Kawamleh, Suzanne Against explainability requirements for ethical artificial intelligence in health care. AI and Ethics (2022). https://doi.org/10.1007/s43681-022-00212-1
130. Kingston, John Using artificial intelligence to support compliance with the general data protection regulation. Artificial Intelligence and Law (2017). https://doi.org/10.1007/s10506-017-9206-9
131. Lorè, Filippo; Basile, Pierpaolo; Appice, Annalisa; de Gemmis, Marco; Malerba, Donato; Semeraro, Giovanni An AI framework to support decisions on GDPR compliance. Journal of Intelligent Information Systems (2023). https://doi.org/10.1007/s10844-023-00782-4
132. Kolasa, Katarzyna; Ken Redekop, W.; Berler, Alexander; Zah, Vladimir; Asche, Carl V. Future of Data Analytics in the Era of the General Data Protection Regulation in Europe. PharmacoEconomics (2020). https://doi.org/10.1007/s40273-020-00927-1

133. Lopes, Isabel Maria; Guarda, Teresa; Oliveira, Pedro General Data Protection Regulation in Health Clinics. Journal of Medical Systems (2020). https://doi.org/10.1007/s10916-020-1521-0
134. Staunton, Ciara; Slokenberga, Santa; Mascalzoni, Deborah The GDPR and the research exemption: considerations on the necessary safeguards for research biobanks. European Journal of Human Genetics (2019). https://doi.org/10.1038/s41431-019-0386-5
135. Davey, Matthew G.; O'Donnell, John P.M.; Maher, Elizabeth; McMenamin, Cliona; McAnena, Peter F.; Kerin, Michael J.; Miller, Nicola; Lowery, Aoife J. General data protection regulations (2018) and clinical research: perspectives of patients and doctors in an Irish university teaching hospital. Irish Journal of Medical Science (2021). https://doi.org/10.1007/s11845-021-02789-8
136. Clarke, Niamh; Vale, Gillian; Reeves, Emer P.; Kirwan, Mary; Smith, David; Farrell, Michael; Hurl, Gerard; McElvaney, Noel G. GDPR: an impediment to research?. Irish Journal of Medical Science (2019). https://doi.org/10.1007/s11845-019-01980-2
137. Tantoso, Erwin; Wong, Wing-Cheong; Tay, Wei Hong; Lee, Joanne; Sinha, Swati; Eisenhaber, Birgit; Eisenhaber, Frank Hypocrisy Around Medical Patient Data: Issues of Access for Biomedical Research, Data Quality, Usefulness for the Purpose and Omics Data as Game Changer. Asian Bioethics Review (2019). https://doi.org/10.1007/s41649-019-00085-3
138. Nyrén, Olof; Stenbeck, Magnus; Grönberg, Henrik The European Parliament proposal for the new EU General Data Protection Regulation may severely restrict European epidemiological research. European Journal of Epidemiology (2014). https://doi.org/10.1007/s10654-014-9909-0
139. Zemła-Pacud, Żaneta; Lenarczyk, Gabriela Clinical Trial Data Transparency in the EU: Is the New Clinical Trials Regulation a Game-Changer?. IIC—International Review of Intellectual Property and Competition Law (2023). https://doi.org/10.1007/s40319-023-01329-4
140. Vukovic, Jakov; Ivankovic, Damir; Habl, Claudia; Dimnjakovic, Jelena Enablers and barriers to the secondary use of health data in Europe: general data protection regulation perspective. Archives of Public Health (2022). https://doi.org/10.1186/s13690-022-00866-7
141. Truyers, Carla; Kellen, Eliane; Arbyn, Marc; Trommelmans, Leen; Nys, Herman; Hensen, Karen; Aertgeerts, Bert; Bartholomeeusen, Stefaan; Hansson, Mats; Buntinx, Frank The use of human tissue in epidemiological research; ethical and legal considerations in two biobanks in Belgium. Medicine, Health Care and Philosophy (2009). https://doi.org/10.1007/s11019-009-9230-y
142. de Carvalho Junior, Marcelo Antonio; Bandiera-Paiva, Paulo Strengthen Electronic Health Records System (EHR-S) Access-Control to Cope with GDPR Explicit Consent. Journal of Medical Systems (2020). https://doi.org/10.1007/s10916-020-01631-5
143. Godyn, Mateusz; Kedziora, Michal; Ren, Yingying; Liu, Yongxin; Song, Houbing Herbert Analysis of solutions for a blockchain compliance with GDPR. Scientific Reports (2022). https://doi.org/10.1038/s41598-022-19341-y
144. Wallace, Rachel; Greene, Elaine Survey of NCHDs in Ireland to assess their views and opinions in relation to participation in health research and the impact of new Irish data protection regulations. Irish Journal of Medical Science (2020). https://doi.org/10.1007/s11845-020-021 85-8
145. Quinn, Paul Research under the GDPR—a level playing field for public and private sector research?. Life Sciences, Society and Policy (2021). https://doi.org/10.1186/s40504-021-001 11-z
146. Brodin, Martin A Framework for GDPR Compliance for Small- and Medium-Sized Enterprises. European Journal for Security Research (2019). https://doi.org/10.1007/s41125-019-00042-z
147. Rydzewska-Siemiątkowska, Joanna Deontic Modality in the GDPR Based Finnish Privacy Notices in the Light of the Transparency Principle. International Journal for the Semiotics of Law—Revue internationale de Sémiotique juridique (2022). https://doi.org/10.1007/s11196-022-09939-w

Chapter 3
Bringing Privacy and Data Protection Issues Closer to Home

3.1 Introduction by the Editor

The Data Act in the European Union positions itself at the cusp of several transitions. These transitions include the growth of Internet of Things (IoT), the exponential processing and sharing of personal data and the process of creating a fair and equitable framework of data sharing.

Recital 5, of the Act states:

"This Regulation ensures that users of a connected product or related service in the Union can access, in a timely manner, the data generated by the use of that connected product or related service and that those users can use the data, including by sharing them with third parties of their choice. It imposes the obligation on data holders to make data available to users and third parties of the user's choice in certain circumstances. It also ensures that data holders make data available to data recipients in the Union under fair, reasonable and non-discriminatory terms and conditions and in a transparent manner".[1] The fundamental idea is to set a framework that protects personal data, make data subjects aware of their rights and set the way personal data are processed in an IoT environment. Recital 5 shifts the control to the data subjects and they get to decide the future and the fate of their personal data. Data subjects also get to control the personal data that they generate by using the connected devices. Further, there is a need to create an equitable framework so that the accessibility of such data does not become a cause for concern. The Act assuages several privacy concerns that originate from the uncertainties surrounding instantaneous processing of personal data through multiple interconnected devices.

The chapter is set in the context where more and more devices are getting interconnected in an environment, which is generally considered as private space. Several of

[1] -Regulation (EU) 2023/2854 of the European Parliament and of the Council of 13 December 2023 on harmonised rules on fair access to and use of data and amending Regulation (EU) 2017/2394 and Directive (EU) 2020/1828 (Data Act) (Text with EEA relevance) *OJ L, 2023/2854, 22.12.2023.*

I. Gupta (ed.), *Operationalizing Expectations and Mapping Challenges of Information Privacy and Data Protection Measures*, https://doi.org/10.1007/978-981-99-7371-2_3

the connected appliances in a home environment are providing a foundation to smart homes.[2] There are important privacy apprehensions that may slow down the rise of interconnected devices at homes. Towards that end, human behaviour seems to be a critical factor that would directly or indirectly influence the use of such devices. The authors derive a conceptual framework by relying upon several models like the Technology Acceptance Model (TAM), the Theory of Planned Behaviour (TPB), Concern for Information Privacy (CFIP), the Internet Users' Information Privacy Concerns (IUIPC) and the Mobile Users' Information Privacy Concerns (MUIPC) model.[3] These models broadly capture perceived risks, apprehensions and the effect that they have on use of interconnected devices. The perceived usefulness of these interconnected devices may be reduced as a result of privacy concerns. The providers of these products and services are therefore required to follow the GDPR norms so that the utility of using these devices is not negatively influenced by privacy apprehensions.[4]

There is also a shift in how tracking happens through mobile phones of data subjects.[5] Mobile phone applications are an integral part of how users communicate with several of the external devices and services that they use. The method of tracking of data subjects has shifted from the use of cookies to the use of platform-based IDs embedded in several of the applications installed on their mobile phones.[6] There are however challenges as to controlling targeted advertisements in a mobile environment.[7] Mobile applications processing sensitive personal information puts the data subjects at a greater risk if the data protection measures are not optimised. They have longer and deeper repercussions on vulnerable individuals.[8]

The concept of ambience intelligence has gained momentum where mobile phones have become the medium of communication.[9] However, the challenges are plenty with several ethical and legal considerations to be fulfilled. The concerns surrounding the collection and use of personal data and the extent of transparency considerations present in the entire ecosystem are some of the issues to be addressed.[10] One has to

---

[2] *Guhr, Nadine; Werth, Oliver; Blacha, Philip Peter Hermann; Breitner, Michael H. Privacy concerns in the smart home context. SN Applied Sciences (2020)* https://doi.org/10.1007/s42452-020-2025-8.

[3] Ibid.

[4] Ibid.

[5] *Sekara, Vedran; Alessandretti, Laura; Mones, Enys; Jonsson, Håkan Temporal and cultural limits of privacy in smartphone app usage. Scientific Reports (2021)* https://doi.org/10.1038/s41598-021-82294-1.

[6] Ibid.

[7] *Ullah, Imdad; Boreli, Roksana; Kanhere, Salil S. Privacy in targeted advertising on mobile devices: a survey. International Journal of Information Security (2022)* https://doi.org/10.1007/s10207-022-00655-x.

[8] *Iwaya, Leonardo Horn; Babar, M. Ali; Rashid, Awais; Wijayarathna, Chamila On the privacy of mental health apps. Empirical Software Engineering (2022)* https://doi.org/10.1007/s10664-022-10236-0.

[9] *Kosta, Eleni; Pitkänen, Olli; Niemelä, Marketta; Kaasinen, Eija Mobile-Centric Ambient Intelligence in Health- and Homecare—Anticipating Ethical and Legal Challenges. Science and Engineering Ethics (2009)* https://doi.org/10.1007/s11948-009-9150-5.

[10] Ibid.

balance the need for innovation with the privacy needs. Research work considered as a part of this chapter points to several such privacy implications while highlighting the manifold advantages.[11] There is an increased risk of surveillance and profiling of individuals as a result of extensive processing, and further makes personal data vulnerable to several external threats. The technology and the processing of personal data that such use of a technology entails sits uncomfortably with several of the data protection principles.[12] The range of data processing may end up not fulfilling the requirements of having a specific purpose of data processing, and the data minimisation principle. Specificity leads to a specific purpose and creates a basis for data processing. The basis has to be legally enforceable with consent as an option giving data subjects an opportunity to freely exercise their choice.[13] Choices exercised by data subjects are premised on "…fundamental rights and freedoms… [contributing] to the accomplishment of an area of freedom, security and justice…"[14] While giving the control back to the data subject, the idea is to boost innovation and not to impede technological progress.[15] One of the legal considerations is to bring back the safety net through the application of data protection by design, and data protection by default encompassing "…measures… of minimising the processing of personal data, pseudonymising personal data as soon as possible, transparency with regard to the functions and processing of personal data, enabling the data subject to monitor the data processing, enabling the controller to create and improve security features"[16] The idea of data protection by design and data protection by default has been captured under Article 25 of the GDPR.[17] This framework reduces uncertainties for the data subject and the data controller. It reduces data privacy risks associated with a product or a service when it has been marketed. The risk assessment happens at the time of designing a product or a service with default settings geared towards fulfilling the data minimisation principle. The data controller fulfils the requirement of data protection principles. This framework would face challenges when a particular data controller merely uses the technology in question and is not the producer of such technology, thereby having lesser control at the design stage of the product or the service in question.[18]

Out of the concerns mentioned above, the risks associated with the overall well-being of the children are the greatest. It is rather difficult for parents to constantly

---

[11] *De Hert, Paul; Gutwirth, Serge; Moscibroda, Anna; Wright, David; González Fuster, Gloria Legal safeguards for privacy and data protection in ambient intelligence. Personal and Ubiquitous Computing (2008)* https://doi.org/10.1007/s00779-008-0211-6.

[12] Ibid.

[13] Recital 32 and Article 7, Regulation (EU) 2016/679 of the European Parliament and of the Council of 27 April 2016 on the protection of natural persons with regard to the processing of personal data and on the free movement of such data, and repealing Directive 95/46/EC (General Data Protection Regulation) (Text with EEA relevance) *OJ L 119, 4.5.2016, pp. 1–88.*

[14] Recital 2, ibid.

[15] Recitals 2 and 6, ibid.

[16] Recital 78, ibid.

[17] Article 25, ibid.

[18] Recital 78, ibid.

monitor and set protocol measures on devices to protect children from online threats. Here the use of machine learning technology and its readings can help differentiating between an adult and an under-age.[19] Operationalising personal data pertaining to children has been a contentious issue and can happen optimally upon suitable intervention of parents and data controllers.[20] The adoption of the design and default approach discussed earlier can also minimise the risks associated with the use of smart connected toys that children have access to at their homes.[21]

3.2 Machine Generated Summaries

Disclaimer: The summaries in this chapter were generated from Springer Nature publications using extractive AI auto-summarization: An extraction-based summarizer aims to identify the most important sentences of a text using an algorithm and uses those original sentences to create the auto-summary (unlike generative AI). As the constituted sentences are machine selected, they may not fully reflect the body of the work, so we strongly advise that the original content is read and cited. The auto generated summaries were curated by the editor to meet Springer Nature publication standards. To cite this content, please refer to the original publications.

Machine generated keywords: app, mobile, scenario, ict, usage, mobile device, threat, legal, user, intelligence, safeguard, smart, phone, track, mobile phone

3.2.1 Privacy Concerns in the Smart Home Context [1]

This is a machine-generated summary of:

Guhr, Nadine; Werth, Oliver; Blacha, Philip Peter Hermann; Breitner, Michael H.: Privacy concerns in the smart home context [1]

Published in: SN Applied Sciences (2020)

Link to original: https://doi.org/10.1007/s42452-020-2025-8

---

19 *Zaccagnino, Rocco; Capo, Carmine; Guarino, Alfonso; Lettieri, Nicola; Malandrino, Delfina Techno-regulation and intelligent safeguards. Multimedia Tools and Applications (2021)* https://doi.org/10.1007/s11042-020-10446-y.

20 Article 8, Regulation (EU) 2016/679 of the European Parliament and of the Council of 27 April 2016 on the protection of natural persons with regard to the processing of personal data and on the free movement of such data, and repealing Directive 95/46/EC (General Data Protection Regulation) (Text with EEA relevance) *OJ L 119, 4.5.2016, pp. 1–88.*

21 *Yankson, Benjamin Continuous improvement process (CIP)-based privacy-preserving framework for smart connected toys. International Journal of Information Security (2021)* https://doi.org/10.1007/s10207-020-00535-2.

If you want to cite the papers, please refer to the original.

For technical reasons we could not place the page where the original quote is coming from.

Abstract-Summary

"The role of privacy has remained largely unexplored in the smart home (SH) usage context."

"This study addresses this gap in literature: to what extend users' concerns for information privacy influence the intended SH usage."

"This study takes an interactional psychology perspective and links the dimensions of the privacy concerns to the intention to use SH devices."

"This study contributes to the literature on privacy, and SH usage by exploring how and why privacy concerns influence participants' intended SH usage."

"The findings, derived from the empirical study emphasize the importance of addressing privacy concerns because they are capable of directly and indirectly influencing the intended SH usage."

Introduction

"Domestic appliances are mutually connected via information and communication technology, resulting in the emergence of smart homes (SH)."

"We aim to identify relevant privacy concerns of potential users in connection with the use of SH devices and to investigate whether these have a significant influence on the acceptance of this particular technology."

"We aim to provide deeper insight into the issue by answering the following research question: How do users' concerns for information privacy influence the intention to use SH devices?"

"We discuss the concept of privacy concerns for the application of SH devices and instruments measuring information privacy concerns and user acceptance."

"Hypotheses are derived and a conceptual model is developed that links these privacy concerns to the intention to use SH devices taking into account critical influencing factors of the technology acceptance model (TAM) and the theory of planned behavior (TPB) as antecedents for to intention to use."

Theoretical Foundation

"Other studies explored privacy hazards directly connected to the usage of an IoT based SH because of the networks between mobile devices of the users and SH infrastructure, like doors or windows (e.g., [2])."

"The most prominent models in current research literature are the Concern for Information Privacy (CFIP), the Internet Users' Information Privacy Concerns (IUIPC) and the Mobile Users' Information Privacy Concerns (MUIPC) model, which serve as one of two theoretical pillars for our research model."

"The model MUIPC measures the privacy concerns of mobile users based on the following three dimensions: perceived intrusion, perceived surveillance, and secondary use of personal information [3]."

"The TAM and the TPB were extensively used to examine the influence of privacy concerns on behavioral intention and were used as a second theoretical pillar in this study of SH usage and privacy concerns."

Hypothesis Generation and Research Model

"The proposed research model aims to examine the way privacy concerns influence the use of SH devices."

"The basic premise is that the more users see their privacy endangered, the lower the intention to use SH devices."

"It is hypothesized that the PU influences the intention to use SH devices—with the perceived privacy concerns as its precursor."

"Privacy concerns are negatively associated with the perceived usefulness of SH devices."

"We propose the following hypothesis: Perceived usefulness is positively associated with the intention to use SH devices."

"We propose the following hypothesis: Perceived behavioral control is positively associated with the intention to use SH devices."

"We propose the following hypothesis: Subjective norms are positively associated with the intention to use SH devices."

"The following hypothesis is proposed: Privacy concerns have a negative impact on the intention to use SH devices."

Research Methods

"The survey questionnaire was distributed via social media platforms, e.g., Facebook, Xing, and LinkedIn in order to recruit participants through announcements."

"The posting provided a cover letter, which stated the objective of the study and also included the link to the survey questionnaire as well as the explanation of the data handling procedure."

"common method variance (CMV)] ex ante in the research design stage (e.g., counterbalancing the order of the questions, presenting the survey items in a random manner, and adapting the measurement items for the independent and dependent constructs from different previously validated studies) and ex post after the research has been conducted."

"Given our care in survey design and based on our analysis, CMV in our model is not likely (compare Online Appendix C)."

"For our survey instrument, we adapted items from existing research literature for all latent constructs wherever possible."

Data Analysis and Results

"To maximum likelihood (ML)-based covariance-based SEM approaches, the evaluation of the structural and measurement model in PLS-SEM is based on a number of nonparametric evaluation criteria [4, 5] and PLS-SEM is more suitable for unestablished models or preliminary model building [6]."

"Measurement validation and model testing were conducted using a two-step approach with the SmartPLS/-Version 3.2.8 software [7]."

"The two-step approach involves the structural model and a separate assessment of the measurement models [4]."

"The hypotheses, we evaluated the validity and reliability of the construct measures."

"We assessed the reflective measurement models with regard to their validity and internal consistency reliability."

"The values for all the constructs in our model except the second-order constructs range from 0.102 to 0.829."

"The effect size f^2 is, in addition to the values of the path coefficients another measure of the substantial effect of exogenous latent variables on the latent endogenous variable and is calculated according to [8], to check for practical significance."

Discussion

"The evaluation of results shows that the developed research model is a meaningful instrument for the investigation influence of user privacy concerns' on behavioral intention to use SH systems."

"We elaborated the influence of different variables (e.g., privacy concerns, PU, PBC, and SN) on the decision to use SH technology."

"To the extent that the privacy concerns is an important factor influencing the behavioral intention to use SH systems, it is also important for service provider that these privacy concerns reduces the perceived usefulness of these systems and users' perceived behavioral control."

"From the practical perspective (i.e., considering government regulations), SH provider should react to users' privacy concerns to comply with government regulations such as the GDPR to avoid any legal complications through the use of SH devices and the associated personal data."

Conclusions

"We assessed the influence of privacy concerns on the acceptance of SH devices."

"We proposed a model based on MUIPC, IUIPC, TBP and TAM to better represent critical factors influencing the intention to use SH devices."

"New technological options as SH applications can have numerous advantages but also challenges like privacy concerns."

3.2.2 Temporal and Cultural Limits of Privacy in Smartphone App Usage [9]

This is a machine-generated summary of:

Sekara, Vedran; Alessandretti, Laura; Mones, Enys; Jonsson, Håkan: Temporal and cultural limits of privacy in smartphone app usage [9]

Published in: Scientific Reports (2021)

Link to original: https://doi.org/10.1038/s41598-021-82294-1

If you want to cite the papers, please refer to the original.

For technical reasons we could not place the page where the original quote is coming from.

Abstract-Summary

"We show that behaviour captured in the form of application usage data collected from smartphones is highly unique even in large datasets encompassing millions of individuals."

"This makes behaviour-based re-identification of users across datasets possible."

"We study 12 months of data from 3.5 million people from 33 countries and show that although four apps are enough to uniquely re-identify 91.2% of individuals using a simple strategy based on public information, there are considerable seasonal and cultural variations in re-identification rates."

"We find that people have more unique app-fingerprints during summer months making it easier to re-identify them."

"Further, we find significant variations in uniqueness across countries, and reveal that American users are the easiest to re-identify, while Finns have the least unique app-fingerprints."

Introduction

"With the standard methods based on cookies for identifying customers not being used in smartphones, combined with the rising usage of ad-blockers among users [10], companies, advertisers, and so-called data brokers are now using smartphone apps to identify and track individuals."

"Identifiers such as the Android Advertisement ID are one example of IDs embedded in app, but they do not allow data brokers to track users across multiple applications or devices, and they can even be reset by users."

"An example of this is TalkingData, China's largest third-party mobile data platform, which collects and sells app usage data of more than 750 million smartphone users [11]."

"Previous work (on datasets of limited size—approximately 50.000 users and without accounting for geographic spread) has shown that 99.4% of smartphone users have unique app usage patterns and established the viability of using apps as markers of human identity, similar in application to fingerprints in forensic science [12]."

Results

"We evaluate the likelihood of identifying individuals within smartphone app usage data using a dataset that spans 12 months (1 Feb 2016 to 31 Jan 2017) and encompassing 3.5 million people from 33 countries using in total 1.1 million unique apps (approximately 40% of all apps on Google Play [13])."

"Given an individual's app-fingerprint, unicity quantifies the number of apps needed to uniquely re-identify that person; the fewer apps we need the more unique a person is and vice versa."

"Given a dataset of app-fingerprints and apps i, j and k, a person u is uniquely identifiable if they, and only they, in the dataset have used apps i, j and k, i.e. matching the fingerprint of user u. In our dataset we evaluate uniqueness as the percentage of individuals we can re-identify using n number of apps."

"The median number of apps, which is a proxy for the size of people's app-fingerprints, is the only informative variable and it is positively associated with country uniqueness values."

Discussion

"These results opens up for a larger discussion around the impact and portability of policies, as a policy intended to protect users within one country might not provide the same protection for individuals living in a different country."

"From our model, we expect knowledge about the properties of a country app ecosystems, especially the distributions of popularity, to lead to better informed policies for protecting user privacy."

"This study was performed using app usage data collected from Android phones from a single vendor only."

"Further, we have no reason to believe that app usage behaviour and uniqueness are fundamentally different for individuals using iOS devices compared to Android users."

"iOS has measures in place to protect users from apps that collect app usage for fingerprinting purposes, but these have been shown to be inadequate [14]."

Methods

"We use a dataset that spans 12 months, from Feb. 1st 2016 to Feb. 1st 2017, and contains monthly app-fingerprints for 3,556,083 individuals with pseudonymized app and user identifiers."

"The number of unique apps in the dataset is 1,129,110, and each individual in the dataset uses at least 3 apps per month."

"Due to privacy concerns raw data cannot be shared publicly, however, a data sample of app-fingerprints from 20,000 randomly selected individual, including the hashed id of the apps used in a given month can be found at https://doi.org/10.11583/DTU.13650797.v1."

"To estimate the uniqueness of app-fingerprints, we apply the unicity framework [15] on s samples of 10,000 randomly selected individuals."

"For each individual we select n apps (without replacement) from their app-fingerprint."

"The person is then said to be unique if they are the only individual in the dataset whose app-fingerprint contains those apps."

3.2.3 Mobile-Centric Ambient Intelligence in Health- and Homecare—Anticipating Ethical and Legal Challenges [16]

This is a machine-generated summary of:

Kosta, Eleni; Pitkänen, Olli; Niemelä, Marketta; Kaasinen, Eija: Mobile-Centric Ambient Intelligence in Health- and Homecare—Anticipating Ethical and Legal Challenges [16]

Published in: Science and Engineering Ethics (2009)

Link to original: https://doi.org/10.1007/s11948-009-9150-5

If you want to cite the papers, please refer to the original.

For technical reasons we could not place the page where the original quote is coming from.

Abstract-Summary

"Ambient Intelligence provides the potential for vast and varied applications, bringing with it both promise and peril."

"The development of Ambient Intelligence applications poses a number of ethical and legal concerns."

"The MINAmI (Micro-Nano integrated platform for transverse Ambient Intelligence applications) FP6 research project aims at creating core technologies for mobile device based Ambient Intelligence services."

"The paper concludes with some thoughts on drafting ethical guidelines for Ambient Intelligence applications."

Introduction

"Ethical issues are not just related to individual applications."

"To general ethics research, there are lots of studies on ethical issues related to ICT, and how the ethical values should be considered in designing computerized systems."

"The ethical-legal analysis described in this paper has been in connection to the development of a new mobile phone based architecture for ambient intelligence applications."

"The development work has been carried out in the frame of an EC-funded research project, MINAmI. The ambient intelligence vision of the project refers to an ambient intelligence environment, where the user has her personal mobile device as a medium of communication."

"The aim has been to ensure that the mobile architecture to be developed will support, encourage and even force ethically sustainable applications."

"As we have been working with the ethical guidelines parallel to the architecture design, we aim to have the guidelines ready before commercial application developments start."

Mobile-Centric Ambient Intelligence

"The MINAmI project aims at accelerating the adoption of Ambient Intelligence applications that will be used in the everyday life of the user."

"MINAmI continues MIMOSA work by enriching the MIMOSA platform with additional technologies enabling more functionalities in the applications that are provided to the user on her personal mobile phone."

"Mobile devices provide trusted intelligent user interfaces and wireless gateways to sensors, networks of sensors, local networks and the Internet."

"This design approach makes possible to affect basic design decisions that may have effects on user acceptance and ethical issues regarding all the applications that will use the MINAmI ambient intelligence infrastructure [17]."

"The information, applications or services provided by the system are read or activated by touching or scanning tags and sensors with the mobile device, providing natural interaction methods to the user."

Ethical-Legal Assessment Process

"Ethical issues related to the forthcoming applications and the platform itself are studied in the MINAmI project in a dedicated work package."

"The MINAmI project has built on the ethical considerations recognised in the MIMOSA project, but external experts have also been called to assist the project researchers."

"For ethical assessment, the project has nominated an Ethical Advisory Board that consists of eleven external experts representing different fields of ethics (Niemelä [18])."

"The aim of the MINAmI ethical assessment work is to create, parallel to the architectural solutions and demonstration applications, ethical guidelines that would guide forthcoming application developers and platform implementers into ethically sustainable solutions."

MINAmI Scenarios and Legal-Ethical Assessment

"The MINAmI demonstrator scenarios entail the processing of personal data, in most cases data of the users of the MINAmI-enabled mobile device."

"The ethical issues in this scenario relate to the reliability of the system (both technical and functional), protection of data, and autonomy of the patient."

"When user-centric ambient intelligence applications, like the MINAmI demonstrator scenarios, utilise information about the user, processing of personal data takes place."

"All of the MINAmI scenarios deal with information relating to an identified or identifiable natural person, usually the user, and thus with the processing of personal data."

"The processing of personal data is allowed only under the grounds mentioned in Article 7 of the Data Protection Directive and shall be respected when the processing of personal data is taking place in ambient intelligence applications."

Towards Ethical Guidelines

"The scenario describing the basic mobile interaction concept, the "Movie trailer downloading from a memory tag", raised ethical issues on privacy and user autonomy if information from the user can be collected in relation to a service use and if the user must involuntary receive information or services (cf.e-mail spam)."

"The MINAmI ethical guidelines are intended to those who design applications and services that utilise MINAmI technology and for those who implement the MINAmI platform itself."

"The guidelines should be based on ethical principles that highlight user needs, but are not too individual-centric, taking into account e.g. societal and environmental interests."

"It is not always necessary to consider the ethical principles of informational privacy when implementing an ambient intelligence application, but obeying the law usually ensures also rather high ethical standards."

"They give guidance on ethical issues that should be taken into account when designing applications and services that utilise MINAmI platform for mobile-centric Ambient Intelligence."

Conclusions

"The experiences of the MINAmI project show that external experts can provide valuable insights to a legal and ethical assessment."

"The comments and the recommendations of the Ethical Advisory Board have influenced the MINAmI scenarios and have led to a second revised set of demonstrator scenarios."

"It has become also clear from the experience of the project that drafting general ethical guidelines to be followed in ambient intelligence services and applications is difficult."

3.2.4 Legal Safeguards for Privacy and Data Protection in Ambient Intelligence [19]

This is a machine-generated summary of:

De Hert, Paul; Gutwirth, Serge; Moscibroda, Anna; Wright, David; González Fuster, Gloria: Legal safeguards for privacy and data protection in ambient intelligence [19]

Published in: Personal and Ubiquitous Computing (2008)

Link to original: https://doi.org/10.1007/s00779-008-0211-6

If you want to cite the papers, please refer to the original.

For technical reasons we could not place the page where the original quote is coming from.

Abstract-Summary

"It constructed four "dark scenarios" showing possible negative implications of AmI, notably for privacy protection."

"Legal analysis of the depicted futures showed the shortcomings of the current legal framework in being able to provide adequate privacy protection in the AmI environment."

"The authors, building upon their involvement in SWAMI research as well as the further advancement of EU privacy analysis, identify various outstanding issues regarding the legal framework that still need to be resolved in order to deal with AmI in an equitable and efficacious way."

Introduction: AmI and Privacy

"Ambient intelligence (AmI) will undoubtedly bring substantial economic and social benefits to European citizens and industry, but they will come alloyed with many risks."

"Even if it is not possible (or desirable for that matter) to bury new technologies, such as those that form the architecture of AmI, it behoves us to anticipate the threats and vulnerabilities in order to prevent them from overwhelming the many desirable features and advantages that AmI will yield."

"Such was the attitude of the SWAMI researchers who took a precautionary but prospective approach towards AmI. SWAMI, the acronym for Safeguards in a World of Ambient Intelligence, was a policy-oriented research project launched within the European Commission's Sixth Framework Programme."

"The project focused on the social, economic, legal, technological and ethical issues arising from AmI with particular regard for privacy, trust, security and identity."

"It then proposes AmI-specific safeguards for privacy."

Dark Scenarios

"In order to identify and understand the possible implications of the technologies that are being used to construct an AmI world, SWAMI researchers collaborated with various stakeholders in developing four dark scenarios showing technology that does not work or that works in an unexpected way."

"The aim of focusing on such situations was to identify and highlight possible adverse impacts of and risks in AmI. The four dark scenarios encompass individual–societal and private–public concerns."

"These concerns formed two scenario axes which helped to reduce the virtually infinite number of possible futures that could be envisaged."

"Following is a thumbnail sketch of each scenario:"

Privacy Threats

"The ensuing analysis of each of the scenarios revealed various risks, threats and vulnerabilities posed by AmI in relation to privacy, trust, security, identity and inclusion, among which were greatly increased surveillance and monitoring, a deepening of the digital divide, identity theft, malicious attacks and so on [20]."

"The massive collection of data by the AmI technologies that populate the intelligent environment enables extensive profiling, which in turn is necessary to deliver the benefits promised by AmI. This extensive profiling is made possible by the availability and exchange of data between various systems, devices and databases (and consequently between different spheres of one's life)."

"The knowledge derived from AmI can create information asymmetries between those surveilled and those surveilling."

"Although RFID technology can provide significant benefits for tracking shipments, inventory management, sales and market analysis, its identification, profiling and monitoring capabilities have raised concerns, particularly with regard to its tagging of personal items."

The legal framework and AmI: lacunae and weaknesses in privacy and data protection law

"Among the conclusions of the analysis were, that AmI can effectively put the individual's privacy into jeopardy and that the existing framework appeared to offer insufficient protection of privacy and personal data."

"The fact that AmI needs as many data as possible to achieve its full potential clearly clashes with some of the main principles of data protection law, namely the data minimisation principle (collecting as little data as necessary for a given purpose) and the purpose specification principle (the collected information can only be used for the purpose defined at the moment of data collection)."

"AmI also causes major problems for the consent principle in data protection law; currently, the unambiguous consent of the data subject is the main factor making a processing of personal data legitimate."

"AmI forces us to reconsider our understanding of current privacy and data protection law It forces us to seek more flexible ways of articulating AmI requirements and concerns related to privacy and data protection, and the implied values of autonomy, self-determination and liberty."

A Legal framework for AmI: specific safeguards for privacy and data protection

"By default, privacy law protects the opacity of the individual, while data protection, also by default, calls for transparency of the processor of personal data."

"We draw attention to the important distinction between opacity and transparency tools in privacy and data protection."

"A key issue in ambient intelligence is how to ensure that data collection and data processing are transparent to the data subject and how to ensure a proper balance between the information provided to data subjects and the information taken from them."

"Complementary to privacy-enhancing technologies (PETs), which aim at controlling the dissemination of data, transparency-enhancing technologies (TETs) [21] could contribute to information exchange and management."

"The individual should always be informed of the presence of tags and readers, the purposes for which data are collected and processed, who is the responsible controller, whether the data (and what kind of data) are stored, the means to access and rectify data, and whether the data will be made available to third parties."

Conclusions

"The SWAMI consortium identified various threats and vulnerabilities affecting privacy and data protection in AmI, including a number of weaknesses in the existing legal framework."

"A new approach to privacy and data protection is needed, based on control and responsibility rather than on restriction and prohibition."

"This article presented a few examples of legal safeguards against the loss of privacy."

"We discussed some possibilities for improving the transparency of the processing of AmI-generated data and for improving the control of such by the user."

"Specific safeguards for RFID were presented as an example of a precautionary approach toward a particular AmI technological application."

3.2.5 *Techno-Regulation and Intelligent Safeguards [22]*

This is a machine-generated summary of:

Zaccagnino, Rocco; Capo, Carmine; Guarino, Alfonso; Lettieri, Nicola; Malandrino, Delfina: Techno-regulation and intelligent safeguards [22]

Published in: Multimedia Tools and Applications (2021)

Link to original: https://doi.org/10.1007/s11042-020-10446-y

If you want to cite the papers, please refer to the original.

For technical reasons we could not place the page where the original quote is coming from.

Abstract-Summary

"The techno-regulation paradigm is inspiring new ways to safeguard legally interests by means of tools allowing to hamper breaches of law In this paper, we focus on the exposure of individuals to specific online threats when interacting with smartphones."

"We propose a novel techno-regulatory approach exploiting machine learning techniques to provide safeguards against threats online."

"We study a set of touch-based gestures to distinguish between underages or adults who is accessing a smartphone, and so to guarantee protection."

"We experimented both single-view and multi-view learning techniques to find the best combination of touch-gestures able of distinguishing between adults and underages."

Introduction

"By starting with the observation that underages and adults perform commonly used touch-gestures in a different way on mobile devices, we developed an Android app in order to collect, in an experimental study involving 147 participants, more than 9000 heterogeneous touch-based gestures."

"We carried out several experiments to find, by exploiting machine learning techniques like in [23–27], the best combination of touch-based gestures capable of distinguishing between adults and underages."

"We study a set of touch-based gestures, to determine whether it is possible to distinguish who is accessing a smartphone, i.e., underage or adult, to guarantee protection."

"We experimented both single-view and multi-view learning techniques to find the best combination of touch-gestures capable of distinguishing between adults and underages."

Related Work

"Differently from these works, we used only the information extracted from touch gesture data."

"Other works available in literature are based on the analysis of touch-based gestures performed on smartphones, arguing that user information could be extracted and used to control smartphones' interactions."

"In [28], the authors describe a novel multi-touch gesture based authentication technique, defined on a set of five-finger touch gestures."

"The authors collected touch data from users interacting with a smartphone by using only up-down and left-right scroll gestures."

"In [29], authors argued that touch-based gestures on touchscreen devices constitute a privacy threat."

"In [30], authors present a technique to classify the users' age group from touch gestures."

"In our work, we do not consider touch gesture windows but combinations of single gestures performed by the same participant."

Child Protection: Why?

"Beyond motivational issues, one thing that clearly emerges by the related works is the need for cross-disciplinary experimental activities leading to more and more efficient techno-regulatory approaches binding regulatory priorities with opportunities provided by technological safeguards."

"In this direction, we propose a techno regulated-based approach that exploits machine learning techniques to classify individuals, specifically underages, while they interact with technology, with the final goal of protecting them against specific online threats."

"The explosion of information and communication technology, as emphasized in the 2015 edition of Guidelines for Industry on Child Online Protection released by International Telecommunication Union (ITU) and by UNICEF, has created unprecedented opportunities for children and young people to communicate and access information but, at the same time, significant challenges to children's safety [31]."

"Parents do not always understand the potential risks their children may encounter since they often underestimate teenagers' exposure to sexual content or overestimate it due to mass media messages [32, 33]."

Our Approach for Age Detection

"The "intermediate integration" consists in performing a features selection for each type of gesture [34, 35] (single-view), and then by combining the features selected; thus, for each individual, we combine (concatenate) such features in order to obtain the samples for the integrated datasets; the advantages of such a technique are: (a) the heterogeneous nature of the gestures' features can be better used by separating the data, (b) the size of the output is reduced, and (c) the separate extraction of features for different type of gestures implements the divide-et-impera principle, reducing the complexity of the operations."

"In the multi-view learning setting, when combining three gestures with the intermediate integration (scrollD_swipeL_pinch), the accuracy increases up to 88% (our best result) and the ROC AUC score reaches almost the same result (0.92) as in the single-view setting."

Discussion

"Looking at the results, we can conclude that scroll down is the type of touch gesture that best allows us to classify among underages and adults, with a ROC AUC score of 0.93 and an accuracy score of 86%, followed by pinch-to-zoom, swipe left, and swipe right."

"This shows that there is a significant variation between underages and adults in the running of touch gestures like swipes and scrolls."

"The touch gesture triple combining scroll down, pinch-to-zoom, and swipe left, with the intermediate integration technique, showed a ROC AUC score of 0.92 and an accuracy score of 88%."

"This means that the way an underage/adult performs a scroll gesture is correlated with how he performs a swipe or a pinch-to-zoom gesture."

Conclusion

"We studied touch gestures to derive whether it is possible to distinguish who is accessing a mobile device (underages or adults) to provide safeguards against threats online."

"Our result is a new regulatory approach that exploits machine learning techniques to provide automatic safeguards against threats online."

"The idea is to relieve parents from the challenging task of configuring protecting tools and constantly monitoring children and trusting an automatic mechanism that identifies the user accessing a smartphone and provides the right protection."

"In terms of techno-regulation solutions, we aim at integrating our approach inside the AI4C app, with a parental control system designed to support in various ways the enforcement of the safeguards resulting from the normative framework in the field of online child protection."

3.2.6 Continuous Improvement Process (CIP)-Based Privacy-Preserving Framework for Smart Connected Toys [36]

This is a machine-generated summary of:

Yankson, Benjamin: Continuous improvement process (CIP)-based privacy-preserving framework for smart connected toys [36]

Published in: International Journal of Information Security (2021)

Link to original: https://doi.org/10.1007/s10207-020-00535-2

If you want to cite the papers, please refer to the original.

For technical reasons we could not place the page where the original quote is coming from.

Abstract-Summary

"This work provides an overview of the SCT privacy problem and the landscape based on previous work to address privacy and safety concerns."

"We further investigate technical and legislative-related privacy issues in SCT and present a confirmatory study on developers' privacy views based on the privacy calculus theory."

"We present an abstract continuous improvement process-based privacy preservation framework using Plan-Do-Check-Act, which can be adopted and used during the SCT design to minimize privacy breaches and serve as a foundation for the deployment of continuous privacy control improvement in a complex SCT development context."

Introduction

"The patent describes the devices as an "anthropomorphic device" in the form of a doll or toy with embedded motors that would enable it to both move their heads and limbs and appear to open and close their eyes [37], and connected to a home network allowing for collection, processing, transmission and storage of PII, interaction data, and context data, which raises serious privacy, data security, and child safety concerns that can be minimized through a well-defined privacy preservation framework as presented in this work."

"The importance protection of data such as PII is a good indication that SCTs have unique privacy requirements considerations, which include specific needs for child privacy, the security of the mobile device (which may connect with other physical toy components), and the trust relationship between the child and the SCT."

Background

"The rise of interest in children's data by hackers results from lax security mechanisms in devices like SCTs (including other children platforms), making the child's PII more vulnerable to breaches and theft [38]."

"While there have been many efforts by governments and international organizations, such as the United Nations Children's Fund (UNICEF), and legislation such as COPPA in the USA, which encourages the protection of children data online, there is currently no limited number of actual implementable privacy preservation framework for SCT which can easily help SCT designers to be fully compliant to such regulations [39, 40] In most jurisdictions across the globe, children are protected by international regulations such as the United Nations Convention on the Rights of the Child (CRC), which protects children from all forms of violence, exploitation, abuse, and discrimination."

Related Work

"Current research on the privacy of smart connected toys (SCTs) and provide details on unique research work, which contributes significantly to SCT-related privacy works."

"On SCT data privacy mechanism, Jones et al. [41], Druga et al. [42], McReynolds et al. [43], Valente and Cardenas [44], Mahmoud et al. [45], and Steie et al. [46] all present research works directly on the privacy of different SCTs, including CogniToys' Dino, Mattel's Hello Barbie, and My Friend Cayla."

"All the previous works under privacy mechanism to address child safety do a fair job of providing some solution, but we enhance it by developing an abstract privacy preservation framework based on PDCA, which can be easily implemented to address privacy-related safety in a full SCT development and use cycle, with the capability of minimizing any probability of a privacy breach when the device is in use by a child."

"Due to a lack of proper security and privacy mechanism, which renders SCT vulnerable, this opens up opportunities for a hacker to hack into IoT-related devices such as SCT, baby monitors, which can view children in their most vulnerable periods [47]."

Experimental Design and Method

"The participants were given a 15-minutes' demonstration of our Zenbo communication to show how Zenbo (SCT) could collect sensitive information Due to privacy concerns, we did not collect any personally identifiable information from our participants."

"There are existing works [48–50] look into the study information privacy violation, such as SCT data privacy from different angles, by adopting various information privacy theories to conduct a study on the user information privacy problem."

"We consider these theorems as essential to support privacy research by helping to extrapolate user information privacy concerns and user response during the collection and storage of SCT-related user information stored with the SCT manufacturer."

"This model depicts the value gained from a trade-off between privacy risk and benefit in cases where SCT users are required to provide information."

Relevant Results and Discussion

"We assess if there is a correlation between participant age and their respected views of the seriousness of privacy preservation relating to children's information and SCT."

"Further, we collected other related information, such as participants with one or more children within their immediate family circle."

"Because of the information provided to the participants about location privacy concerns, we hypothesize that the participants with children in their family circle will select Petri net as the preferred data modeling option."

"We are also interested in understanding whether there is an impact choice participant will make based on gender with respect to protecting children's information."

"We expected that based on Privacy Education Training and Awareness (PETA) provided as part of the study, the majority of the participants would see a link between privacy breach and safety of children."

PDCA-Based Privacy Preservation Framework for SCT

"The purpose of the conceptual model is to provide an opportunity to clearly understand the data flow between the various aspects and functions of the smart toy in order to help determine appropriate privacy controls to develop in order to minimize the privacy violation risk during design phase violation."

"In the "Plan" phase, we start by identifying the possible features of the SCT, identifying the type of data and support privacy laws required for compliance, identifying all context of use and possible threats and vulnerabilities, and finally, the protection based on sound privacy principle such as SPIAT privacy principles [51]."

"The different phases of the PDCA framework can then be repeated until set for the SCT maintains a privacy protection base."

"The privacy risk minimization techniques provision by PDCA-based privacy preservation framework, as discussed in the different phases, includes a focus on categories of threats, such as surveillance, secondary use, disclosure, exclusion, and over-collection of children's information without consent."

Conclusion

"We demonstrated the various privacy issues plaguing the SCT industry and the privacy risk SCT poses to our children as a result of a lack of serious consideration to privacy surrounding the collection, processing, and storage of children's information."

"We further demonstrate the exploration of the following research questions: (1) Is there any privacy preservation framework available that the toy industry can adopt to address the concern of SCT data collection, processing, and storage? (2) What do we (parent, government and SCT Industry) know about children's data privacy? (3) Where do we learn about privacy, specifically PII and context?"

"This PDCA-based privacy preservation framework can be adopted and implemented by the SCT industry to help minimize the possibility of privacy violations in SCTs and alleviate the safety concerns of SCT users."

3.2.7 Privacy in Targeted Advertising on Mobile Devices: A Survey [52]

This is a machine-generated summary of:

Ullah, Imdad; Boreli, Roksana; Kanhere, Salil S.: Privacy in targeted advertising on mobile devices: a survey [52]

Published in: International Journal of Information Security (2022)

Link to original: https://doi.org/10.1007/s10207-022-00655-x

If you want to cite the papers, please refer to the original.

For technical reasons we could not place the page where the original quote is coming from.

Abstract-Summary

"The advertising and analytics companies collect, aggregate, process, and trade a vast amount of users' personal data, which has prompted serious privacy concerns among both individuals and organisations."

"This article presents a comprehensive survey of the privacy risks and proposed solutions for targeted advertising in a mobile environment."

"We outline details of the information flow between the advertising platform and ad/analytics networks, the profiling process, the measurement analysis of targeted advertising based on user's interests and profiling context, and the ads delivery process, for both in-app and in-browser targeted ads; we also include an overview of data sharing and tracking technologies."

"We discuss challenges in preserving the mobile user's privacy that include threats related to private information extraction and exchange among various advertising entities, privacy threats from third-party tracking, re-identification of private information and associated privacy risks."

Introduction

"The collection and use of personal data pose serious threats to the privacy of users [53–58], when websites or apps indicating sensitive information are used as the basis for profiling, e.g. a gaming app showing a gambling problem."

"In a stronger push towards increased user's privacy control over the collection and use of their data, Apple has recently introduced the Identification for Advertisers (IDFA) opt-in overhaul in iOS 14.5, which will have a significant impact on targeted ads and mobile ad/data attribution."

"Mobile users do want to view relevant (interest-based) ads, provided that their information is not exposed to the outside world including the advertising companies."

"We then describe, for both in-app (note that we interchangeably use 'mobile' and 'in-app') and in-browser targeted ads: the user profiling process, data collection, and tracking mechanism, the ad delivery process and the process of ad characterisation."

Methodology

"We consider representative works for both surveys of privacy in related fields to mobile advertising and in privacy technologies and systems applicable to our focus area."

"Previous literature surveys related to our work consider broader areas of Personalization and privacy [59], Online advertising [60], the privacy-personalisation trade-offs [61] and Online behavioural advertising [62]."

"We consider the following conditions to select and include articles in our survey: (1) The published work must be within the domain of advertising systems, (2) the ad system may be browser-based or app-based, (3) the research article may have addressed part of a private/secure advertising system, e.g. private profiling, (4) the research article may be related to performance measurements, advertising measurements or traffic analysis, etc., (5) we also consider supporting articles to elaborate a particular concept or theory, and (6) we include conference papers, journals, books, early access articles, magazines, and survey articles only."

The Mobile Advertising Network

"A user profile comprises a number of interests, that indicates the use of related apps, e.g. sports, business, etc., constructed by e.g. Google Advertising network for Mobile (AdMob) and Flurry [63] (note that the latter is only visible to app developers)."

"The advertising companies, e.g. Google, profile users based on the information they add to their Google account, data collected from other advertisers that partner with Google, and its estimation of user's interests based on mobile apps and websites that agree to show Google ads."

"Google presents a comprehensive set of demographic targeting options for search campaigns, ads display, etc The demographics D are typically grouped in diverse categories, with specific options, such as age-ranges, e.g. '18–24', '25–34', '35–44', '45–54', '55–64', '65 or more', and gender e.g. 'Male', 'Female', besides other options, e.g. location, household income, etc. Recall that the user profiling results when the user device interacts with Google analytics via AdMob SDK [54] for various activities."

Operations of the Advertising System

"We identify the workflow of a mobile app requesting a Google AdMob ad and the triggered actions resulting from e.g. a user click (we note that other advertising networks, such as Flurry, use different approaches/messages to request ads and to report ad clicks)."

"We note that there are notable differences in collecting and analysing the mobile and in-browser user's ad/data traffic for the ad delivery mechanism, in order to target users."

"There are a number of reasons why the existing in-browser [64–73] ads characterisation approaches on targeted advertisements cannot be directly applied to the evaluation of in-app targeted ads: First, there may be various factors for in-app ads targeting that go beyond what is considered for in-browser ads, e.g. the context of mobile apps installed on the user device, their utilisation behaviour (e.g. heavy gamers may receive specific ads)."

"The MAdScope [74] and [75] collects data from a number of apps, probes the ad network to characterise its targeting mechanism and reports the targeted advertising using profiles of specific interests and preferences."

Privacy in Mobile Advertising: Challenges

"We note that in all these scenarios, generally the user does not oppose the profiling process and is willing to receive behavioural services, e.g. targeted ads, on selected topics of interest, but does not wish for specific parts of their profile (attributes), based on the apps contexts (s)he considers private, to be known to the associated analytics companies or any third party, or to be used for personalised services."

"The profiling attributes that are sensitive to the users are considered private information, e.g. political or religious views, sexual orientation, etc. Several works investigate the mobile targeted ad traffic from the point of view of privacy and security concerns."

"The user profile, i.e. information collected or determined by these two analytics services, consists of personally identifiable information including the unique device ID, user demographics, their interests inferred from the app usage, etc. A crucial technique to extract user profiles from the analytics services (we mainly target Google and Flurry analytics services) is to first impersonate the victim's identity."

Privacy in Mobile Ads: Solutions

"Other works based on obfuscation techniques, e.g. text it differential privacy, obscure user preferences; however, it may result in lower accuracy of targeted ads and hence result in lower revenues."

"In the scenario of an advertising system, recall that the user privacy is mainly breached for their context, i.e. specific use of mobile apps from an app category, and their profiling interests along with the ads targeting based on these interests."

"An important parameter is 'impact on targeted ads' as a results of achieving user privacy with various techniques e.g. Crypto-based techniques (such as PIR), Blockchain and Data masking and generalisation techniques will have no impact on targeted ads, alternatively, the Differential privacy, Obfuscation and Randomisation will have an impact on targeted ads and can be adjusted to achieve various trade-offs, i.e. 'low-relevant vs. high-relevant interest-based ads', as is also evident in [54, 55]."

Open Research Issues

"The RTB necessitates that A &A companies also work with publishers to help manage their relationship for ad exchange (in addition to user's data tracking) and to optimise the ad placement (i.e. targeted ads) and bidding on advertiser's behalf."

"In order to reduce the complexity of the advertising systems, we envision several additional areas of research: devising processing-sensitive frameworks, limiting the direction-redirection of requests among A &A entities, opening up the user data exchange processes within the ad platform, identifying new privacy threats, and devising new protection mechanisms."

"The applicability of this private billing model can be verified in the online advertising system, which would require changes on both the user and ad system side."

"This framework provides: a private profiling mechanism, privately requesting ads from the advertising system, the billing mechanisms for ads monetisation, uploading ads to the cloud system, various types of transactions to enable advertising operations in Blockchain-based network, and access policy over cloud system for accessing various resources (such as ads, mobile user profiles)."

Conclusion

"We provide a comprehensive overview of privacy issues and solutions in mobile-based targeted advertising systems."

"Considering the proposed and in some cases used (in the industry) privacy-preserving approaches, we have found that it is extremely difficult to provide user privacy in a way that enables greater user control of their private data and to simultaneously reduce the financial impact of introducing privacy mechanisms, without significantly changing the advertising ecosystems and their operations."

"To aid further development of privacy-enabled ad systems, we also identify open research issues that need to be solved in future work."

3.2.8 On the Privacy of Mental Health Apps [76]

This is a machine-generated summary of:

Iwaya, Leonardo Horn; Babar, M. Ali; Rashid, Awais; Wijayarathna, Chamila: On the privacy of mental health apps [76]

Published in: Empirical Software Engineering (2022)

Link to original: https://doi.org/10.1007/s10664-022-10236-0

If you want to cite the papers, please refer to the original.

For technical reasons we could not place the page where the original quote is coming from.

Abstract-Summary

"An increasing number of mental health services are now offered through mobile health (mHealth) systems, such as in mobile applications (apps)."

"Although there is an unprecedented growth in the adoption of mental health services, partly due to the COVID-19 pandemic, concerns about data privacy risks due to security breaches are also increasing."

"Whilst some studies have analyzed mHealth apps from different angles, including security, there is relatively little evidence for data privacy issues that may exist in mHealth apps used for mental health services, whose recipients can be particularly vulnerable."

"This paper reports an empirical study aimed at systematically identifying and understanding data privacy incorporated in mental health apps."

"We analyzed 27 top-ranked mental health apps from Google Play Store."

"Based on the empirical findings of this study, we provide recommendations to be considered by different stakeholders of mHealth apps in general and apps developers in particular."

"We conclude that while developers ought to be more knowledgeable in considering and addressing privacy issues, users and health professionals can also play a role by demanding privacy-friendly apps."

Introduction

"The real or perceived security risks leading to data privacy compromises are particularly concerning for mental health apps because they deal with highly sensitive data, in contrast to other general mHealth apps, e.g., for fitness and wellness."

"The above-mentioned mHealth apps' data privacy concerns warrant evidence-based inquiries for improved understanding and actionable measures as there is a paucity of empirical research on understanding the full range of privacy threats that manifest in mental health apps; the existing research has only focused on privacy policy analysis (O'Loughlin et al. [77]; Powell et al. [78]; Robillard et al. [79]; Rosenfeld et al. [80]), or 3rd-party data sharing (Huckvale et al. [81])."

"Unawareness is also related to such threats, given that apps do not explain (e.g., in the privacy policy) the risks posed by targeted advertising on people experiencing mental problems and the risk of re-identification and disclosure of mental health conditions (e.g., anxiety, depression)."

"The findings highlight the prevalence of data privacy problems among the top-ranked mental health apps."

Background

"We are mainly interested in the LINDDUN threat taxonomy, which can be used as a standard reference for discussing privacy threats: Linkability: an adversary can link two items of interest (IOI) without knowing the identity of the data subject(s) involved (e.g., service providers are able to link data coming from different apps about the same data subject)."

"Shipp and Blasco [82] also looked into menstrual apps in order to show that developers often fail to consider menstruation and sex data as especially sensitive, mentioning only common pieces of personal data (e.g., name, email) in their privacy policies."

"Other studies have privileged the "invisible" aspects of mHealth apps' security and privacy, e.g., using pentesting tools to analyze the apps' code, network traffic, logs, and generated data (He et al. [83]; Papageorgiou et al. [84]; Hussain et al. [85]; LaMalva and Schmeelk [86])."

Methodology

"We performed a detailed analysis of the logs and apps' data files that were obtained from MobSF dynamic analysis."

"We analyzed Logcat logs and Frida API monitor logs generated during the dynamic analysis process to identify whether or not these logs reveal personal information of a user, reveal apps' behavior, usage, and activities, reveal tokens and credentials used by the app, or reveal the details of web traffic, parameters and Post values."

"As the final step of the dynamic analysis process, we analyzed the data generated by each app (i.e., files and databases) to see whether or not an app had insecurely stored any user's sensitive data."

"The Dynamic Security Analysis was also crucial for gathering various data generated by the apps, especially in terms of data flows (e.g., network traffic and system logs) and data stored (e.g., files and databases)."

Results

"Typical examples are anonymous letters written by the same person, web pages visited by the same user, entries in two databases related to the same person, people related by a friendship link, etc. Such linkability threats are revealed during the dynamic analysis of an apps when network traffic was manually inspected."

"Identifiability of log-in used (I_e1) and contextual data (I_df2) were the most common types of threat found in the 18 apps that went through dynamic analysis."

"Software developers cannot easily address such threats, considering that existing apps would have to provide relatively advanced privacy controls, such as using covert channels and anonymous communication."

"We found 15 apps that stored users' personal information (e.g. email, password, address) in files or databases."

"A type of threat concerns providing too much personal data (U_1), which can be linked to the list of permissions required by the apps to run."

Discussion

"Based on that, this section discusses the following concerning topics: (1) privacy impacts of mental health apps; (2) apps' permissions and data access; (3) apps' security testing and coding; (4) Privacy Impact Assessments and responsible disclosure; (5) privacy policies; and, (6) recommendations."

"Our analysis identified vulnerabilities related to all seven Android app vulnerability categories (i.e., cryptography API, inter-component communication, networking, permission, data storage, system processes, and web API) presented by Ranganath and Mitra [87]."

"Some issues identified in this privacy analysis would have a lower impact in a general Android app (e.g., WhatsApp, Twitter, Netflix apps)."

"It would be recommended that app developers ask for permissions "in context", i.e., when the user starts to interact with the feature that requires it."

"Companies should review their apps' privacy policy and, most importantly, change the apps to honestly inform users, recording their consent to collect and process data."

Limitations

"It may not be possible for researchers doing empirical studies to consider all sorts of tools available due to scope and resources limitations."

"Apart from tools such as Drozer, Qualys SSL, CLAUDETTE and PrivacyCheck, we relied on MobSF for various static and dynamic analyses since this framework integrates many tools to provide a broad coverage of penetration tests."

"We limited our analysis to the files flagged by MobSF."

"Even though using these wrapper classes and util methods in security contexts would lead to a security vulnerability, our analysis did not investigate such usages as it would increase the complexity and resources required for the study."

"Regarding the analysis of the privacy policies, we relied on two AI-based tools: (1) CLAUDETTE, to identify unfair clauses; and, (2) PrivacyCheck, to calculate user's control and GDPR compliance scores."

Conclusion

"Given the users of mental health apps are usually facing psychological issues such as depression, anxiety and stress, the detrimental impact of an app's data privacy breach can have more significant negative impact on users."

"We decided to empirically study the data privacy of mental health apps."

"This study has provided us with sufficient empirical evidence to assert that mHealth apps in general and also mental health apps in particular ought to be developed by following a privacy by design paradigm (Cavoukian et al. [88]; Gürses et al. [89])."

"This study has also enabled us to surmise that apart from developers, other stakeholders can also play important roles in ensuring data privacy in mHealth apps."

"We expect that these recommendations will enable all the key stakeholders, particularly the apps developers, to play their respective parts in order to ensure the privacy of the data of mHealth apps."

References

1. Guhr, Nadine; Werth, Oliver; Blacha, Philip Peter Hermann; Breitner, Michael H. Privacy concerns in the smart home context. SN Applied Sciences (2020). https://doi.org/10.1007/s42452-020-2025-8
2. Geneiatakis D, Kounelis I, Neisse R, Nai-Fovino I, Steri G, Baldini G (2017) Security and privacy issues for an IoT based smart home. In: Proceedings of the 40th international convention on information and communication technology, electronics and microelectronics
3. Xu H, Gupta S, Rosson MB, Carroll JM (2012a) Measuring mobile users' concerns for information privacy. In: Proceedings of the 33rd international conference on information systems
4. Petter S, Straub D, Rai A (2007) Specifying formative constructs in information systems research. Manag Inf Syst Q 31(4):623–656
5. Edwards JR (2001) Multidimensional constructs in organizational behavior research: an integrative analytical framework. Organ Res Methods 4(2):144–192
6. Himmel S, Ziefle M (2016) Smart home medical technologies: users' requirements for conditional acceptance. i-Com 15(1):39–50
7. Richter NF, Cepeda G, Roldán JL, Ringle CM (2016) European management research using partial least squares structural equation modeling (PLS-SEM). Eur Manag J 34(6):589–597
8. Farrell AM (2010) Insufficient discriminant validity: a comment on Bove, Pervan, Beatty, and Shiu (2009). J Bus Res 63(3):324–327
9. Sekara, Vedran; Alessandretti, Laura; Mones, Enys; Jonsson, Håkan Temporal and cultural limits of privacy in smartphone app usage. Scientific Reports (2021). https://doi.org/10.1038/s41598-021-82294-1
10. PageFair. The state of the blocked web—2017 global adblock report (2017). https://pagefair.com/blog/2017/adblockreport/
11. Kaggle. TalkingData Mobile User Demographics (accessed 2020-10-01); https://www.kaggle.com/c/talkingdata-mobile-user-demographics (2016)
12. Achara, J. P., Acs, G. & Castelluccia, C. On the unicity of smartphone applications. In Proceedings of the 14th ACM Workshop on Privacy in the Electronic Society 27–36 (ACM, 2015)
13. AppBrain. Google Play statistics (accessed 2017-04-18); https://www.appbrain.com/stats (2017)
14. Gulyás, G. G., Acs, G. & Castelluccia, C. Near-optimal fingerprinting with constraints. Proc. Priv. Enhanc. Technol. 2016, 470–487 (2016)
15. De Montjoye, Y.-A., Hidalgo, C. A., Verleysen, M. & Blondel, V. D. Unique in the crowd: The privacy bounds of human mobility. Sci. Rep. 3, 1376 (2013)
16. Kosta, Eleni; Pitkänen, Olli; Niemelä, Marketta; Kaasinen, Eija Mobile-Centric Ambient Intelligence in Health- and Homecare—Anticipating Ethical and Legal Challenges. Science and Engineering Ethics (2009). https://doi.org/10.1007/s11948-009-9150-5
17. Kaasinen, E., Ermolov, V., Niemelä, M., Tuomisto, T., & Välkkynen, P. (2006a). Identifying user requirements for a mobile terminal centric ubiquitous computing architecture. In FUMCA 2006: System support for future mobile computing applications. Workshop at Ubicomp 2006. http://ieeexplore.ieee.org/stamp/stamp.jsp?arnumber=04021039. Accessed May 07 2008
18. Niemelä, M. (2006, September 20–22). MIMOSA/MINAmI: Building mobile-centric ambient intelligence. In AmI.d conference, Sophia Antipolis

19. De Hert, Paul; Gutwirth, Serge; Moscibroda, Anna; Wright, David; González Fuster, Gloria Legal safeguards for privacy and data protection in ambient intelligence. Personal and Ubiquitous Computing (2008). https://doi.org/10.1007/s00779-008-0211-6
20. Wright D, Gutwirth S, Friedewald M, Punie Y, Vildjiounaite E (eds) (2008) Safeguards in a world of ambient intelligence, Springer Press, Dordrecht, p 291
21. Hildebrandt M, Koops BJ (eds) A vision of ambient law, FIDIS (Future of Identity in the Information Society) D7.9, version as of 15.11.2007 (http://www.fidis.net)
22. Zaccagnino, Rocco; Capo, Carmine; Guarino, Alfonso; Lettieri, Nicola; Malandrino, Delfina Techno-regulation and intelligent safeguards. Multimedia Tools and Applications (2021). https://doi.org/10.1007/s11042-020-10446-y
23. Cosimato A, De Prisco R, Guarino A, Lettieri N, Malandrino D, Sorrentino G, Zaccagnino R (2019) The conundrum of success in music: playing it or talking about it? IEEE Access 7:123289–123298. https://doi.org/10.1109/ACCESS.2019.2937743
24. Cozza F, Guarino A, Isernia F, Malandrino D, Rapuano A, Schiavone R, Zaccagnino R (2020) Hybrid and lightweight detection of third party tracking: design, implementation, and evaluation. Comput Netw 167(106):993
25. Klempous R, Nikodem J, Jacak W, Chaczko Z (2013) Advanced methods and applications in computational intelligence, vol 6. Springer Science & Business Media
26. Narudin F A, Feizollah A, Anuar N B, Gani A (2016) Evaluation of machine learning classifiers for mobile malware detection. Soft Comput 20(1):343–357
27. Rzecki K, Pławiak P, Niedźwiecki M, Sośnicki T, Leśkow J, Ciesielski M (2017) Person recognition based on touch screen gestures using computational intelligence methods. Inf Sci 415:70–84
28. Sae-Bae N, Ahmed K, Isbister K, Memon N (2012) Biometric-rich gestures: a novel approach to authentication on multi-touch devices. In: Proceedings of the SIGCHI conference on human factors in computing systems. ACM, pp 977–986
29. Masood R, Zhao B. ZH, Asghar HJ, Kaafar MA (2018) Touch and you're trapp (ck) ed: quantifying the uniqueness of touch gestures for tracking. Proc Privacy Enhanc Technol 2018(2):122–142
30. Vatavu RD, Anthony L, Brown Q (2015) Child or adult? Inferring Smartphone users' age group from touch measurements alone. In: IFIP conference on human-computer interaction. Springer, pp 1–9
31. Choo BJ, Joo HJ, Kim BS, Cha SH, Myung DH (2019) Using multimedia based big data for child abuse prevention system (based on the establishment of e-child welfare support system). Multimed Tools Appl 78(20):28805–28814
32. Mitchell KJ, Finkelhor D, Wolak J (2005) Protecting youth online: family use of filtering and blocking software. Child Abuse Neglect 29(7):753–765
33. Richardson CR, Resnick PJ, Hansen DL, Derry HA, Rideout VJ (2002) Does pornography-blocking software block access to health information on the internet? JAMA 288(22):2887–2894
34. Li Y, Ngom A (2015) Data integration in machine learning. In: 2015 IEEE international conference on bioinformatics and biomedicine (BIBM). IEEE, pp 1665–1671
35. Li Y, Wu FX, Ngom A (2016) A review on machine learning principles for multi-view biological data integration. Brief Bioinform 19(2):325–340
36. Yankson, Benjamin Continuous improvement process (CIP)-based privacy-preserving framework for smart connected toys. International Journal of Information Security (2021). https://doi.org/10.1007/s10207-020-00535-2
37. Smart2Zero.: Google patents connected smart toys that listen, speak, Smart2.0. (2019). https://www.smart2zero.com/news/google-patents-connected-smart-toys-listen-speak. Accessed 19 May 2019
38. Federal Trade Commission: Children's online privacy protection rule: a six-step compliance plan for your business. https://www.ftc.gov/tips-advice/business-center/guidance/childrens-online-privacy-protection-rule-six/-step-compliance. Accessed 30 Jul 2019

39. Hung, P.C.K.: Mobile Services for Toy Computing, The Springer International Series on Applications and Trends in Computer Science. Springer International Publishing, Berlin (2015)
40. Lin, P., Yankson, B., Lu, Z., Hung, P.C.K.: Children privacy identification system in LINE Chatbot for smart toys. In: 2019 IEEE 12th International Conference on Cloud Computing (Cloud), Milan, Italy, pp. 86–90 (2019)
41. Jones, M.L., Meurer, K.: Can (and should) Hello Barbie Keep a Secret?, Ethics in engineering, science and technology (ETHICS). In: 2016 IEEE International Symposium on Ethics in Engineering pp. 1–6, (2016)
42. Druga, S., Williams, R., Park, H.W., Breazeal, C.: How smart are the smart toys?: Children and parents' agent interaction and intelligence attribution. In: Proceedings of the 17th ACM Conference on Interaction Design and Children, p. 231, (2018)
43. McReynolds, E., Hubbard, S., Lau, T., Saraf, A., Cakmak, M., Roesner, F.: Toys that listen: a study of parents, children, and internet-connected toys. In: Proceedings of the 2017 CHI Conference on Human Factors in Computing Systems, p. 519, (2017)
44. Valente, J., Cardenas, A.: Security and privacy in smart toys. In: Proceedings of the 2017 Workshop on Internet of Things Security and Privacy, p. 19, (2017)
45. Mahmoud, M., Hossen, M.Z., Barakat, H., Mannan, M., Youssef, A.: Towards a comprehensive analytical framework for smart toy privacy practices. In: Proceedings of the 7th Workshop on Socio-Technical Aspects in Security and Trust, p. 64, (2018)
46. Strei, J., Kenny, O., Das, S., Leeth, A., Camp, L.J.: Who's watching your child? Exploring home security risks with smart toy bears, internet-of-things design and implementation (IoTDI). In: 2018 IEEE/ACM 3rd International Conference on, p. 285, (2018)
47. BBC UK.: German parents told to destroy Cayla dolls over hacking fears. (2017). http://www.bbc.com/news/world-europe-39002142
48. Dinev, T., Hart, P.: An extended privacy calculus model for e-commerce transactions. Inf. Syst. Res. 17(1), 61–80 (2006)
49. Rensel, A.D., Abbas, J.M., Rao, H.R.: Private transactions in public places: an exploration of the impact of the computer environment on public transactional web siteuse. J. Assoc. Inf. Syst. 7(1), 19–51 (2006)
50. Li, Y.: Theories in online information privacy research: a critical review and an integrated framework. Decis. Support Syst. 54, 471–481 (2019)
51. Yankson, B., Iqbal, F., Hung, P.C.K.: Privacy Preservation Framework for Smart Connected Toys. Computing in Smart Toys, pp. 149–164. Springer, Berlin (2017)
52. Ullah, Imdad; Boreli, Roksana; Kanhere, Salil S. Privacy in targeted advertising on mobile devices: a survey. International Journal of Information Security (2022). https://doi.org/10.1007/s10207-022-00655-x
53. Ullah, I., Sarwar, B. G., Boreli, R., Kanhere, S. S., Katzenbeisser, S., Hollick, M.: Enabling privacy preserving mobile advertising via private information retrieval. 2017 IEEE 42nd Conference on Local Computer Networks (LCN), pp. 347–355 (2017)
54. Ullah, I., Boreli, R., Kanhere, S.S., Chawla, S., Ahanger, T.A., Tariq, U.: Protecting private attributes in app based mobile user profiling. IEEE Access 8, 143818–143836 (2020)
55. Chen, T., Ullah, I., Kaafar, M. A., Boreli, R.: Information leakage through mobile analytics service. ACM HotMobile 15th International workshop on mobile computing systems and applications, (2014)
56. Mamais, S.: Privacy-preserving and fraud-resistant targeted advertising for mobile devices. PhD thesis, Cardiff University, https://orca.cardiff.ac.uk/id/eprint/125897/1/2019mamaisssphd.pdf (2019)
57. Liu, Y., Simpson, A.A.: Privacy-preserving targeted mobile advertising: requirements, design and a prototype implementation. Softw. Pract. Exp. 46(12), 1657–1684 (2016)
58. Wang, Y., Genc, E., Peng, G.: Aiming the mobile targets in a cross-cultural context: effects of trust, privacy concerns, and attitude. Int. J. Hum. Comput. Interact. 36(3), 227–238 (2020)
59. Toch, E., Wang, Y., Cranor, L.F.: Personalization and privacy: a survey of privacy risks and remedies in personalization-based systems. User Model. User Adapt. Interact. 22(1), 203–220 (2012)

60. Estrada-Jiménez, J., Parra-Arnau, J., Rodríguez-Hoyos, A., Forné, J.: Online advertising: analysis of privacy threats and protection approaches. Comput. Commun. 100, 32–51 (2017)
61. Kaaniche, N., Laurent, M., Belguith, S.: Privacy enhancing technologies for solving the privacy-personalization paradox: taxonomy and survey. J. Netw. Comput. Appl. 171, 102807 (2020)
62. Boerman, S.C., Kruikemeier, S., Zuiderveen Borgesius, F.J.: Online behavioral advertising: a literature review and research agenda. J. Advert. 46(3), 363–376 (2017)
63. Flurry advertisers, publishers, and analytics http://www.flurry.com (2022)
64. Castelluccia, C., Kaafar, M.-A., Tran, M.-D.: "Betrayed by your ads!," Springer, Privacy Enhancing Technologies (PETs), pp. 1–17 (2012)
65. Yan, J., Liu, N., Wang, G., Zhang, W., Jiang, Y., Chen, Z.: How much can behavioral targeting help online advertising? Proceedings of the ACM 18th international conference on World wide web, pp. 261–270, (2009)
66. Guha, S., Cheng, B., Reznichenko, A., Haddadi, H., Francis, P.: Privad: Rearchitecting online advertising for privacy. Proceedings of Hot Topics in Networking (HotNets) (2009)
67. Toubiana, V., Narayanan, A., Boneh, D., Nissenbaum, H., Barocas, S.: Adnostic: Privacy preserving targeted advertising. Proceedings Network and Distributed System Symposium (2010)
68. Balebako, R., Leon, P., Shay, R., Ur, B., Wang, Y., Cranor, L.: Measuring the effectiveness of privacy tools for limiting behavioral advertising. Web 2.0 Security and Privacy Workshop, (2012)
69. Wills, C. E., Tatar, C.: Understanding what they do with what they know. Proceedings of the 2012 ACM Workshop on Privacy in the Electronic Society, pp. 13–18 (2012)
70. Goldfarb, A., Tucker, C.: Online display advertising: targeting and obtrusiveness. Market. Sci. 30, 389–404 (2011)
71. Farahat, A., Bailey, M. C.: How effective is targeted advertising? Proceedings of the ACM 21st international conference on World Wide Web, pp. 111–120 (2012)
72. Evans, D.S.: The online advertising industry: economics, evolution, and privacy. J. Econ. Perspect. 23, 37–60 (2009)
73. Barford, P., Canadi, I., Krushevskaja, D., Ma, Q., Muthukrishnan, S.: Adscape: harvesting and analyzing online display ads. Proceedings of the ACM 23rd international conference on World wide web, pp. 597–608 (2014)
74. Nath, S.: Madscope: characterizing mobile in-app targeted ads. Proceedings of the 13th ACM annual international conference on mobile systems, applications, and services, pp. 59–73 (2015)
75. Ullah, I., Boreli, R., Kaafar, M. A., Kanhere, S. S.: Characterising user targeting for in-app mobile ads. 2014 IEEE Conference on computer communications workshops (INFOCOM WKSHPS), pp. 547–552, (2014)
76. Iwaya, Leonardo Horn; Babar, M. Ali; Rashid, Awais; Wijayarathna, Chamila On the privacy of mental health apps. Empirical Software Engineering (2022). https://doi.org/10.1007/s10664-022-10236-0
77. O'Loughlin K, Neary M, Adkins EC, Schueller SM (2019) Reviewing the data security and privacy policies of mobile apps for depression. Internet Interventions 15:110–115
78. Powell AC, Singh P, Torous J (2018) The complexity of mental health app privacy policies: a potential barrier to privacy. JMIR Mhealth and Uhealth 6(7):e158
79. Robillard JM, Feng TL, Sporn AB, Lai JA, Lo C, Ta M, Nadler R (2019) Availability, readability, and content of privacy policies and terms of agreements of mental health apps. Internet Interventions 17:100243
80. Rosenfeld L, Torous J, Vahia IV (2017) Data security and privacy in apps for dementia: An analysis of existing privacy policies. The American Journal of Geriatric Psychiatry 25(8):873–877. use of Technology in Geriatric Mental Health
81. Huckvale K, Torous J, Larsen ME (2019) Assessment of the data sharing and privacy practices of smartphone apps for depression and smoking cessation. JAMA Network Open 2(4):e192542–e192542
82. Shipp L, Blasco J (2020) How private is your period?: a systematic analysis of menstrual app privacy policies. Proc Priv Enhancing Technol 2020 (4):491–510

83. He D, Naveed M, Gunter CA, Nahrstedt K (2014) Security concerns in Android mhealth apps. AMIA Annual Symposium proceedings AMIA Symposium 2014:645—654
84. Papageorgiou A, Strigkos M, Politou E, Alepis E, Solanas A, Patsakis C (2018) Security and privacy analysis of mobile health applications: the alarming state of practice. IEEE Access 6:9390–9403
85. Hussain M, Al-Haiqi A, Zaidan AA, Zaidan BB, Kiah M, Iqbal S, Iqbal S, Abdulnabi M (2018) A security framework for mhealth apps on Android platform. Computers & Security 75:191–217
86. LaMalva G, Schmeelk S (2020) MobSF: Mobile health care Android applications through the lens of open source static analysis. In: 2020 IEEE MIT Undergraduate Research Technology Conference (URTC), IEEE, pp 1–4
87. Ranganath VP, Mitra J (2020) Are free Android app security analysis tools effective in detecting known vulnerabilities? Empir Softw Eng 25(1):178–219
88. Cavoukian A, et al. (2009) Privacy by design: The 7 foundational principles. Information and privacy commissioner of Ontario, Canada 5:12
89. Gürses S, Troncoso C, Diaz C (2011) Engineering privacy by design. Computers, Privacy & Data Protection 14(3):25

Chapter 4
Risks, Threats and Way-Outs to Protecting Privacy in Tech-Environment: Finding Technical Solutions to Safeguard Privacy of Individuals

4.1 Introduction by the Editor

In an environment where rampant use of technology raises several privacy risks, suitable technological interventions are required to protect the interests of those who are using such technology. In the process, those interventions may raise more privacy questions and therefore, must be calibrated to assuage the concerns of the data subjects. The sheer volume of personal data processing that happens in a data dependency environment reduces the degree of control[1] that data subjects should have over their data. There are several technical tools like encryption, pseudonymization,[2] use of synthetic data[3] and use of multi-party computation in business-to-business data transfer[4] that have been suggested to mitigate some of the concerns. However, it is difficult to find an optimal solution or several of such optimal solutions in a data dependency environment.

From a legal perspective, anonymisation has the force of law, which does not allow identification of a natural person, although using anoymised data from multiple

---

[1] *Maaser, Michael; Langendörfer, Peter Privacy from Promises to Protection. Mobile Networks and Applications (2008)* https://doi.org/10.1007/s11036-008-0116-7.

[2] Recital 26, 28 and 29, Regulation (EU) 2016/679 of the European Parliament and of the Council of 27 April 2016 on the protection of natural persons with regard to the processing of personal data and on the free movement of such data, and repealing Directive 95/46/EC (General Data Protection Regulation) (Text with EEA relevance) *OJ L 119, 4.5.2016, pp. 1–88.*

[3] *Thambawita, Vajira; Isaksen, Jonas L.; Hicks, Steven A.; Ghouse, Jonas; Ahlberg, Gustav; Linneberg, Allan; Grarup, Niels; Ellervik, Christina; Olesen, Morten Salling; Hansen, Torben; Graff, Claus; Holstein-Rathlou, Niels-Henrik; Strümke, Inga; Hammer, Hugo L.; Maleckar, Mary M.; Halvorsen, Pål; Riegler, Michael A.; Kanters, Jørgen K. DeepFake electrocardiograms using generative adversarial networks are the beginning of the end for privacy issues in medicine. Scientific Reports (2021)* https://doi.org/10.1038/s41598-02101295-2.

[4] *Agahari, Wirawan; Ofe, Hosea; de Reuver, Mark It is not (only) about privacy: How multi-party computation redefines control, trust, and risk in data sharing. Electronic Markets (2022)* https://doi.org/10.1007/s12525-022-00572-w.

I. Gupta (ed.), *Operationalizing Expectations and Mapping Challenges of Information Privacy and Data Protection Measures*, https://doi.org/10.1007/978-981-99-7371-2_4

datasets run the risk of reidentifying natural persons.[5] GDPR leaves anonymized data and those processing such data outside the scope of the legal framework.[6] Sometimes, anonymized data may not be useful to serve the objective of data collection that a data controller follows. Research suggests about using techniques that can make sense of encrypted data without breaching confidentiality of data subjects.[7]

However, it is imperative to introduce technical mechanisms that help data subjects take informed decisions.[8] It also makes the idea of informed consent relevant and worthy in an environment where it is increasingly becoming difficult to exercise control over personal data.[9] While introduction of technical mechanisms are useful, perception of privacy concerns that individuals have must be considered. Several technologies that individuals seem to be comfortable with may continue to record and track them on a regular basis.[10] Therefore, there is no place for information asymmetry concerning technologies that are geared to making lives easier, and they should be processing personal data within limits and should increase transparency in the process.[11]

Tracking provides an option through which data subjects receive personalized curated information. Such information often comes in exchange for privacy unless positive technical interventions using high-level data encryption restores balance.[12] Information about individuals obtained either directly or using multiple anonymized datasets can lead to a situation of predictive analytics.[13] Such a situation leads to infringement of predictive privacy where data controllers derive "…a prediction about individuals or profiling groups from data that were collected from many other users of networked digital services".[14]

---

[5] *Maritsch, Friedrich; Cil, Ingeborg; McKinnon, Colin; Potash, Jesse; Baumgartner, Nicole; Philippon, Valérie; Pavlova, Borislava G. Data privacy protection in scientific publications: process implementation at a pharmaceutical company. BMC Medical Ethics (2022)* https://doi.org/10.1186/s12910-02200804-w.

[6] Recital 26, Regulation (EU) 2016/679 of the European Parliament and of the Council of 27 April 2016 on the protection of natural persons with regard to the processing of personal data and on the free movement of such data, and repealing Directive 95/46/EC (General Data Protection Regulation) (Text with EEA relevance) *OJ L 119, 4.5.2016, pp. 1–88.*

[7] *Susilo, Willy; Win, Khin Than Security and Access of Health Research Data. Journal of Medical Systems (2007)* https://doi.org/10.1007/s10916-006-9035-y.

[8] *Ackerman, Mark S. Privacy in pervasive environments: next generation labeling protocols. Personal and Ubiquitous Computing (2004)* https://doi.org/10.1007/s00779-004-0305-8.

[9] Ibid.

[10] *Nguyen, David H.; Hayes, Gillian R. Information privacy in institutional and end-user tracking and recording technologies. Personal and Ubiquitous Computing (2009)* https://doi.org/10.1007/s00779-009-0229-4.

[11] Ibid.

[12] *El-Ansari, Anas; Beni-Hssane, Abderrahim; Saadi, Mostafa; El Fissaoui, Mohamed PAPIR: privacy-aware personalized information retrieval. Journal of Ambient Intelligence and Humanized Computing (2021)* https://doi.org/10.1007/s12652-020-02736-y.

[13] *Mühlhoff, Rainer Predictive privacy: towards an applied ethics of data analytics. Ethics and Information Technology (2021)* https://doi.org/10.1007/s10676-021-09606-x.

[14] Ibid.

The other issue is with the ecosystem in which privacy information is exchanged with data subjects. It gives a little room to negotiate or allow data subjects to exercise free choices. Technical interventions in this regard are useful, thereby bringing about a greater degree of transparency.[15] There is a separate concern about operationalizing notice and consent requirement with its primary task of notifying and informing data subjects about the consequences of personal data processing. Interventions centered around the use of technology and its application can go a long way in improving the situation with the use and operationalisation of the notice and consent framework.[16] The workings of the notice and consent framework rely on the effectiveness of privacy policies that data controllers draft to convey privacy-related information to data subjects. Research shows that "…an excellent privacy policy can be effective in reducing privacy concern, it is insufficient when highly sensitive information is being collected".[17] There are suggestions about automatically generating privacy policies from the source code of software compared to privacy policies that are drafted in a standalone mode.[18] There are several instances where technology have and can play a pivotal role in protecting the privacy of data subjects. The above used instances brought about a tangible change. However, the concerns surrounding such technical interventions must be equally addressed.

The use of blockchain technology has been proposed to provide a technical solution that prevents misuse of personal data and increase data reliability for the end-users.[19] The process of verification of records at different stages of professional and academic life of individuals has undergone substantial changes. Use of blockchains is one such method which enhances and preserves the privacy of individuals compared to the previously used methods.[20] Certain measures can be undertaken that make blockchain technology yet more effective and efficient towards fulfilling the privacy needs of data subjects.[21]

---

[15] *Filipczuk, Dorota; Baarslag, Tim; Gerding, Enrico H.; schraefel, m. c. Automated privacy negotiations with preference uncertainty. Autonomous Agents and Multi-Agent Systems (2022)* https://doi.org/10.1007/s10458-022-09579-1.

[16] *Vidhani, Kumar; Banahatti, Vijayanand; Mirza, Hussain Privacy self management in enterprise. CSI Transactions on ICT (2022)* https://doi.org/10.1007/s40012-022-00347-w.

[17] *Lwin, May; Wirtz, Jochen; Williams, Jerome D. Consumer online privacy concerns and responses: a power–responsibility equilibrium perspective. Journal of the Academy of Marketing Science (2007)* 10.1007/s11747-006–0003-3.

[18] *Hjerppe, Kalle; Ruohonen, Jukka; Leppänen, Ville Extracting LPL privacy policy purposes from annotated web service source code. Software and Systems Modeling (2022)* https://doi.org/10.1007/s10270-022-00998-y.

[19] *Pericàs-Gornals, Rosa; Mut-Puigserver, Macià; Payeras-Capellà, M. Magdalena Highly private blockchain-based management system for digital COVID-19 certificates. International Journal of Information Security (2022)* https://doi.org/10.1007/s10207-022-00598-3.

[20] *Dewangan, Narendra K; Chandrakar, Preeti; Kumari, Saru; Rodrigues, Joel J.P.C. Enhanced privacy-preserving in student certificate management in blockchain and interplanetary file system. Multimedia Tools and Applications (2022)* https://doi.org/10.1007/s11042-022-13915-8.

[21] *Bansod, Smita; Ragha, Lata Challenges in making blockchain privacy compliant for the digital world: some measures. Sādhanā (2022)* https://doi.org/10.1007/s12046-02201931-1.

One must understand that any use of technology is a work in progress and may require effective customization before it optimally protects the privacy of individuals.[22] The idea of managing personal sensitive data[23] at the time of health crisis has been severely tested during the COVID-19 times. Contact-tracing applications were used to prevent the spread of the disease. However, all these applications required sharing of personal sensitive data, and at times the enormity of the crisis led to the over-processing of such data.[24] The COVID-19 crisis taught us how the risk-assessment exercise concerning the processing of personal sensitive data is equally important.[25] It also showcased the importance of keeping health records of individuals portable.

More portable devices carrying personal health records will be used in the future, but there are lots of inbuilt privacy concerns embedded with each such use.[26] It is rather important to intervene at the design stage to stop unwarranted data leaks[27] alongside the readiness of effective digital forensic services,[28] the use of blockchain technology and other technological interventions to address such threats.[29] Further protection must be taken when these personal data records are transferred to the cloud.[30] The act of transferring personal sensitive data to a cloud server possess certain tangible risk to the privacy of individuals.[31] It is important to assess the

---

[22] *Noam, Oded; Rottenstreich, Ori Realizing privacy aspects in blockchain networks. Annals of Telecommunications (2021)* https://doi.org/10.1007/s12243-021-00861-z.

[23] See generally to ascertain sensitivity of personal data, *Hekel, Rastislav; Budis, Jaroslav; Kucharik, Marcel; Radvanszky, Jan; Pös, Zuzana; Szemes, Tomas Privacy-preserving storage of sequenced genomic data. BMC Genomics (2021)* https://doi.org/10.1186/s12864-021-07996-2.

[24] *Sowmiya, B.; Abhijith, V.S.; Sudersan, S.; Sakthi Jaya Sundar, R.; Thangavel, M.; Varalakshmi, P. A Survey on Security and Privacy Issues in Contact Tracing Application of Covid-19. SN Computer Science (2021)* https://doi.org/10.1007/s42979-021-00520-z.

[25] Ibid.

[26] *Huang, Lu-Chou; Chu, Huei-Chung; Lien, Chung-Yueh; Hsiao, Chia-Hung; Kao, Tsair Embedding a Hiding Function in a Portable Electronic Health Record for Privacy Preservation. Journal of Medical Systems (2008)* https://doi.org/10.1007/s10916-008-9243-8.

[27] Ibid.

[28] *Chernyshev, Maxim; Zeadally, Sherali; Baig, Zubair Healthcare Data Breaches: Implications for Digital Forensic Readiness. Journal of Medical Systems (2018)* https://doi.org/10.1007/s10916-018-1123-2.

[29] *Vidhya, S.; Kalaivani, V. A blockchain based secure and privacy aware medical data sharing using smart contract and encryption scheme. Peer-to-Peer Networking and Applications (2023)* https://doi.org/10.1007/s12083-023-01449-1.

[30] *Kanwal, Tehsin; Anjum, Adeel; Khan, Abid Privacy preservation in e-health cloud: taxonomy, privacy requirements, feasibility analysis, and opportunities. Cluster Computing (2020)* https://doi.org/10.1007/s10586-020-03106-1.

[31] *Anand, K.; Vijayaraj, A.; Vijay Anand, M. Privacy preserving framework using Gaussian mutation based firebug optimization in cloud computing. The Journal of Supercomputing (2022)* https://doi.org/10.1007/s11227-021-04173-w.

nature of such susceptibility, and the measures undertaken must protect personal data from different intrusions.[32]

As technology progresses to reduce human interventions, there are situations where physiological sensitive personal data are collected. There are enormous concerns as to the further use of such data and the protection of privacy of individuals in an ecosystem where connectedness of different devices has become the new normal.[33]

Identifying the location of data subjects is an opportunity that can prove useful for streamlining many tasks, yet the practice may give rise to many privacy concerns.[34] Such data can encourage both tracking and generating subsequent reports concerning the habits of natural persons. Therefore, to offset such concerns, location services must provide additional safeguards to protect personal autonomy and data safety.[35]

The use of RFID tags can decide ownership issues in a secure way. However, the process of communication through RFID tags can also identify a natural person.[36] Similar issues exist with the devices that are connected to the IoT networks. The use of privacy enhancing technology is a possible way out in a situation where multiple devices are integrated into one single device. Data minimization is a key factor in this regard, thereby avoiding over processing of personal data.[37] Process of automated data anonymization has also been suggested to protect privacy and confidentiality of users.[38] In relation to the IoTs and their operational privacy optimisation needs, several privacy frameworks have been adopted. There are divergent opinions as to their efficacy.[39]

---

[32] Ibid; See, *Li, Jin; Xiao, Wei; Zhang, Chong Data security crisis in universities: identification of key factors affecting data breach incidents. Humanities and Social Sciences Communications (2023)* https://doi.org/10.1057/s41599-023-01757-0.

[33] *Wang, Meng; Qin, Yalin; Liu, Jiaojiao; Li, Weidong Identifying personal physiological data risks to the Internet of Everything: the case of facial data breach risks. Humanities and Social Sciences Communications (2023)* https://doi.org/10.1057/s41599-023-01673-3.

[34] *Krumm, John A survey of computational location privacy. Personal and Ubiquitous Computing (2008)* https://doi.org/10.1007/s00779-008-0212-5.

[35] Ibid; See, *Schnell, Rainer; Klingwort, Jonas; Farrow, James M. Locational privacy-preserving distance computations with intersecting sets of randomly labeled grid points. International Journal of Health Geographics (2021)* https://doi.org/10.1186/s12942-021-00268-y.

[36] *Langheinrich, Marc A survey of RFID privacy approaches. Personal and Ubiquitous Computing (2008)* https://doi.org/10.1007/s00779-008-0213-4.

[37] *Staudemeyer, Ralf C.; Pöhls, Henrich C.; Wójcik, Marcin What it takes to boost Internet of Things privacy beyond encryption with unobservable communication: a survey and lessons learned from the first implementation of DC-net. Journal of Reliable Intelligent Environments (2019)* https://doi.org/10.1007/s40860-019-00075-0.

[38] *Ren, Wang; Tong, Xin; Du, Jing; Wang, Na; Li, Shancang; Min, Geyong; Zhao, Zhiwei Privacy Enhancing Techniques in the Internet of Things Using Data Anonymisation. Information Systems Frontiers (2021)* https://doi.org/10.1007/s10796-021-10116-w.

[39] *Pamarthi, Satyanarayana; Narmadha, R. Adaptive Key Management-Based Cryptographic Algorithm for Privacy Preservation in Wireless Mobile Adhoc Networks for IoT Applications. Wireless Personal Communications (2021)* https://doi.org/10.1007/s11277-021-09360-9.

Privacy setting goals should be made part of the internal governance structure of any organization.[40] It gives a heads-up to optimize business processes and shows the impact that privacy requirements have over business processes. Often the idea of keeping the privacy requirements in digital applications in an embedded way is a localised discussion and not necessarily involving all stakeholders in the process.[41] Research shows while taking privacy decisions a dual approach should be adopted by using the "…security-oriented requirement engineering methodologies and privacy enhancing technologies".[42] Also, how these privacy enhancing technologies function and the promises that they make to protect privacy must be tested.[43] Other suggestions include using a privacy-criteria method at an early stage of development of a software.[44]

There is a greater need to adopt human-centred design practices in processes that imbibe complex integration of several components.[45] These design practices reflect intrinsic and personal needs of data subjects.[46] Further on, there must exist co-operation amongst multiple organisations to set up multiple intrusion detection systems to prevent privacy breach, and in that regard a privacy-preserving alert system can help.[47]

There are several uses of already processed data. By way of process mining, the event logs of processed data provide valuable information.[48] Therefore, research

---

[40] *Kalloniatis, Christos; Kavakli, Evangelia; Gritzalis, Stefanos Addressing privacy requirements in system design: the PriS method. Requirements Engineering (2008)* https://doi.org/10.1007/s00766-008-0067-3.

[41] *Ansari, Md Tarique Jamal; Baz, Abdullah; Alhakami, Hosam; Alhakami, Wajdi; Kumar, Rajeev; Khan, Raees Ahmad P-STORE: Extension of STORE Methodology to Elicit Privacy Requirements. Arabian Journal for Science and Engineering (2021)* https://doi.org/10.1007/s13369-021-05476-z.

[42] Ibid.

[43] *Pullonen, Pille; Tom, Jake; Matulevičius, Raimundas; Toots, Aivo Privacy-enhanced BPMN: enabling data privacy analysis in business processes models. Software and Systems Modeling (2019)* https://doi.org/10.1007/s10270-019-00718-z.

[44] *Peixoto, Mariana; Silva, Carla; Araújo, João; Gorschek, Tony; Vasconcelos, Alexandre; Vilela, Jéssyka Evaluating a privacy requirements specification method by using a mixed-method approach: results and lessons learned. Requirements Engineering (2022)* https://doi.org/10.1007/s00766-022-00388-2.

[45] *Carew, Peter J.; Stapleton, Larry; Byrne, Gabriel J. Implications of an ethic of privacy for human-centred systems engineering. AI & SOCIETY (2007)* https://doi.org/10.1007/s00146-007-0149-7.

[46] Ibid.

[47] *Spathoulas, Georgios; Theodoridis, Georgios; Damiris, Georgios-Paraskevas Using homomorphic encryption for privacy-preserving clustering of intrusion detection alerts. International Journal of Information Security (2020)* https://doi.org/10.1007/s10207-020-00506-7.

[48] *Mannhardt, Felix; Koschmider, Agnes; Baracaldo, Nathalie; Weidlich, Matthias; Michael, Judith Privacy-Preserving Process Mining. Business & Information Systems Engineering (2019)* https://doi.org/10.1007/s12599-019-00613-3; See, *Tavani, Herman T. KDD, data mining, and the challenge for normative privacy. Ethics and Information Technology (1999)* https://doi.org/10.1023/a:1010051717305.

suggests the way to protect privacy of event logs containing personal data while balancing the need of process mining.[49]

In terms of using new technology, Article 35 of the GDPR refers to the idea of data protection impact assessment exercise catering towards mitigating the data privacy risks.[50] Research suggests that measuring security risk through a certain set of parameters is not the same as measuring privacy risks.[51] Therefore, these parameters must be drawn separately to have correct readings of the risks following the perspective of the data subjects involved.[52] The risk assessment exercise could be a complex process where processing involves personal sensitive data comprising medical records.[53]

The overarching idea is to give the control and sense of personal autonomy back to the data subjects. Ultimately, the data subjects should be able to choose between data controllers based on the processes they adopt towards effective protection of personal data. Here the use of independent trusted third parties have been suggested to increase the control that data subjects have over their personal data with an additional option to monitor the use of such data.[54] A privacy-enhancing dashboard that manages records of personal data processing, thereby maintaining transparency about the associated data privacy risks is desirable.[55]

4.2 Machine Generated Summaries

Disclaimer: The summaries in this chapter were generated from Springer Nature publications using extractive AI auto-summarization: An extraction-based summarizer aims to identify the most important sentences of a text using an algorithm and uses those original sentences to create the auto-summary (unlike generative AI). As the constituted sentences are machine selected, they may not fully reflect the

---

[49] Ibid.

[50] Article 35, Regulation (EU) 2016/679 of the European Parliament and of the Council of 27 April 2016 on the protection of natural persons with regard to the processing of personal data and on the free movement of such data, and repealing Directive 95/46/EC (General Data Protection Regulation) (Text with EEA relevance) *OJ L 119, 4.5.2016, pp. 1–88.*

[51] *Hart, Stephen; Ferrara, Anna Lisa; Paci, Federica Fuzzy-based approach to assess and prioritize privacy risks. Soft Computing (2019)* https://doi.org/10.1007/s00500-019-03986-5.

[52] Ibid.

[53] *Holub, Petr; Müller, Heimo; Bíl, Tomáš; Pireddu, Luca; Plass, Markus; Prasser, Fabian; Schlünder, Irene; Zatloukal, Kurt; Nenutil, Rudolf; Brázdil, Tomáš Privacy risks of whole-slide image sharing in digital pathology. Nature Communications (2023)* https://doi.org/10.1038/s41467-023-37991-y.

[54] *Mitropoulos, Dimitris; Sotiropoulos, Thodoris; Koutsovasilis, Nikos; Spinellis, Diomidis PDGuard: an architecture for the control and secure processing of personal data. International Journal of Information Security (2019)* https://doi.org/10.1007/s10207-019-00468-5.

[55] *Silva, Paulo; Casaleiro, Rui; Simões, Paulo; Antunes, Nuno; Curado, Marilia; Monteiro, Edmundo Risk Management and Privacy Violation Detection in the PoSeID-on Data Privacy Platform. SN Computer Science (2020)* https://doi.org/10.1007/s42979-020-00198-9.

body of the work, so we strongly advise that the original content is read and cited. The auto generated summaries were curated by the editor to meet Springer Nature publication standards. To cite this content, please refer to the original publications. Machine generated keywords: blockchain, digital, attack, file, user, privacy requirement, provider, method, encryption, IoT, event, cloud, identifier, datum breach, patient.

4.2.1 *Identifying Personal Physiological Data Risks to the Internet of Everything: The Case of Facial Data Breach Risks [232]*

This is a machine-generated summary of:

Wang, Meng; Qin, Yalin; Liu, Jiaojiao; Li, Weidong: Identifying personal physiological data risks to the Internet of Everything: the case of facial data breach risks [232].

Published in: Humanities and Social Sciences Communications (2023).

Link to original: https://doi.org/10.1057/s41599-023-01673-3.

If you want to cite the papers, please refer to the original.

For technical reasons we could not place the page where the original quote is coming from.

Abstract-Summary

"Personal physiological data is the digital representation of physical features that identify individuals in the Internet of Everything environment."

"As facial recognition technology has become prevalent and smarter over time, facial data associated with critical personal information poses a potential security and privacy risk of being leaked in the Internet of Everything application platform."

"Based on the risks identified, we then listed intermediate events and basic events according to the causal logic, and drew a complete fault tree diagram of facial data breaches."

"The study determined that personal factors, data management and supervision absence are the three intermediate events."

"We anticipate that this study will explain the manageability and traceability of personal physiological data during its lifecycle."

"This study contributes to an understanding of what risks physiological data faces in order to inform individuals of how to manage their data carefully and to guide management parties on how to formulate robust policies and regulations that can ensure data security."

Introduction

"The IoE era is aimed at connecting everything, and physiological data is easily accessible due to a large number of digital collecting devices around us with high-risk factors for data breaches."

"The identification of personal physiological data risks is the main task of current work and is also an important prerequisite to solve the security and privacy problems of personal physiological data."

"By collecting, summarising and studying cases of facial data breaches, we can summarise the causes of personal physiological data breaches through the fault tree analysis method and provide insight into personal physiological data security."

"The IoE context connecting everything leads to more risk sources for data breaches."

"Based on these myriad considerations, this study uses facial data breach accidents as the research objects and adopts fault tree analysis to systematically and comprehensively investigate the risks of physiological data breaches in line with the data lifecycle."

The generation and application of personal physiological data in the era of IoE

"The IoE is a massive, complex network ecosystem consisting of various elements such as objects, digital devices, individuals, enterprises, governments and data resources through the support of digital platforms and digital processes (Li, 1)."

"Personal physiological data has been a crucial resource for data collection in the IoE environment."

"The development of face recognition technology and human body recognition technology has allowed increasing amounts of facial data and human body data to be perceived and collected."

"With the innovation of emotion recognition system and emotion computing technology, some new application platforms are trying to scientifically monitor individual emotions (surprised, happy, sad, angry, calm), psychological conditions and ideas by analysing the facial data or voice [2, 3]."

"Medical and health wearable devices generally have built-in sensors for measuring human health that can automatically collect important data such as blood pressure, pulse, heart rate and body temperature."

Risks to personal physiological data in the IoE

"Worse still, in the era of the IoE, technology companies, digital media platforms and the government—powered by intelligent collection equipment and information technology—are collecting physiological data on people's digital daily lives in invisible ways."

"Physiological data is sensitive personal data and can face unprecedented threats once leaked [4]."

"These threats abound in personal physiological data and may carry dire consequences for individuals, enterprises and nations [5]."

"For individuals, physiological data breaches can cause privacy violations, property loss, genetic discrimination, security threats, and serious emotional and mental harm [6–10]."

"On the national level, the impact of a physiological data breach can be far-reaching."

"US credit enterprise Equifax was attacked by hackers who breached the data of 143 million US citizens, posing a serious threat to US national security."

The application and uniqueness of facial data

"As a special category of personal physiological data, facial data is a highly sensitive form of personal sensitive information."

"Facial data has assumed as an increasingly vital role in daily life due to the ubiquitous presence of face recognition technology, which is characterised by high acceptability, collectability and universality [11–13]."

"Facial data can be collected imperceptibly and used widely compared with fingerprint, palmprint or iris data due to non-contact and wide application of face recognition technology, thus facing more breach risks."

"Facial data is a special kind of personal sensitive data related with privacy, which not only covers the personality interests of the 'face', such as personal portrait, reputation and human dignity, but also reflects the additional economic value of data."

"The identification of personal physiological data risks, especially facial data risks, is the main task of current work."

Current research on risk identification

"When identifying personal data breach risks, previous scholars have mainly adopted empirical and positive research approaches, and empirical research is more used comparatively."

"Scholars have also dedicated research efforts to the identification of personal physiological data breach risks [14–16]."

"Previous studies have determined that the risks of physiological data breaches mainly come from indirect attacks and direct attacks [17–19]."

"Existing research on the risks of personal data breaches has established a basic risk identification reference and listed several risk sources, but the research also has deficiencies."

"Facial data, as a special kind of personal physiological data, is widely used and has high breach risks."

"There is little research on the identification of personal physiological data breach risks, especially facial data breach risks."

Methods

"In order to identify and enumerate the risks leading to facial data breach and to list the casual events in the fault tree diagram, we collect cases from the China National Knowledge Internet (CNKI) newspaper database, China Search (chinaso.com), Baidu News and Google News using the keywords 'facial data breach', 'facial data accident', 'facial data abuse', 'face recognition' and 'facial data'."

"This paper starts from the top event and then determines the intermediate events, basic events and the logical relationship between them level by level based on the risks of facial data breach cases identified via the aspects of direct cause and expert analysis."

"We simplify the fault tree diagram of facial data breach by Boolean algebra as follows: After step-by-step analysis, the bottom combination which cannot be decomposed constitutes the MCS, or the most basic path to the top event."

"Principle, the structural importance analysis of the basic event of a facial data breach is carried out."

Discussion

"We identify potential and traditional breach risks of personal physiological data and assess the degree of importance of risk factors based on the fault tree diagram of facial data breach cases."

"The results indicated that 'Lack of laws and regulations' (X_{43}), 'Immature face recognition technology' (X_7), 'Personal greed for small gains' (X_1), 'Incautious about the related products' (X_2), 'Individuals lack deletion consciousness' (X_3), 'Individuals use the simple password' (X_4), 'Individuals upload facial data for entertainment actively and casually' (X_5) and 'Individuals download unapproved apps' (X_6) are the main factors leading to a facial data breach."

"In the context of the IoE, the risk of personal physiological data breach is usually long-term and hidden, while the immediate interest temptation is direct, individuals usually give up their right to privacy control in order to obtain more personalised services."

"We also found the hidden risk sources of personal physiological data breaches in the IoE terminals and cloud."

Conclusion

"The identification of personal physiological data risks is key to ensuring privacy and data security in the IoE environment."

"This study provides an important practical approach to revealing the occurrence and evolution mechanism of personal physiological data breaches and to implementing privacy protection measures by taking facial data breaches as a research object."

"Few studies have used systematic and objective methods to study and explain personal physiological data breach risks and their logical relationship."

"It is of practical significance for improving the ability of risk adaptive governance and supporting the manageability and traceability of physiological data from data collection to data destruction. (3) In terms of theory, this study explores the plight of physiological data in the IoE era, providing a reference for privacy protection under the new situation."

4.2.2 Challenges in Making Blockchain Privacy Compliant for the Digital World: Some Measures [233]

This is a machine-generated summary of:

Bansod, Smita; Ragha, Lata: Challenges in making blockchain privacy compliant for the digital world: some measures [233]

Published in: Sādhanā (2022).

Link to original: https://doi.org/10.1007/s12046-022-01931-1.

If you want to cite the papers, please refer to the original.

For technical reasons we could not place the page where the original quote is coming from.

Abstract-Summary

"Privacy has been made a critical parameter in Data Protection Regulations leading to the search for such a privacy compliant system which is also resilient to attacks."

"A detailed analysis of the Blockchain technology, which is becoming popular for secure applications in the finance sector, indicates that there are several challenges relating to user identity, transaction linkability, crypto-keys management, data privacy, usability, interoperability, and post- quantum compliance of privacy regulations which need to be resolved before its widespread adoption."

"This paper discusses the development flow of some of the privacy enhancement mechanisms like ZKPs, SMPC, Ring signatures, Mixing, Homomorphic Encryption and quantum resilient computing, bringing out their features and lacunae."

"There is a detailed discussion of the privacy mechanisms adopted by blockchain platforms like ZCash, Zerocoin, Hyperledger, Wanchain, Coin Party, Monero, Cryptomate, MixCoin, Coinshuffle, PICNIC and New Hope."

Introduction

"The enormous increase in online interactions between governments, businesses and individual citizens demanded a high level of protection and privacy of all data used in these interactions."

"Since data protection and privacy are critical elements of any online interaction, Blockchain technology is considered to have the features to revolutionize the online interactions by bringing in transparency across all the users by revamping currently existing processes and bringing in efficiency and adding value."

"There are many questions relating to Data Privacy that need to be analysed, discussed, debated and researched in order to provide convincing solutions complying with the Data Protection Regulations so that Blockchain technology could fit in as the preferred system."

"This paper discusses the features of Blockchain technology with a focus on the different mechanisms as well as platforms that need to be studied further to find acceptable solutions to the privacy preservation and security attack issues."

Motivation and contribution

"The draft proposes to mandate strict privacy requirements for personal data in respect of digital applications."

"The expert committee set up to review the provisions of the draft bill has recommended that the digital economy should aim to benefit the citizens by strengthening the privacy, safety and security aspects of all the digital transactions."

"There is a growing demand that data privacy should be given the status of a fundamental human right."

"On reviewing the latest literature, it is apparent that there are several gaps in the present privacy, safety and security aspects of blockchain applications."

Brief background

"Sensitive applications, like User identity management (location related), Patients' data (health record, medicines, state of mind, physical condition, behaviour pattern), Financial data, Educational data or any Business data should have special data privacy protection. (3) Privacy protection Regulations—the Indian Scene: There are various laws relating to data privacy applicable to different countries like the General Data Protection Regulation (GDPR) in Europe, Freedom of Information Act 2000 (FIA) in the United States of America, Data Protection Act 1998 (DPA) in the UK, Personal Information Protection Act (PIP) in South Korea and the Personal Data Protection Act 2012 (PDPA) in Malaysia."

"In India, Data Privacy is presently governed by the rules under the Information Technology Act, 2000."

"With the intention of bringing about a comprehensive overhaul of the existing data privacy regime, the Government proposed the Personal Data Protection Bill [2] in the year 2019, which is currently being examined by the Joint Parliamentary Committee and is expected to be presented before the Parliament very soon."

Basic terminologies

"A summary of the basic features of this technology is given as: (1) Distributed Database With the distributed database, there is no need for any intermediary to verify the records by any user as the complete history is available. (2) Mesh connected network Each peer can communicate with another peer directly and broadcast the information to all nodes. (3) Transparency The unique alpha-numeric address identifies each user or node and transactions take place between these with pseudonymous addresses."

"This will ensure anonymity of users and transactions if required. (4) Irreversibility of Records The algorithms deployed in this technology enable the system to store all entered data without any provision for deletion or manipulation."

"Blockchain technology is vulnerable to other kinds of attacks, namely Cyber Crimes, Double spending, Privacy leakage in transaction, 51% attack on PoW, Private key security, Smart contract frauds, and Price manipulation operations [3]."

Issues in Blockchain relating to privacy preservation

"Onion or Garlic cast routing is designed to be highly resistant to wide range of attacks while ensuring a high level of anonymity but these are still vulnerable to different types of attacks such as timing analysis. (c) Key management or wallet management is another big challenge in Blockchain technology."

"The various issues related to different Key management methods and Recovery schemes are discussed in paper [4]. (d) Data privacy One of the features of Blockchain is permanency of all information entered into the systems."

"Quantum resistant cryptography or Post Quantum Cryptography is an area that requires continuous research to remain ahead of the attackers. (h) Compliance It may be seen from the above discussion that several features of Blockchain, which are termed as advantages of the technology, are known to be non-compliant with the provisions of the Personal Privacy Protection regulations."

Blockchain security

"The paper [5] mentions a four layered structure namely network, consensus, replicated state machine/transaction and application layers with details of attacks and risk assessments of the model."

"The Network layer is the layer which is most vulnerable to attacks."

"While validating the blocks, the Blockchain network generates the fork due to misalignment of different miners, which metamorphoses into Stale Blocks or Orphaned Blocks leading to Sybil or selfish mining attacks by assigning several identifiers to the same node."

"The Consensus Layer protocols are vulnerable to generalized attacks like Majority attack or 51% attack, Double-spending attack, Finney attack, Time-Validation Attack, Block Withholding Attack by (selfish miner to create double spending attack), Classical block withholding attack to harm pool operator in PoW,

Pool Specific Attack, Memory Pool (mempool) attack, Consensus delay attack, Time-jacking Attack, Time validation attack, and Counter block withholding attack [5, 6]."

"Replicated State Machine Layer or Transaction layer is vulnerable to user identity, transaction data confidentiality, integrity, authentication, availability threats and related attacks."

Mechanisms for privacy preservation

"A mixing transaction privacy enhancement scheme [7] is constructed by using aggregate signature and group signature algorithms of a set of mixing peers."

"In Secure Multi Party Computation two or more parties of different data owners do not share their private data but agree to compute some joint function to ensure privacy (parties acquire only the output of the computation) and correctness (output of the computation is correct)."

"Various techniques, such as, Shamir Secret Sharing, Honest-majority MPC with secret sharing (BGW protocol), Threshold cryptography, and Dishonest-majority MPC are being considered to address the secure Multi party communication issues."

"As Blockchain does not ensure privacy and anonymity, the various algorithms are proposed using ring signature concept."

"The mixing scheme in paper [8] provides privacy using signature protocol without involvement of third party and extra transaction fees."

"Encryption algorithms, like quantum key distribution (QKD), which are created using quantum computing, may provide strong security to Blockchain based applications."

Blockchain platforms/applications based on privacy mechanisms

"ZCash is vulnerable to any of the attacks like Sapling Wood-Chipper, Remote Side-Channel on Anonymous Transactions and Allowed Infinite Counterfeit [9]."

"It includes the following three mechanisms: (i) Threshold secret-sharing technology, (ii) Multi-party computing for the locked account management, and (iii) The ring signatures and one-time accounts-based privacy security system for smart contract token transactions [10]."

"On an elliptic curve, bit-length leakage during scalar multiplication is possible, which might enable functional recovery of the long-term private key which leads to timing attack [11]."

"To maintain external anonymity, Mixcoin uses a central mixing server (or trusted third party) to mix transaction addresses and provide mixing services to users."

"TLS fortune attack due to quantum computing can be resolved by Picnic's quantum-resistant key exchange and signature algorithms."

"Erdem Alkim, Leo Ducas, Thomas Poppelmann, and Peter Schwabe developed NewHope [12], a key-agreement protocol designed to resist quantum machine attacks."

Limitations

"The above analysis clearly shows the need for further study which is required to determine the relative strengths and disadvantages of the various PETs, create new PETs or enhance the efficacy of existing ones, and consider the challenges to PET deployment in the online marketplace."

"Individuals often need to be made aware of the role of PETs so that they can make an informed choice of the technique to be adopted in order to secure their information on the internet and benefit the most from these techniques."

"The high computational power and the low usability continue to be the major issues in this area making these techniques quite expensive."

"PET algorithms need to be protected from black market as these could become the root cause for Silk Road or Darknet market."

Proposed privacy enhanced architecture

"The sequence of operations is discussed below: (1) User sends Request to Issuer(E-Government) for DID. (2) Issuer asks for PII which require to provide DID. (3) User provides only authenticated essential information i.e. ZKP enabled identity to issuer. (4) Issuer provides identity certificates in the form of X0, X1,,... Xn to User which is in segregated information format signed by Issuer. (5) Also hash values H(X0), H(X1) ...H(Xn) are generated by Issuer for verification purpose which can be stored on the blockchain. (6) User stores the identity certificates on personal database in Symmetric homomorphic encrypted format. (7) User also stores the encrypted form data on the blockchain. (8) The User who has the Identity certificates, can request for services to the Service provider, as and when required. (9) Service provider demands Identity data Xa, Xc or any other information. (10) User provides required information as Xa, Xc. (11) Service Provider verifies the Hash values and digital signature on the certification. (12) On successful verification, Service Provider provides services to User."

Conclusion

"This paper identified the technological requirements of data privacy and protection in general for the online transactions for the contactless working of today's digital world."

"The suitability of Blockchain technology to meet the requirements of security and privacy was examined and the aspects which require further research and study in order to make Blockchain technology future proof have been analysed in detail."

"The need for efficient crypto-privacy algorithms which will make Blockchain technology more effective and enable building of quantum resistant ledgers has been emphasized."

"It is believed that these measures will lead to more wide spread use of Blockchain in the digital world fully compliant with the privacy regulations."

4.2.3 Risk Management and Privacy Violation Detection in the PoSeID-On Data Privacy Platform [234]

This is a machine-generated summary of:

Silva, Paulo; Casaleiro, Rui; Simões, Paulo; Antunes, Nuno; Curado, Marilia; Monteiro, Edmundo: Risk Management and Privacy Violation Detection in the PoSeID-on Data Privacy Platform [234].

Published in: SN Computer Science (2020).

Link to original: https://doi.org/10.1007/s42979-020-00198-9.

If you want to cite the papers, please refer to the original.

For technical reasons we could not place the page where the original quote is coming from.

Abstract-Summary

"As personal data establishes itself as one of the main resources of our digital society, ways of controlling, monitoring, managing and securing personally identifiable information have become crucial."

"We present and discuss PoSeID-on's risk management and personal data analyser strategies, which support core functions of the data privacy platform being developed."

Introduction

"Willingly or unwillingly, people are continuously feeding data processors and/or data controllers with items of personal information and, in most cases, user control over who does what with someone's personal data is not practical or inexistent, despite EU's General Data Protection Regulation (GDPR) [13]."

"The PoSeID-on H2020 European Project is developing a platform for ensuring protection of and user control over Personally Identifiable Information (PII), in environments where multiple data processors deal with data pertaining to a potentially high number of data subjects."

"Two key functions of this platform are Risk Management and Personal Data Analysis."

"The Risk Management Module (RMM) and the Personal Data Analyser (PDA), together with privacy-enhanced dashboard (PED), aim to empower data subjects and reduce the unnecessary exposure of their PII."

"The RMM and PDA autonomously monitor such interactions and warn the data subjects of potentiation privacy risks."

The PoSeID-on Platform

"The goal of the PoSeID-on Project ("Protection and control of Secured Information by means of a privacy-enhanced dashboard") [14] is to develop a system for personal data protection, in line with GDPR with respect to digital security."

"The project aims to design, implement and validate a privacy-enhancing dashboard for personal data protection, a platform that manages all the personal data transactions between a data subject (owner of personal data) and private or public entities acting as data controllers or data processors."

"All relevant information shall be made available to users via a user-friendly web dashboard that allows them to track PII, manage PII access permissions, and view the risk level stemming from their data exposure."

"The PoSeID-on solution is based on technologies such as blockchain, smart contracts, and cloud computing, enabling users to manage personal data and data access authorizations in an easy, secure and auditable way."

Risk Management

"In order to integrate Drain into the RMM, a Java implementation of the algorithm was developed, in order to fully integrate it in the Spark Streaming [15] pipeline; feature extraction—once logs haven been parsed, each log is assigned to one of the log templates derived from the Drain parsing step it is then necessary to create numerical feature vectors which will be provided to the machine learning models performing the anomaly detection."

"anomaly detection—for the anomaly detection step, in initial phases of the RMM deployment, unsupervised learning models are more favourable due to the lack of operational data, such as K-Means and Principal Component Analysis, which do not require previous labelled data."

"The batch layer manages a master dataset which is used for historic risk analysis, taking advantage of having a large collection of data, which time span will depend on how long the RMM is allowed to retain data, provided the data subject has given explicit consent for this purpose."

Privacy Violation Detection

"The requests arriving from the message bus to the request processor are processed and dispatched to other components according to the data type (e.g. direct messages, RPC inputs, PDF, TXT, or other types of structured information)."

"The current version of the PDA (developed in Python version 3) provides a functional NLP pipeline capable of identifying named entities such as persons' names, cities, countries, locations, birth dates and others."

"The current performance metrics available (although not final) show that one instance of the PDA is capable of processing at least 100 messages per second without engaging the NLP pipeline, and at least 3 messages per second when the NLP pipeline is called."

Experimental Results

"The parameters that suffered variations during testing were the ones related to the pipeline and anomaly detection approach, more specifically: whether or not feature vector overlap was considered for training when overlapping is used, the length of the window is larger than the length of the slide, which means that some values in the window are new (the ones that correspond to the most recent slide) and the remaining ones are reused from the previous window; what percentage of points a cluster must have to be considered anomalous since small clusters are hypothesized to contain anomalies, a threshold must be set to define what is a small cluster."

Conclusion

"We have presented and discussed two crucial modules of the PoSeID-on personal data protection platform."

"The risk management module is used for evaluating and managing privacy and operational risks within the PoSeID-on system, through machine learning-based analysis of operational logs and PII exchanges, and managing a risk score associated with each data processor."

"To detect and prevent anomalies and misbehaved transactions, the Personal data analyser is used to monitor personal data transactions and related warnings generated by the blockchain platform."

"The PoSeID-on platform is now under intense study and evaluation, and this will continue in the immediate future, with the aim of improving the efficacy and efficiency of the risk management and privacy violation detection functionality."

4.2.4 *Privacy Risks of Whole-Slide Image Sharing in Digital Pathology [235]*

This is a machine-generated summary of:

Holub, Petr; Müller, Heimo; Bíl, Tomáš; Pireddu, Luca; Plass, Markus; Prasser, Fabian; Schlünder, Irene; Zatloukal, Kurt; Nenutil, Rudolf; Brázdil, Tomáš: Privacy risks of whole-slide image sharing in digital pathology [235].

Published in: Nature Communications (2023).

Link to original: https://doi.org/10.1038/s41467-023-37991-y.

If you want to cite the papers, please refer to the original.

For technical reasons we could not place the page where the original quote is coming from.

Abstract-Summary

"We develop a model for privacy risk analysis for whole-slide images which focuses primarily on identity disclosure attacks, as these are the most important from a regulatory perspective."

"We introduce a taxonomy of whole-slide images with respect to privacy risks and mathematical model for risk assessment and design."

"We develop guidelines for risk assessment and recommendations for low-risk sharing of whole-slide image data."

Introduction

"The privacy risks related to WSI sharing have not yet been systematically explored and the practice of sharing is extremely heterogeneous: from the above-mentioned approaches considering WSIs low-risk data and sharing them as open data sets, to the opposite extreme considering them as sensitive as other clinical data and sharing them only as a part of pseudonymized data sets under contracts compliant with the applicable data protection laws, such as the General Data Protection Regulation (GDPR) in European countries."

"At a first glance, it would appear unfeasible to identify the source individual from a WSI alone, which uniquely refers to the biological material that has already been removed from the patient's body and therefore histological slides and WSIs produced from this tissue cannot be directly reproduced from the patient."

Results

"Assume that an attacker possesses a background knowledge consisting of a data set of patients indexed by a PatientID (for simplicity, we assume that PatientID is unique) and including WSI data linked to the patients."

"If the linking attack is successful, the attacker can link this additional data to the correct patient, thus expanding his knowledge about the individual."

"In the problem statement in the previous paragraph, this work focuses on the analysis of the likelihood dimension for linking disclosure attacks, where the WSIs can act as a key for linking different data sets—i.e., linking the WSIs in background knowledge to the set of WSI probes, possibly associated with additional data."

"The results show that WSIs of the same tissue acquired using different scanners are very similar and can be used in a successful WSI linking attack."

Discussion

"Subsections we define the relevant terminology, discuss risk assessment aspects, and propose two sets of guidelines: the first for releasing WSI as de facto anonymous data sets, and the second for releasing pseudonymized (or even identified) data set under a contract defining appropriate technical and organizational safeguards."

"The following technical and organizational measures shall be considered before releasing WSI data in de facto anonymous data sets."

"Rationale: This follows from ability to do bit-by-bit match of identical WSIs or more effectively by comparing hashes of WSIs as discussed in Characterization of privacy risks considered. (b) If the same WSI is released into a de facto anonymous data set and other non-open data sets, gaining information from the de facto anonymous data set is technically trivial for the recipients of the non-open data set and must be considered from the risk assessment perspective."

Methods

"The above metric can also be generalized in several ways. (a) One may consider randomized attacks, where the attacker no longer deterministically assigns probes to patients but may instead employ randomness in the choice: from random or fixed assignments of probes to patients without considering the content of each probe (probably not really useful except very small data sets with very privacy-threatening background knowledge) to any type of randomized algorithm."

"We consider attackers assigning WSI probes to a patient's WSIs using similarity measures on features extracted from WSIs using common deep learning methods."

"The attack analyses focused primarily on linking complete WSIs and we performed experiments with the following data sets based on the WSI spatiotemporal hierarchy. (1) [Based on (H-1)] Consists of 28 slides from different patients with prostate, breast, or colorectal cancer provided by MMCI, which were scanned on Pannoramic® MIDI and Pannoramic® 250 Flash III scanners by 3DHistech (Budapest, Hungary) at a resolution of 0.172 μm/px (WSI sizes: 18.3 Gpx for Flash scanner and 23.3 Gpx for MIDI scanner)."

4.2.5 Privacy Preservation in e-health Cloud: Taxonomy, Privacy Requirements, Feasibility Analysis, and Opportunities [236]

This is a machine-generated summary of:

Kanwal, Tehsin; Anjum, Adeel; Khan, Abid: Privacy preservation in e-health cloud: taxonomy, privacy requirements, feasibility analysis, and opportunities [236].

Published in: Cluster Computing (2020).

Link to original: https://doi.org/10.1007/s10586-020-03106-1.

If you want to cite the papers, please refer to the original.

For technical reasons we could not place the page where the original quote is coming from.

Abstract-Summary

"Electronic health records (EHRs) are increasingly employed to maintain, store and share varied types of patient data."

"With the increasing cost and scarcity of healthcare services, healthcare organizations feel at ease in outsourcing these services to cloud-based EHRs."

"Regardless of its increased effectiveness and growing eagerness in its adoption, not much care is being employed to the privacy issues that might arise."

"Our work aims at answering three major questions: firstly, how privacy models and privacy techniques correlate with each other, secondly, how we can fix the privacy-utility-trade off by using different combinations of privacy models and privacy techniques and lastly, what are the most relevant privacy techniques that can be adapted to achieve privacy of EHR on cloud."

Introduction

"In cloud-based EHR, at one end, the dissemination of patient's data is greatly beneficial but on the other hand, it must be performed so immaculately that patient's privacy ought to be preserved."

"The privacy techniques that are secure in legacy systems for EHR systems might be susceptible to different kinds of attacks when introduced for EHR on the cloud. (3) Most of the privacy techniques and models that are applicable to EHRs in the cloud-based environment or standard computing environment for privacy models are almost same, though the settings change overall general mechanisms of underlying privacy techniques remain unchanged."

"The privacy concerns are frequently faced by the data owners and practitioners [16, 17], despite the benefits of cloud-based EHR services."

Privacy preservation in cloud-based EHRs

"We categorize privacy techniques for cloud-based EHRs into cryptographic, cryptographic policy aware hybrid and privacy-aware anonymity-based techniques."

"Patients' data privacy is claimed to be achieved by encrypted EHR, using the SKE hybrid cloud."

"Hybrid schemes are a combination of cryptographic and access control based techniques to preserve the privacy of EHR at a cloud."

"CAM-Access of EHRs with privacy and auditability Solutions claim to provide protection in the mobile healthcare system at private cloud."

"To preserve the privacy of cloud-based EHR, a hybrid approach is used that combines CP-ABE and k-Anonymity for anonymization."

"Privacy-preserving techniques for cloud-based EHRs based on the cloud deployment model will be analysed in this subsection."

"We have observed that most of the privacy preservation work is performed at the public cloud, whereas the hybrid cloud has been limitedly used for EHRs data privacy protection."

"Privacy-preserving anonymity based techniques are applied to the hybrid cloud [18–20] and public cloud [21]."

A comprehensive overview: privacy preserving techniques

"Data swapping techniques can be applied to for anonymization of numerical attributes and categorical sensitive attributes. (c) Synthetic data generation In this method, a mathematical model is constructed that uses the original data to generate synthetic records."

"Privacy of set-valued data also called transactional data, is preserved in [22, 23] by applying full subtree generalization."

"The focus here is to rearrange traditional privacy techniques to handle big data scalability and anonymization problem [24–26]."

"Discussion, we emphasize that to manage the expanding requirements from data receivers, some privacy models and techniques have been proposed which consider certain scenarios against the structure of underlying data types, the possibilities of privacy threats inferences, etc. If we are looking to achieve an enhanced balance between privacy and utility, the answer lies in achieving the best combination of privacy preserving model and techniques."

Privacy preserving requirements for cloud-based EHR

"Unlinkability can be properly achieved by several privacy preserving models (k-anonymity, t-closeness etc....) and techniques E.g. "Generalization and Suppression" etc. Data integrity means that health data should remain accurate and consistent when it is stored in the cloud."

"Data utilization means that EHRs data remains useful after applying Standard Privacy rules like GDPR Compliance and other security and privacy protection measures [27]."

"Some other requirements like relevance, patient's identifiable & sensitive information disclosure is bound to be addressed when we use privacy techniques to preserve patients' EHRs data."

"Some patient-centric requirements like patient consent and information ownership are not affected by privacy preserving techniques as these requirements directly relate to patients or information owner while the other requirements need to be effectively addressed by privacy techniques to achieve the best trade-off between privacy and utility for outsourced data."

Comparison: privacy-aware anonymity-based techniques against privacy preserving requirements

"For all privacy aware techniques, the general trend is that where privacy is satisfied, data utilization is low."

"In privacy techniques like permutation, perturbation, condensation, randomization and differential privacy, data are modified in different ways, so the data utilization is low."

"In privacy techniques like Anatomy, Slicing, and Angel, there is a general trend that highlights high privacy level and high data utilization."

"Data integrity is satisfied in the case of generalization, angel, slicing and differential privacy but it is not satisfied in the case of suppression."

"Privacy techniques like anatomy, angel, slicing and differential privacy are the most prominent techniques that satisfied maximum privacy aware requirements that are e-health specific."

"For above-mentioned techniques, there is a general trend that shows high privacy level and data utilization."

Conclusion and future work

"In the said perspective, we conducted a comprehensive research to do an in-depth analysis of privacy preserving techniques in e-health cloud."

"An effort is made to determine what is the actual need for preserving the privacy of EHRs, stored at the cloud?"

"The main theme of our work is to highlight the need for privacy preserving techniques when we outsource EHRs data to the cloud."

"We provided a deep comparative analysis of privacy techniques based on the privacy preserving requirements."

"As future work, we are keen on exploring the most innovative ideas and features in privacy-preserving techniques and models in terms of EHRs data confidentiality in cloud scenarios."

4.2.6 KDD, Data Mining, and the Challenge for Normative Privacy [237]

This is a machine-generated summary of:

Tavani, Herman T.: KDD, data mining, and the challenge for normative privacy [237].

Published in: Ethics and Information Technology (1999).

Link to original: https://doi.org/10.1023/a:1010051717305.

If you want to cite the papers, please refer to the original.

For technical reasons we could not place the page where the original quote is coming from.

Abstract-Summary

"The present study examines certain challenges that KDD (Knowledge Discovery in Databases) in general and data mining in particular pose for normative privacy and public policy."

"Whereas the main purpose of the earlier essay was to show what those specific privacy concerns are and to describe how exactly those concerns have been introduced by the use of certain KDD and data-mining techniques, the present study questions whether the use of those techniques necessarily violates the privacy of individuals."

"The implications of that privacy theory for a data-mining policy are also considered."

4.2.7 Privacy Preserving Framework Using Gaussian Mutation Based Firebug Optimization in Cloud Computing [238]

This is a machine-generated summary of:

Anand, K.; Vijayaraj, A.; Vijay Anand, M.: Privacy preserving framework using Gaussian mutation based firebug optimization in cloud computing [238].

Published in: The Journal of Supercomputing (2022).

Link to original: https://doi.org/10.1007/s11227-021-04173-w.

If you want to cite the papers, please refer to the original.

For technical reasons we could not place the page where the original quote is coming from.

Abstract-Summary

"Data confidentiality and data privacy are still considered as significant challenges while sharing sensitive data."

"The cloud storage servers based on unencrypted data are susceptible to both external and internal attacks established by strangers or untrustworthy cloud service providers."

"Since the medical data are sensitive, the risk based on privacy enhances at the moment of subcontracting entity medical records to the cloud."

"The significant intention of the proposed approach involves securing and preserving sensitive healthcare data."

"Data hiding and data restoration operations are considered as two significant operations of the proposed framework."

"An optimal key is generated in the data hiding operation."

Introduction

"Data transmission using the cloud faces various risks regarding privacy and security management."

"In this, data storage security incorporates data confidential related problems when the data gets saved in the cloud centre, while data processing security incorporates the problems related to data confidential runtime."

"Cloud computing platforms are growing rapidly in the health field for the secure storage of sensitive health records in cloud services [28]."

"The location-based data such as time, important sign readings are transferred to privacy providers for granting high quality services so that the patients can be saved in case of an emergency situation."

"The sensitive data are authenticated and authorized in the cloud for secure data preservation."

"The significant intention of the proposed approach involves in securing and preserving the sensitive healthcare data."

"The data hiding and data restoration operations are employed to perform an effective privacy preserving based healthcare design."

Review of related works

"Wu and others [29] presented the edge cloud computing paradigm based private random decision tree (PRDT) framework for better privacy preservation and data utilization process."

"Abdo and others [30] developed the mobile health care monitoring system based on cloud computing along with location privacy preservation approach."

"The performance of the approach was evaluated in terms of reliability, scalability, efficiency, privacy and security, while the analysis result proved it managed the trade-off between data utility and location privacy."

"Yang and others [31] implemented the stateless cloud auditing technique intended for privacy preserving process in non-manager dynamic group data."

"Jayaram and others [32] suggested the secure edge cloud-based healthcare system (SECHS) for efficient prediction and privacy preservation process."

"This technique was performed by three steps, namely data registering, authenticating and accessing processes and the evaluation result showed that it had enhanced privacy and decreased computation complexity but reduced data integrity."

Problem statement

"Even though, the cloud technology comprises immense benefits, data confidentiality and data privacy are still considered as a significant challenge during the utilization of cloud storage services."

"The cloud storage servers based on unencrypted data are susceptible to both external and internal attacks established by strangers or untrustworthy or dubious cloud service providers."

"Since the medical data are sensitive, the risk based on privacy enhances at the moment of subcontracting entity medical records to the cloud."

"Medical purpose sensitive data are involved mostly in the cloud service system."

Proposed approach

"An optimal key is generated in data hiding operation."

"An optimal key is generated using a Gaussian mutation-based firebug optimization (GM-FBO) algorithm."

"The original hidden sensitive data are recovered by the receiver by using an inverse optimal key which comes under restoration operation."

"The significant intention of the proposed approach involves securing and preserving the sensitive healthcare data, and the main objective of this paper is to accomplish an optimal key and it can be done by employing a hybrid algorithm named GM-FBO algorithm."

"The Firebug swarm optimization algorithm along with Gaussian mutation operation and cross over operation of a genetic algorithm is proposed to conquer the drawbacks like poor convergence rate and minimum global search capability of the original FBO algorithm."

"The successful restoration of original data takes place only if the optimal key generated using GM-FBO is accurate."

Experimental results and analysis

"It is proven that the proposed methodology offers low fitness which is effective when generating an optimal key with the minimum length for different medical datasets."

"The optimal key obtained using the proposed GM-FBO algorithm is the best key obtained for the secure generation process."

"The proposed GM-FBO algorithm offers optimal outcomes for the three datasets."

"In KPA attack analysis, the proposed methodology offers improved performance of 8.5%, 5.9%, 6.9%, 4.5%, 2.9%, and 1.9% when compared to the state-of-the-art techniques such as SKMA-SC [33], PRDT-DRU [34], J-SSO [35], Improved honeypot [28], and SECHS [36]."

"For dataset-I, the proposed GM-FBO offers an improvement of 33.25, 71.25, 74.65, and 74.65% when compared to the GA, PSO, FA, and WOA algorithms."

"The performance outcomes of the three datasets using our proposed methodology show that our model's restoration process is beneficial when compared to the conventional techniques."

Conclusion

"This paper presents a Gaussian Mutation based Firebug optimization (GM-FBO) algorithm for improving the security in cloud-based healthcare applications."

"The GM-FBO algorithm is used to offer secure data storage and restoration operations."

"The convergence efficiency of the proposed GM-FBO algorithm is analysed by comparing it with different heuristic algorithms such as GA, PSO, FA, and WOA."

"The simulation results show that the GM-FBO algorithm offers maximum privacy with minimal fitness value."

4.2.8 PAPIR: Privacy-Aware Personalized Information Retrieval [239]

This is a machine-generated summary of:

El-Ansari, Anas; Beni-Hssane, Abderrahim; Saadi, Mostafa; El Fissaoui, Mohamed: PAPIR: privacy-aware personalized information retrieval [239].

Published in: Journal of Ambient Intelligence and Humanized Computing (2021).

Link to original: https://doi.org/10.1007/s12652-020-02736-y.

If you want to cite the papers, please refer to the original.

For technical reasons we could not place the page where the original quote is coming from.

Abstract-Summary

“The problem of information overload on the Internet increased the need for personalized information retrieval (PIR) systems capable of providing information that corresponds to the user interests.”

“Although, for most people, the word personalization comes with trust issues and privacy concerns.”

“Since giving the user a personalized browsing experience usually comes at the cost of his privacy.”

“We propose a new model for privacy protection in PIR systems.”

“Our model aims at achieving a trade-off between the personalization quality and the privacy risk, to keep the latter under control.”

“We have studied the assets and drawbacks of the existing profile-based PIR structures, from a privacy protection perspective, along with the possible privacy threats in this field in a threat modeling approach.”

“While privacy protection is ensured during the personalization process, by taking into consideration the user’s privacy requirements, and through encryption.”

Introduction

“For those reasons, users need intelligent personalized web applications that simplify information access and content discovery, based on user preferences, and delivers services in a most valuable and convenient way.”

“These systems offer users personalized answers about services, products, and information.”

“The second challenge is the privacy protection problem, since giving the user a personalized browsing experience comes at the cost of his privacy.”

“Most former research works in the field focused on improving the personalization quality disregarding user privacy.”

"The purposes of our research, in general, are first to enhance the personalization quality by using accurate profiles capable of reflecting the user's changing interests."

"To ensure user's privacy by protecting his sensitive data on the client-side, through the Internet channel, and more importantly on the server-side where most privacy risks come (data misuse, leakage, etc.)."

"PAPIR is a model that aims at improving both personalization quality and privacy protection."

Related work

"Zhu et al. [55] tried to address the privacy protection problems in personalized systems by protecting user identification using techniques such as the pseudo-identity, the group identity, no identity, and no personal information."

"Authors in Liu et al. [56] presented a hybrid approach for privacy-preserving recommender systems by combining DP with RP to offer more privacy protection."

"Both DP and RP techniques are designed for aggregate data privacy ignoring each user's privacy requirements and they both decrease the personalization accuracy."

"As an individual privacy (IP) solution, Shou et al. [57] claims that better results can be achieved with privacy guarantee if the personalization is only performed based on less sensitive user data."

"Wu et al. [58] proposed a framework for the privacy protection in book search based on the idea of constructing a group of plausible fake queries for each user query to cover up the sensitive subjects behind users' queries."

Privacy in personalized IR systems

"Based on the user profile's storage (on the client-side or the server-side) and use for personalization."

"If the user gives his permission, the search system will hold all the personally identifiable data possibly available on the server."

"The user profile is still on the client-side, but the server also participates in search personalization."

"In Shou et al. [57], the contextual information sent to the server is a generalized profile that specifies the user search preferences without exposing the sensitive data in the user profile."

"Exploiting a user's device to store sensitive data (user profile) by personalized IR applications, introduces risks."

"Since most personalized systems including personalized search engines and social media store the collected massive user data on their servers making it the first target for attackers."

"The server is compromised, and the attacker can collect data sent by the user, then guess the original user profile as in the second scenario."

The proposed PAPIR model

"Helps preserve user privacy by minimizing the risk of guessing the user's whole profile from the query vector sent to the server."

"In our query expansion algorithm, the terms we add to the user query are extracted from the user profile reflecting his interests and needs."

"Selecting terms (to add) is done by computing the similarity between the query vector and each topic in the user profile (STP) taxonomy to extract related topics of interest."

"The following algorithm describes the query expansion process: We compute the cosine similarity metric between the query vector Q and each concept's vector C in the STP as follows:A high cosine value indicates that the query is closely related to the topic of interest in the user profile."

"In the encrypted search process, once the user query is prepared and expanded as a vector, the encryption agent encodes and encrypts this query vector using the public key pk."

Proof of concept implementation

"Homomorphically encrypted search is also assured by the encryption program implemented on the server-side based on the SEAL [59] library."

"To test the distributed computation effect on the performance of our PAPIR model, we implemented a 2nd version of the previously mentioned encryption program using the SparkFHE project on our local server."

"This component includes a query expansion feature (Algorithm 1) to personalize and expand the user query based on his profile."

"Encryption agent is responsible for protecting user data confidentiality in different states: At-rest, In-transit, and In-use."

"We implement the user profile protection At-rest using the AES algorithm with a 256-bit key, offering strong protection with lower computational requirements."

"To protect the user personalized query vector both In-transit and In-use along with the search results, we used homomorphic encryption that showed promising results in one of our previous works Makkaoui et al. [60]."

Experiments and results discussion

"We analyze the communication and computation cost for a single search operation both on the client-side and server-side."

"Algorithm 1 computes the similarity between the query vector with user profile concepts' vectors of the same dimension."

"Our model aims to enhance the search accuracy and results' relevance for each user based on his profile while preserving the user's privacy."

"Use different system versions to perform search operations and provide feedback on the search result and the user profile."

"Initial experiments prove that our model significantly improves the accuracy and relevance of the search results, as shown in the above table."

"By mapping the accumulated search queries for every user with the dataset and user profile."

"The initial experiments proved the feasibility and efficiency of our model for privacy protection in personalized information retrieval systems."

Conclusion and future work

"The work presented in this paper focuses on protecting the user's privacy by protecting the sensitive data in his profile."

"The use of homomorphic encryption as an extra protection layer reinforces the user's privacy protection, which is still one of the major issues in the field of personalized IR systems."

"A personalized IR system must ensure privacy protection to earn the user's trust."

"Only a minority of users will use it, to whom the personalized experience is more important than their privacy."

4.2.9 Predictive Privacy: Towards an Applied Ethics of Data Analytics [240]

This is a machine-generated summary of:

Mühlhoff, Rainer: Predictive privacy: towards an applied ethics of data analytics [240].

Published in: Ethics and Information Technology (2021).

Link to original: https://doi.org/10.1007/s10676-021-09606-x.

If you want to cite the papers, please refer to the original.

For technical reasons we could not place the page where the original quote is coming from.

Abstract-Summary

"This paper stresses the severe ethical and data protection implications of predictive analytics if it is used to predict sensitive information about single individuals or treat individuals differently based on the data many unrelated individuals provided."

"To tackle these concerns in an applied ethics, first, the paper introduces the concept of "predictive privacy" to formulate an ethical principle protecting individuals and groups against differential treatment based on Machine Learning and Big Data analytics."

"The paper sheds light on what is qualitatively new in the way predictive analytics challenges ethical principles such as human dignity and the (liberal) notion of individual privacy."

"The paper summarizes that data protection in the age of predictive analytics is a collective matter as we face situations where an individual's (or group's) privacy is violated using data other individuals provide about themselves, possibly even anonymously."

Introduction

"One major application is the algorithmic prediction of human behavior, or human "fate", if you will: Predictive Analytics (PA) leverages large behavioral data sets to classify individuals according to future risks, economic developments or expected costs and utility, as predicted from data correlations [61–63]."

"Other concerns relate to transparency and explainability of PA: in the common depiction as "black boxes", many ML and PA algorithms do not provide the necessary output to "justify" their decisions (or predictions) to a satisfying extent, making it ethically and legally questionable whether life-changing consequences can and should be based on such output [64]."

"This paper will take a more detailed look at the data processing cycle of predictive systems in order to shape an applied ethics of PA that includes but also goes beyond the hot-spot problems of bias and explainability."

Predictive systems and predictive privacy

"The ethical approach of this paper addresses the twofold challenge regarding privacy and data protection due to the possibility of predicting sensitive information about individuals from proxy data through lateral comparisons to the data of many "data donors"."

"A violation of predictive privacy is not committed by stealing or leaking information from someone's private sphere, but by deriving a prediction about individuals or profiling groups from data that were collected from many other users of networked digital services."

"To pinpoint this ethical concern, I will give a formal definition of the negative of predictive privacy—its violation: An individual's or group's predictive privacy is violated if sensitive information about that person or group is predicted against their will or without their knowledge on the basis of data of many other individuals, provided that these predictions lead to decisions that affect anyone's social, economic, psychological, physical,... well-being or freedom."

Ethical evaluation of predictive systems: a step-by-step guide

"In order to assess the ethical implications of predictive systems and precisely locate the violation of predictive privacy in these complex assemblages, I propose the following minimal model of the data processing cycle of predictive systems."

"Promises about anonymisation of training data will make many users feel that the data processing is harmless, while they do not understand that they are contributing their data to a model that allows the prediction of sensitive information about other users, including those who would not consent to processing of their information."

"Further ethical questions arise if the predictive model is trained from data that is being re-purposed from other contexts."

Collective ethical concerns: from predictive privacy to unfair bias and discrimination

"Extending on the data processing minimal model presented in the previous section, I will now discuss ethical requirements to PA operators that contribute to reducing unfair bias and discrimination."

"This potentially results in a perpetuation of existing discriminatory patterns and social inequalities when predictive systems are deployed to treat these clusters differently in terms of access to resources, information, education, etc. It is important to notice that this kind of unfair bias has, in general, no technological solution [65, 66]."

"I will exclusively address Type A bias, as I intend to discuss ethical imperatives to PA operators to reduce unfair bias."

"A key ethical demand to counter the problem of Type A bias (systematic misprediction) is the constant verification and re-training of predictive models in the prediction phase through the implementation of feedback loops [61]."

Conclusion and outlook

"I maintain that protecting predictive privacy is an inevitable part of making sense of and reasserting human dignity in the age of Big Data and AI by extending the scope of privacy to include the harm we help inflict to others through the data traces we leave behind."

"Normatively, predictive privacy sits at the intersection of ethics, data protection and anti-discrimination, based on the ethical intuition that persons are wronged if they are judged by behavioral comparisons with others on facts they would not want to reveal about themselves [67]."

"Raising awareness of predictive privacy in our societies means challenging the liberal presupposition underlying the Western notion of "privacy", according to which each data subject should be able to decide for themselves what data they want to disclose about themselves."

"And perhaps most importantly, an ethical awareness of predictive privacy as a fundamental value among a large majority of data subjects is urgently needed as precondition for any successful regulation."

4.2.10 *It is not (Only) About Privacy: How Multi-party Computation Redefines Control, Trust, and Risk in Data Sharing [241]*

This is a machine-generated summary of:

Agahari, Wirawan; Ofe, Hosea; de Reuver, Mark: It is not (only) about privacy: How multi-party computation redefines control, trust, and risk in data sharing [241].

Published in: Electronic Markets (2022).

Link to original: https://doi.org/10.1007/s12525-022-00572-w.

If you want to cite the papers, please refer to the original.

For technical reasons we could not place the page where the original quote is coming from.

Abstract-Summary

"This paper investigates if MPC affects known antecedents for data sharing decisions: control, trust, and risks."

"Through 23 qualitative interviews in the automotive industry, we find that MPC (1) enables new ways of technology-based control, (2) reduces the need for inter-organizational trust, and (3) prevents losing competitive advantage due to data leakage."

"MPC also creates the need to trust technology and introduces new risks of data misuse."

"Our findings show that known antecedents of data sharing should be specified differently with MPC in place."

Introduction

"MPC could create new sources of control, new forms of trust, and new types of risk that were not known in existing data sharing approaches."

"This paper aims to investigate the impact of MPC on perceived control, trust, and perceived risks in the context of data sharing by firms, in order to understand how MPC affects known antecedents of firms' data sharing decisions."

"Based on the research objective and setting, the research question for this paper is: What are the impacts of multi-party computation (MPC) on perceived control, trust, and perceived risks in data sharing by firms through data marketplaces?"

"This study is among the first to show that MPC challenges what we know about the key antecedents of data sharing decisions: perceived control, trust, and perceived risks."

"We set a basis to extend existing theory on data sharing antecedents to the emerging context of MPC."

Background

"Although the three streams are distinct and focus on three different concepts of control, trust, and perceived risks, they cannot provide a comprehensive understanding of why firms share data with other firms when viewed separately."

"We also investigate the impact of MPC in enhancing owners' control in data sharing."

"The extant literature suggests that control, trust, and perceived risks are vital aspects that affect firms' willingness to share data."

"Reinforcing various control mechanisms is important in data sharing since it could reduce risks and establish trust between data owners and data users."

"With MPC in place, we expect that perceived control might still be relevant to understanding firms' data sharing decisions in data marketplaces."

"Perceived control over data is more relevant while sharing through data marketplaces that use MPC."

Research approach

"While much research has been done on control, trust, and perceived risk as antecedents of business-to-business data sharing, the impact of MPC has not yet been studied empirically."

"To ensure that interviewees had the same understanding of data marketplaces and MPC, we developed a short presentation explaining (1) the definition of data marketplaces and their use cases; (2) how MPC works and its use cases; and (3) how MPC can be implemented in data marketplaces."

"We moved to the second part of the interview, starting with introducing interviewees to MPC and its possible use case in data marketplaces."

"It is only added during the axial coding because it explains conditions under which MPC impacts control, inter-organizational trust, and perceived risks in data sharing."

"In the selective coding phase, all codes were re-examined to establish an understanding of how they are related to the main topic of the impact of MPC on control, inter-organizational trust, and perceived risks in data sharing."

Findings

"Technology-based control is more relevant while sharing through data marketplaces that use MPC."

"The more control you have of the data, the more you control how it can be used [and] the less you have to trust in the partner not cheating on you. (A06) [D]oing [data sharing in data marketplaces] based on an MPC algorithm where the OEMs keep control of their data and not giving away their data to other parties gives them trust to actually collaborate because they do not have that much more to lose anymore. (A15) Strikingly, some interviewees argued that MPC is not just about reducing the relevance of inter-organizational trust but increasing the importance of trust in technology."

"Trust in technology is more relevant while sharing through data marketplaces that use MPC."

"Competitiveness risk is less relevant while sharing through data marketplaces that use MPC."

Discussion

"We found that MPC changes the relevance of control, trust, and risk in sharing through data marketplaces, particularly in the automotive industry."

"These factors should be specified in new ways to understand firms' data-sharing decisions through MPC-based data marketplaces, namely technology-based control, trust in technology, and data misuse risk."

"We found that trust in other actors involved and competitiveness risk, which was relevant in the current data sharing situations, are less relevant with MPC in place."

"While MPC lowers the risks for data owners, it creates new risks for data users in revealing sensitive information while sharing through data marketplaces."

"We find that the impact of MPC on perceived control, trust, and perceived risks in data sharing depends on three conditions."

"Our findings imply that scholars interested in researching business-to-business data sharing should specify the concepts of control, trust, and perceived risks in a new way while considering the impact of MPC."

Conclusions

"This paper shows that using MPC in business-to-business data sharing via data marketplaces can provide higher control over data through technology-based control, lower the need for trust in other actors involved, and reduce competitiveness risks."

"This paper contributes to the literature on business-to-business data sharing by being among the first that specifies the concepts of perceived control, trust, and perceived risk into a set of propositions, which holds in the initial phase of MPC before its widespread adoption."

"Future research could perform a follow-up study by exploring the impact of MPC on control, trust, and risks in data sharing based on a working prototype or its real-life implementation, as the perception regarding MPC might differ when it is widely adopted by businesses."

"Scholars could also investigate the differences in the control-, trust-, and risk-related implications of MPC within the context of data-sharing platforms that purely focus on data exchange between partners."

4.2.11 Locational Privacy-Preserving Distance Computations with Intersecting Sets of Randomly Labeled Grid Points [242]

This is a machine-generated summary of:

Schnell, Rainer; Klingwort, Jonas; Farrow, James M.: Locational privacy-preserving distance computations with intersecting sets of randomly labeled grid points [242].

Published in: International Journal of Health Geographics (2021).

Link to original: https://doi.org/10.1186/s12942-021-00268-y.

If you want to cite the papers, please refer to the original.

For technical reasons we could not place the page where the original quote is coming from.

Abstract-Summary

"We introduce and study a recently proposed method for privacy-preserving distance computations which has received little attention in the scientific literature so far."

"The method, which is based on intersecting sets of randomly labeled grid points, is henceforth denoted as ISGP allows calculating the approximate distances between masked spatial data."

"The method allows the computation of distances between locations L when the locations at different points in time t are not known simultaneously."

"ISGP is a new tool for privacy-preserving data handling of geo-referenced data sets in general."

"This technique can be used to include geographical identifiers as additional information for privacy-preserving record-linkage."

"We discussed a new method for privacy-preserving distance computations in microdata."

"The method is highly accurate, fast, has low computational burden, and does not require excessive storage."

Background

"The method introduced here could be used for research applications under privacy legislation such as the General Data Protection Regulation as implemented in different ways among European countries."

"Due to privacy concerns in most countries, survey agencies and official statistics bureaus are often required to separate research data and respondent identifying information [37–39]."

"Different approaches for the masking of spatial data have been suggested in the literature."

"Based on [40], the methods sketched in the review by [41] can be classified into three categories: (1) methods that aggregate spatial points, (2) methods that modify coordinates, and (3) methods that release contextual data only."

"Kerschbaum introduced a distance-preserving pseudonymization technique for timestamps and spatial data."

"The new method allows the computation of distances between locations when the locations at different points in time are not known simultaneously."

Methods

"The regularity of the grid is important, so that identical distances between considered points yield (dependent on the radius) nearly the same number of grid points enclosed by the area of intersection."

"A research group will use these sets of grid points to compute the distances they need for their research (Step 9)."

"The resulting set consists of the random labels of those grid points, which have a distance less than r from the respective point P, Q. Here, the radius (r) is set to 30 km."

"The remaining problem is the computation of the approximated distance given the already computed similarity of the two sets of grid points."

"For a fixed radius, the number of common grid points of the circles around P and Q strongly depends on the number of grid points sampled from the area of intersection."

Results

"The largest absolute relative errors resulted for the nearest hospital with smaller radii and fewer grid points."

"With an increasing number of grid points and increasing radii size, the quality of the approximations increases since the absolute relative errors are decreasing."

"The mean absolute error by radii and the number of grid points for each of the three nearest hospitals are shown."

"The largest mean absolute relative error of about 8% is observed for 50,000 grid points and a radius of 10 km."

"The effect of the number of grid points on the quality of the approximations decreases with the size of the radius."

"Further, the effect of the number of grid points on the quality of the approximations also vanishes with larger distances."

"It should be noted that the variances of errors of approximated distances for fixed radii and fixed numbers of grid points are very small."

Discussion and conclusion

"We have introduced ISGP as a method for the calculation of the distance between masked geographical data."

"All geo-masking methods can be attacked with a graph-theoretical approach, if a distance matrix and restricting additional information is available [42, 43]."

"A further advantage of the described approach is the prospect to use IGSP encoded geographical information for privacy preserving record-linkage (PPRL) applications (for a review, see [44])."

"Bloom-filter encoded IGSP are currently the only PPRL method, which can efficiently utilize geographical information."

"The use of intersecting sets of randomly labeled points permits fast distance approximations with errors below 10 %, where larger errors are due to unsuitable parameter choices."

"A systematic comparative study of the accuracy and privacy of geomasking methods, in general, is lacking in the literature and subject of ongoing research."

4.2.12 Consumer Online Privacy Concerns and Responses: A Power-Responsibility Equilibrium Perspective [243]

This is a machine-generated summary of:

Lwin, May; Wirtz, Jochen; Williams, Jerome D.: Consumer online privacy concerns and responses: a power–responsibility equilibrium perspective [243].

Published in: Journal of the Academy of Marketing Science (2007).

Link to original: https://doi.org/10.1007/s11747-006-0003-3.

Copyright of the summarized publication:

Springer-Verlag 2007.

If you want to cite the papers, please refer to the original.

For technical reasons we could not place the page where the original quote is coming from.

Abstract-Summary

"In our model, consumer privacy concern is the endogenous mediating entity linking business policy and regulatory perceptions to negative online user responses."

"We investigated the moderating role of information sensitivity and congruency on the business policy–concern relationship across three industry contexts."

"Both hypothesized two-way interactions were confirmed, suggesting that a strong business policy is effective in reducing concern when low sensitivity data are gathered, but insufficient in reducing concern for highly sensitive data."

"Concern increased dramatically when sensitive data were collected that were incongruent with the business context."

Introduction

"The same technological advances that have made the Internet a potent marketing tool have also multiplied the threats to user privacy."

"Many users take countermeasures to protect their privacy, including supplying false or fictitious information to a Web site [76], managing the use of cookies [77], and even refusing to purchase from particular Web sites [78]."

"Marketers need to address online consumer privacy concerns as they can undermine a firm's marketing effectiveness, especially given the shift of marketing transactions to the Internet [79]."

"From a policy perspective, governments in many countries have recently given more consideration to consumer privacy policy in their efforts to regulate the Internet [80]."

Model development

"The second stream has concentrated on the ethical, legal, regulatory and public policy factors, and corporate responsibilities concerning online privacy [81–83]."

"The link between perceived levels or quality of privacy policies on consumer privacy concern has received scant empirical attention [82]."

"In the Internet context, many corporate privacy policies do not fully comply with Federal Trade Commission (FTC) guidelines, which in turn has been linked to heightened consumer privacy concerns [83]."

"We propose that consumers' privacy concerns and subsequent actions are affected by consumer perceptions of corporate policies and practices."

"Regulation refers to the perceived regulatory policies of how various government agencies, the other key power holders, devise Internet privacy regulation, and direct and police the use of consumer data."

"We propose that perceived effectiveness of regulatory policies and their enforcement will reduce consumer online privacy concerns."

Consumer responses to privacy concern

"The following hypotheses summarize our discussion: H1: Privacy concern mediates the causal relationship between perceived company privacy policy and protective consumer responses."

"The weaker (less effective) the perceived company policy is, the higher will be the degree of privacy concern, and as a result, the higher will be the likelihood of users engaging in protective response behaviors, including degree of (a) fabrication of personal information, (b) adoption of protection technologies, and (c) withholding from interacting with a Web site."

"H2: Privacy concern mediates the causal relationship between perceived online privacy government regulation and protective consumer responses."

"The weaker (less effective) the perceived government online privacy regulation is, the higher will be the degree of privacy concern, and as a result, the higher will be the likelihood of users engaging in protective response behaviors, including degree of (a) fabrication of personal information, (b) adoption of protection technologies, and (c) withholding from interacting with a Web site."

Extending the base model

"Policy should be highly effective in reducing concern for less sensitive data."

"As sensitivity increases, policy alone will become less effective in reducing concern, which will remain high when data are highly sensitive."

"For highly sensitive information, such as income or health information, a comprehensive policy might be necessary but not quite sufficient to bring concern levels down to a low level, whereas for data with low sensitivity, a good policy might be sufficient."

"H3: The hypothesized causal link in H1 from perceived company policy to privacy concern is moderated by the sensitivity of data being collected."

"We expect congruency to have a stronger effect on concern when information of high rather than low sensitivity is involved."

"H4: At high levels of data sensitivity, incongruency will lead to high levels of concern, whereas at decreasing sensitivity, the effect of incongruency on concern will decline."

Research method

"Three scale items were selected based on Karson's [84] work, who adapted this scale to the Internet context to measure consumer perceptions of a company's online privacy policies and practices."

"No existing scales were found in the literature to measure the perception of legal and regulatory policy in the online privacy context."

"We therefore built on measures used in non-online privacy regulation research [85, 86] to develop three statements for our context."

"Two scale items were adapted from a study on Internet user attitudes about online privacy [87]."

"A further item was adapted from Milne and Boza's (88) research on privacy concerns pertaining to database marketing."

"The final item was taken from a study on situation-specific dimensions of privacy concern [89]."

"Our three-item scale was adapted from measures of the 10th WWW Survey [90] and from an online fabrication behavior scale [91]."

Study 1: testing the base model

"The anticipated main effects were significant for Policy [$F(2,177) = 428.7$, $p < 0.001$] and for Regulation [$F(2,177) = 439.6$, $p < 0.001$] on their respective manipulation checks."

"In sub-model 1, a two-way ANOVA was performed to examine the effects of Policy and Regulation on Concern."

"We found significant main effects for Policy [$F(2,171) = 9.6$, $p < 0.001$] and for Regulation [$F(2,171) = 14.1$, $p < 0.001$]."

"This finding suggests that in spite of the remaining significant effects of Regulation, response behaviors were largely mediated by concern."

"The findings from the three sub-models suggest that Policy and Regulation significantly affect Concern (sub-model 1), as well as Fabricate, Protect and Withhold (sub-model 2)."

"Concern fully mediated the relationship from Policy to the three response behaviors, and largely mediated the relationship from Regulation to the responses (sub-model 3)."

Study 2—information sensitivity and congruency

"Using three services allowed us to examine the robustness of our findings across industry contexts and operationalizations of data lists for the sensitivity and congruency manipulations."

"A particular scenario described an online bank with a highly impressive privacy policy (policy manipulation) asking respondents to disclose highly sensitive financial information (sensitivity manipulation) that is congruent with the industry context such as credit card number, home phone number, bank account details and income (congruency manipulation)."

"Sensitivity and Congruency were manipulated by requesting a battery of five data items from each respondent."

"In our first pre-test, we asked 24 students to rate the sensitivity of each data item (independent of industry), and the congruency of the data item with each of the three industries on a seven-point scale."

"A third pre-test with 30 students confirmed that the six experimental conditions (i.e., the three sensitivity and two congruency manipulations) were perceived as intended across three industries."

Discussion and implications

"This positioning links existing research streams, for example, in examinations of corporate policies and practice versus associating specific privacy concerns to user responses."

"We also demonstrated that privacy concern is a key mediating entity linking both business policy and regulatory perceptions to negative online user responses."

"Our policy–sensitivity interaction shows that although an excellent privacy policy can be effective in reducing privacy concern, it is insufficient when highly sensitive information is being collected."

"Policy efforts to improve privacy protection should be clearly communicated to the public along with the creation of a response outlet for privacy concerns [92]."

"Within the U.S., positive developments include policy makers increasingly turning their attention to privacy legislation that attempts to lessen consumer privacy concern (e.g., California's Database Security Breach Notification Act at the state level, and The Health Insurance Portability and Accountability Act (HIPAA) at the federal level)."

4.2.13 Evaluating a Privacy Requirements Specification Method by Using a Mixed-Method Approach: Results and Lessons Learned [244]

This is a machine-generated summary of:

Peixoto, Mariana; Silva, Carla; Araújo, João; Gorschek, Tony; Vasconcelos, Alexandre; Vilela, Jéssyka: Evaluating a privacy requirements specification method by using a mixed-method approach: results and lessons learned [244].

Published in: Requirements Engineering (2022).

Link to original: https://doi.org/10.1007/s00766-022-00388-2.

If you want to cite the papers, please refer to the original.

For technical reasons we could not place the page where the original quote is coming from.

Abstract-Summary

"Privacy needs to be properly specified, but agile requirements engineering techniques do not explicitly represent privacy requirements and, therefore, are not able to proper analyze such requirements."

"Privacy Criteria Method (PCM), an approach to specify privacy in requirements activities, was proposed to produce more complete and detailed privacy requirements."

"By considering PCM a promising approach to be used in ASD and the importance of empirical evaluation of new methods, we have as objectives: 1 evaluate the ability of PCM to support systems analysts in specifying privacy requirements when used in conjunction with some agile specification methods; and 2 show our lessons learned in conducting empirical research based on an mix-method approach defined to empirically evaluate the suitability of a requirements specification in specifying privacy requirements."

"The requirements specifications following PCM allow to represent privacy aspects, such as user's personal data and the privacy mechanism that can be used to mitigate a privacy risk scenario."

"We also observed that some extra time is necessary to specify privacy requirements with PCM, but it does not imply a greater perceived effort."

"PCM assists in specifying more complete and detailed in relation to traditional techniques used in ASD, which facilitates communication between the requirements analysts and developers."

Introduction

"We could not find any proposal in the direction of supporting agile developers in specifying privacy requirements in the beginning of software development."

"Motivated by this scenario, in previous research, we created an approach, called Privacy Criteria Method (PCM) [46], to specify privacy requirements in ASD that was conceived based on four exploratory studies: (1) In-depth semi-structured interviews aimed at capturing the understanding of privacy by 13 agile developers [47]; (2) a web based survey also aimed at capturing the understanding of privacy by 108 agile developers; (3) a survey study with nine privacy experts to validate privacy concepts. (4) A framework of privacy specification capabilities."

"PCM aims to provide specifications that are more complete and detailed since it addresses five essential quality factors for software requirements specifications in ASD, indicated by Medeiros [48], such as: Automated support The activities of specification and management of requirements are carried out with the support of a specialized tool (we developed PCM tool); Readability The description of the requirements is easy to understand (we have explicit guidelines on how to specify with PCM); Team-oriented The requirements are described using a language directed to the developer."

Background and related work

"We developed PCM to address the demand to specify privacy requirements (complete and detailed) during the earlier stages of software development and following common agile practices [46] and also, to consider the development of any product or service in any context."

"There is a lack of studies evaluating requirements specification approaches regarding their ability to specify privacy requirements in ASD [49]."

"Only two empirical studies to evaluate requirements specification techniques in ASD were identified, but none of them evaluated how agile methods support the specification of privacy requirements."

"From the analysis of these related works, selected in a non-systematic way, we noticed that although there are several studies concerned with privacy requirements, none of them proposes a privacy specification method evaluated in the context of agile software development."

Research methodology

"RQ3: To what extent does PCM provide coverage, applicability, usefulness, and ease of use to specify privacy requirements from system analysts' point of view? (This question aims to collect the opinion of the participants regarding the detailed metrics."

"RQ4: What is the quality (regarding syntactic and semantics) of the privacy requirements specified using PCM? (This question aims to collect information about the quality that may be indicative of a greater detailing of the artifacts)."

"To achieve this goal, we consider: Quantity (number of privacy requirements specifications); Time spent (to specify the requirements); Questions asked during the experiment; The specifications' quality considering Syntactic (Well-formed, Atomic and Minimal) and Semantics (Conceptually sound and Problem-oriented); The perceived effort/use intentions according to well-known questionnaires (NASA-TLX for perceived effort; and TAM instrument for use intentions); Study participants' point of view (regarding PCM's coverage, applicability, usefulness, and ease of use)."

Results and analysis

"There was no difference between G1 and G2 in relation to the perceived effort about the scales MD, PD, TD, P, E, and F. The results indicate that the students of the two different groups (G1 and G2) did not present a difference in the perceived effort when performing the specification activity even though the activity of G1 (they specified US, AC, and PCM) was higher than that of G2 (they specified US and AC)."

"When we observed the standard deviation, we concluded that there was a low dispersion of answers concerning the general average of PCM Ease of Use (0.895); PCM coverage (1.089); PCM applicability (1.263); and PCM usefulness (1.255)."

"The results of these three procedures showed that the participants could obtain a good Syntactic quality of the privacy specifications created with PCM."

Discussion

"Results indicating that the PCM has proper constructs (coverage, applicability, usefulness, and ease of use) for specifying privacy according to the 18 system analysts participating in the feasibility study."

"Results indicating that the specifications produced with PCM were Well-formed, Atomic and Minimal in both studies."

"Students had access to the PCM as an example of specifying privacy requirements."

"Obtain subjects' permission for their participation in the study: The researchers informed the students that they would participate in an empirical study regarding the privacy topic."

"We informed that G1 would have the PCM training and G2 would do other activities not related to the study itself."

"G1 received training on the PCM, already understanding the need to consider the specification of privacy requirements."

"This motivated the students in this group to participate in the study, as they realized that they would have the same knowledge gathered by G1."

Threats to validity

"Participants in the feasibility study also received PCM training and specified the same scenario of the experiment."

"As the Mann–Whitney test indicates the difference between the means and the fact that our study was carried out with a small number of participants, it is necessary to indicate that we have a limitation on the results at this point."

"We observed similar results from the G1 of the experiment and the results obtained in the feasibility study, referring to quality, for example."

"To reduce the threats of a possible difference on the experience regarding requirements specification in both groups, also in the feasibility study, all the participants had the same theoretical and practical classes on how to specify requirements using User Stories."

"We cannot generalize our results to the entire population of requirements engineers, despite the fact that participants of both studies were postgraduate students in computer science with some industry experience."

Conclusion and future work

"We evaluated PCM by performing two empirical studies, a controlled experiment with 32 students and a feasibility study with 18 students."

"During the controlled experiment, we found that PCM leads system analysts to specify more privacy requirements than those who did not use PCM."

"We tried to make the PCM evaluation method as transparent as possible for other researchers to replicate it (and others could learn from our experience)."

"It means that developing a general evaluation method and evaluating the evaluation method defined goes beyond the scope of this research, although we have good evidence that the mix-method defined to evaluate the PCM is generic enough to evaluate any other requirements specification technique."

"Evaluating the mix-method defined in this paper performing other empirical studies with other requirements specification techniques is a future action."

"We intend to conduct case studies to evaluate PCM in companies working with ASD."

4.2.14 Extracting LPL Privacy Policy Purposes from Annotated Web Service Source Code [245]

This is a machine-generated summary of:

Hjerppe, Kalle; Ruohonen, Jukka; Leppänen, Ville: Extracting LPL privacy policy purposes from annotated web service source code [245].

Published in: Software and Systems Modeling (2022).

Link to original: https://doi.org/10.1007/s10270-022-00998-y.

If you want to cite the papers, please refer to the original.

For technical reasons we could not place the page where the original quote is coming from.

Abstract-Summary

"Privacy policies are a mechanism used to inform users of the World Wide Web about the processing of their personal data."

"Privacy policies are documents used, among other things, to inform the data subject about processing of their personal data."

"These are formally represented by privacy languages."

"We present a technique for constructing Layered Privacy Language policy data from web service code bases."

"We model the purposes of processing within web services by extending the privacy language with composition."

"We also present a formal analysis method for generating privacy policy purposes from the source code of web services."

Introduction

"One possible explanation for this is the long-standing gap between research and practice in the domain of privacy and data protection [50]."

"Value can be provided to service providers in terms of tool-assisted privacy policy management, and our work aims to be useful for both."

"A privacy policy for a web service is also a documentation of its behavior concerning personal data processing (along with additional factors, of course)."

"The goal of this work is to formulate a method to extract data from web services in a way that fits into a privacy policy."

"The following two research questions (RQs) are examined: RQ1: How can we model personal data processing for web services using LPL?"

"In terms of RQ1, existing models for privacy policies are extended and augmented in the web application context; in terms of the RQ2, a concrete implementation is presented."

Background and related work

"While some previous research exists for programmable privacy languages [51], the LPL's abstraction level focuses on the modeling of privacy policies."

"There appears to be no previous work for mapping LPL specifically to web services, nor does there seem to exist previous research for grouping purposes."

"Despite these advances, there appears to be no existing tools specifically tailored for generating privacy policies from source code."

"This perspective on web services is suitable for modeling combinations of different services, and the parallel to compositions of layered privacy policies is apparent."

"On this topic there are further related work as well—for example, Diver and Schafer make use of the modeling technique in order to visualize information flows within applications in order to facilitate privacy by design [52]."

Definitions

"Given these preliminaries from the LPL definitions, our goal is to extend the language by defining the relation among composite Purposes in a Policy."

"In the present context, any web service is assumed to be governed by a privacy policy, and that policy can be modeled with LPL."

"In essence: (i) any WS processing personal data is governed by a privacy policy; (ii) the act of processing in LPL is encoded in Purposes; and, therefore, (iii) at least one Purpose governs any web service processing personal data."

"Although any compliant web service processing personal data does so under a privacy policy, this condition does not dictate the practice."

"A further benefit originates from the fact that modeling can be done at either side of the abstraction; at the level of web services; or at the level of privacy policies."

Implementation of approach

"The first step is to use static declarations of personal data processing in annotations of source code."

"We can reason about the structure of a program and the way the program processes personal data, and to use this information to create privacy policy language constructs from the capabilities of a given web service."

"Since the personal data processing purposes coupled to web services are derived from the behavior of services, the next part is to define a way to generate data for privacy policies via static analysis."

"Using both the application entry-point and personal data ORM-annotations, a code base declares how personal data is processed statically."

"To marking a point of a data transfer in source code statically, an annotation instance is parameterized to contain the identifier of the policy it refers to."

Validation: a case study

"To demonstrate the presented method in practical use, the implemented tool was applied to a web service code base of a company."

"The initial analysis of the case web service was done in [53], and the data extraction and analysis were repeated after the extensions presented in this paper, in order to have the full set of policy data."

"The case WS was found to comprise 30 Controller modules, each of which was composed of multiple entry-points (a total of 245 with a range of [1, 33], and an average of 8.2)."

"Following our formulation of the problem, the case WS was viewed as set of component web services (i.e., Controller modules), which in turn are composed of entry-point web services."

"The analysis tool extracts the data for each entry-point, most of which were omitted from these figures for brevity."

Discussion

"This paper provides a method for improved means to express and reason about privacy policies and personal data processing formally."

"Any web service processing personal data must be governed by a privacy policy."

"Having the means to automatically generate data might lower the effort of creating expressive privacy policies."

"The automated method to generate privacy policy data does require a change in practice instead of only implementing technical improvements—but the process can be incremental and iterative."

"The question can be raised whether using source code as the basis for generating privacy policy data is useful in practice, in comparison with other approaches."

"While our approach leaves the final composition of a privacy policy to a practitioner, the automatic data generation facilitates a fine-grained policy."

"Our approach leverages static information encoded in source code files for generating the privacy policy Purposes of web services."

4.2.15 Healthcare Data Breaches: Implications for Digital Forensic Readiness [246]

This is a machine-generated summary of:

Chernyshev, Maxim; Zeadally, Sherali; Baig, Zubair: Healthcare Data Breaches: Implications for Digital Forensic Readiness [246].

Published in: Journal of Medical Systems (2018).

Link to original: https://doi.org/10.1007/s10916-018-1123-2.

If you want to cite the papers, please refer to the original.

For technical reasons we could not place the page where the original quote is coming from.

Abstract-Summary

"While the healthcare industry is undergoing disruptive digital transformation, data breaches involving health information are not usually the result of integration of new technologies."

"Health information is considered to be one of the most attractive targets for cybercriminals due to its inherent sensitivity, but digital investigations of incidents involving health information are often constrained by the lack of the necessary infrastructure forensic readiness."

"Following the analysis of healthcare data breach causes and threats, we describe the associated digital forensic readiness challenges in the context of the most significant incident causes."

Introduction

"Technology-based innovation trends, specifically in the areas of (1) digital health, (2) big data and (3) precision health are instrumental in supporting the delivery of the future healthcare vision [54]."

"The associated threat landscape suggests that data breaches involving health information are not generally the result of sophisticated attacks on contemporary technological building blocks, such as IoT sensors and wearable medical devices [55]."

"Leveraging publicly available healthcare data breach information and industry reports, we examine the associated implications from a digital forensic perspective."

"We identify relevant digital forensic readiness challenges that reflect the unique threat pattern of the healthcare sector."

"Despite widespread availability of healthcare data breach information and statistical analyses provided by industry, we are not aware of other works examining the associated implications with a specific digital forensic focus."

Data breaches in healthcare

"Although other nations do not necessarily have healthcare-specific provisions in their privacy legislation, health information is still usually covered under the broader definition of personal data."

"Speaking, notifications are issued when the data breach poses high risk of harm to affected individuals, which is often the case with health information, given its highly sensitive nature."

"Unlike other definitions of health information, which operate in broader terms, the HIPAA's protected health information (PHI) description also includes a list of 18 elements, which must be removed from the data set so that it is no longer identifiable health information."

"The fact that the sector is facing a unique threat pattern has been recognized by the latest special issue Protected Health Information Data Breach Report (PHIDBR) published by Verizon [55]."

"The 2018 report examines 1368 healthcare data breaches disclosed since the beginning of 2016 and in particular breaches that affect the healthcare industry or involve patients or their health information."

Digital forensic implications

"From a digital forensic infrastructure readiness perspective, investigation of privilege abuse would be supported by the ability to identify and produce forensically sound digital evidence based on the digital artifacts contained in EMR audit logs."

"These challenges include both the requirement specification and implementation issues, and there is an opportunity to introduce intelligent solutions that are able to identify MLEs based on operations of health information and can preserve the associated attributes in a forensically sound manner."

"Numerous solutions and architectures have been proposed to leverage this service delivery model to enable secure cloud-based aggregation and analysis of digital artifacts to improve forensic infrastructure readiness [56–59]."

"Collection of artifacts associated with threats applicable to EMR systems such as hacking can readily be achieved from traditional sources as we have described in Section C. At the same time, the proposed architecture can be utilized for other types of systems that involve access to and operations on potentially sensitive personal information."

Conclusion

"Although human error is widespread, the rate of privilege abuse associated with unauthorized access to health information is alarming."

"Focusing on privilege abuse, we have discussed the implications for infrastructure forensic readiness in the context of EMR systems used by many healthcare services providers to handle health information."

"We found that these systems do not always incorporate a sufficient level of forensic logging required to assist investigations focusing on privilege abuse."

"We have proposed an architecture that incorporates an intelligent real-time artifact identification module which can be deployed alongside the EMS and be integrated into cloud forensic logging service."

4.2.16 A Blockchain Based Secure and Privacy Aware Medical Data Sharing Using Smart Contract and Encryption Scheme [247]

This is a machine-generated summary of:

Vidhya, S.; Kalaivani, V.: A blockchain based secure and privacy aware medical data sharing using smart contract and encryption scheme [247].

Published in: Peer-to-Peer Networking and Applications (2023).

Link to original: https://doi.org/10.1007/s12083-023-01449-1.

If you want to cite the papers, please refer to the original.

For technical reasons we could not place the page where the original quote is coming from.

Abstract-Summary

"With the rapid progress of network information technologies, security in medical data transmission has emerged as a popular research topic."

"Electronic Health Data (EHR) is vulnerable to security breaches, tampering, manipulation, and data leakage."

"This paper presents a Blockchain-based Access Control Scheme (BACS) with Multiple Party Authority (MPA), smart contract, and proxy re encryption to secure the EHR."

"The proposed scheme consists of six phases: the registration phase, appointment phase, data aggregation stage, storage phase, query phase, and validation phase."

"The medical data is encrypted by using a Lightweight Fused Cryptographic (LFC) algorithm with the signature of the patient and doctor to ensure confidentiality and integrity."

"A security analysis is conducted for the proposed method, and the performance is compared with the existing techniques based on encryption time, decryption time, and computation cost."

Introduction

"The fast advancement of information technology has increased the use of electronic data and, as a result, a huge amount of medical information such as disease types, verified reports, medical images, and electronic health records is created every day [60, 61]."

"Medical organizations outsource encrypted EHR to third parties (such as cloud servers) for administration, which not only saves local storage space and reduces computing costs, but also allows for rapid data retrieval [62]."

"The security of EHR can be further enhanced by using blockchain technology which allows for the creation of an open and dispersed online data folder consisting of a list of data configurations that are interconnected to one another [63]."

"To maintain data security and privacy before storing the data in IPFS, a few traditional methods adapt cryptographic methods, such as those based on public key infrastructure and public clouds [64, 65]."

Literature review

"This section provides the literature review of the existing secure medical data transmission methodologies."

"Cheng and others [66] proposed a secure medical storage model using blockchain technology."

"Qiu [67] proposed a secure health data sharing system for cyber-physical systems."

"The goal of this method is to securely share data in healthcare systems."

"A Selective Encryption Algorithm (SEA) united with distribution and fragmentation for storage is intended to secure the message even when both the key and the unrestricted section of electronic health record data on clouds are leaked."

"Fang and others [68] established a secure transmission of medical data."

"The proposed scheme gives a digital watermark connected with their documentation between patient and doctor and provides a new access control model to securely transfer the medical data."

"Akash and others [69] introduced a Healthcare Digital Twin (HDT) model based on blockchain for managing the healthcare data."

Proposed blockchain based secure medical data sharing model

"Physicians examine and treat patients, and generate SHR/ EMR records. (ii) Delegated proxy (DP) The DG gives authorization for the DP to manage and upload the encrypted medical data to the IPFS. (iii) System Administrator (SA) They are primarily in charge of certificate issuing, identity authentication, and user registration.

(iv) IPFS It is a decentralized system that stores EHR of patients and shares it with hospital users and third party users."

"When the IPFS is queried to access the file, it returns the encrypted file and symmetric key to the proxy. (xxii) Data Requesters (DR) The third-party users and hospital users are called DR."

"All transactions may be monitored and regulated in a transparent, traceable, auditable, immutable, and secure way using smart contracts. (ix) Proxy re-encryption server To retrieve data and perform sophisticated processes, proxy re-encryption servers or oracles are needed."

Experimental results analysis

"The proposed LFC encryption algorithm is compared with the benchmark encryption algorithms such as RSA [70], Blowfish [71], AES [72], 3DES [73], DES [74] based on encryption time and decryption time."

"When a healthcare professional needs to access the medical information at the point of service, the time taken to encrypt or decrypt the EHR plays a crucial role in treating the patient quickly as possible."

"The encryption and decryption time of the proposed LFC algorithm is lower than the existing algorithms such as RSA, Blowfish, AES, 3DES, and DES."

"It is also observed that among all the methods, RSA has the highest encryption and decryption times."

"Since the complexity is reduced by the LFC algorithm, the proposed method has low encryption and decryption time."

"The average transaction time of the proposed method is low compared to the existing techniques such as ERFC [75], MCPS [66], GDPR [76], KP-ABE [68], HDT [69], FBASHI [77], MCGV [78], and sc-PBFT [79]."

Conclusion

"The data owner provides a re-encryption key to the oracle to perform proxy re-encryption."

"Access to any document will be denied if the request does not satisfy the delegation time, and the reputation score of the oracle will be reduced if any malicious behavior is recognized."

"The security analysis shows that the proposed method has high privacy, integrity, access control, and is secure against brute force attack, collusion attack, and man in the middle attack."

"The performance analysis of the proposed method demonstrates that the encryption time, decryption time, and computation cost are much lower compared to the existing algorithms."

4.2.17 Privacy-Preserving Storage of Sequenced Genomic Data [248]

This is a machine-generated summary of:

Hekel, Rastislav; Budis, Jaroslav; Kucharik, Marcel; Radvanszky, Jan; Pös, Zuzana; Szemes, Tomas: Privacy-preserving storage of sequenced genomic data [248].

Published in: BMC Genomics (2021).

Link to original: https://doi.org/10.1186/s12864-021-07996-2.

If you want to cite the papers, please refer to the original.

For technical reasons we could not place the page where the original quote is coming from.

Abstract-Summary

"Processing and storing of this data introduce a risk of abuse by potential offenders since the human genome contains sensitive personal information."

"We used a public set of population allele frequencies to mask the personal alleles detected in genomic reads."

"Each personal allele described by the public set is masked by a randomly selected population allele with respect to its frequency."

"Our method masked the personal variants and introduced new variants detected in a personal masked genome."

"The method is reversible and personal alleles in specific genomic regions can be unmasked on demand."

"Our method masks personal alleles within genomic reads while preserving valuable non-sensitive properties of sequenced DNA fragments for further research."

"Personal alleles in the desired genomic regions may be restored and shared with patients, clinics, and researchers."

Background

"The advancements in DNA sequencing technology support increasingly complex and accurate interpretation of genomic data which leads to exposure of sensitive personal information [80–83]."

"Such method stores only the variants in a secure form and discards the original genomic reads or encrypts them, so they can be reanalysed in future [84–89]."

"It is a common practice to confirm uncertain variants by manual examination of the underlying mapped genomic reads (alignments), and specific variants can remain undetected due to their misclassification as sequencing errors [90]."

"Various privacy-preserving solutions for processing genomic data have been proposed."

"We mask personal single nucleotide variation (SNV) alleles within alignments of a sequenced genome while preserving existing alignment data (coverage, quality, etc.)."

"We implemented the proposed methods, validated them with real personal genomic data, and evaluated the reported genomic variation."

Results

"We validated the performance of our methodology with five distinct analyses: (1) the single case study on a selected personal genomic sample, (2) the principal component analysis (PCA) on a set of genomic samples to show the effect of masking on personal alleles, (3) the second PCA analysis on two distinct populations, (4) the comparison of VCF files called on original samples with VCF files called on masked samples, and (5) the comparison between detected pathogenic variants in clinically relevant genes before and after masking."

"The population VCF contains a vast amount of low-frequency alleles which have little chance to be introduced by the masking process into the masked VCF despite considering every variant covered by personal BAM."

Discussion

"Many methods for preserving genomic privacy encrypt genomic data entirely aiming to secure personal variants [84–89]."

"Our method allows an individual to retain full control over his or her digital genome, supporting a dynamic consent approach with limited access only to his or her personal alleles."

"Other methods detect sensitive personal reads using the Bloom filter, which is built from a public database of genomic variation and relies on exact matching through hashing [91, 92]."

"This format provides a solution to privacy control of genomic reads; however, it does not address the issue of secure sharing of retrieved reads."

"Considering that masking population allele frequencies are public, the offender can always tell which genomic positions may be masked, and exploit any rare personal variants not covered by the method."

Conclusions

"Our method masked SNV alleles within genomic alignments and securely preserved them using standard RSA encryption."

"The extent of masking depends on the comprehensiveness of the population allele frequencies used as an input."

"Using masking allele frequencies from the same population as the sample is suitable for hiding an individual within this particular population."

"We can also use masking allele frequencies from a different population to make the encrypted sample resemble a sample from this particular population."

Methods

"The file with masked alleles is decrypted with the associated private key and is processed simultaneously with masked alignments back into personal alignments."

"If an alternative allele replaces either of the personal reference alleles, a new variant can be called at this position in masked alignments and it thus becomes introduced."

"All alleles within masked alignments, or their specific subsets, can be unmasked by the BDIFF file, containing the replaced personal alleles and deleted qualities."

"User has to provide masked alignments and an associated encrypted BDIFF file along with the RSA private key whose public counterpart was used in the BDIFF encryption."

"The holder of the private key that was used to encrypt the BDIFF file can share the alleles described by the BDIFF file and associated masked alignments by re-encrypting the BDIFF file in the desired genomic range."

4.2.18 Security and Access of Health Research Data [249]

This is a machine-generated summary of:

Susilo, Willy; Win, Khin Than: Security and Access of Health Research Data [249].

Published in: Journal of Medical Systems (2007).

Link to original: https://doi.org/10.1007/s10916-006-9035-y.

If you want to cite the papers, please refer to the original.

For technical reasons we could not place the page where the original quote is coming from.

Abstract-Summary

"To achieve this, we have examined available options identified in enhancing information security of health research data."

"It has been noted that there are gaps in information security for health research data."

Introduction

"Availability of health information in the research databases assist in public health research, health services research and clinical research."

"There are concerns of privacy and confidentiality of information in these databases."

"It is essential to maintain the information security of health databases as these databases are rich in information resources."

"These databases contain health data concerns with past and present health status of individuals and identifiable information of individuals."

Background

"Options to enhance information security of health databases have been addressed by various researchers, which include a. use of personal data with consent from the data subjects b. anonymise the data, then use them c. use personal data without explicit consent under a public interest mandate [93] These options will be explored further to determine applicability and practicability for health research data."

"A qualitative study on "patient's consent preferences in the use of their health information for research purposes" conducted by Willison et al. [94] revealed that, "most patients are supportive of the use of their information for research purposes" and "would like to be consulted first on the use of that information-even if the information was de-identified.""

"There were concerns of privacy of health information as de-identified data does not guarantee confidentiality as the anonymity in research databases can be reversed through the disambiguation process [95]."

"If the patient is to be identified, it is important to have the patient's consent for use of this health information."

Method

"The question is how to maximize the use of health data for research purposes without breaching the patient's confidentiality."

"Maximum access of data is required to maintain the confidentiality."

"Various security mechanisms have been implemented to ensure information security of electronic health data."

"To that data encryption is another option to secure the data during transmission to maintain confidentiality."

"Researches [96] indicated that encryption is not a solution as encrypted data cannot be analysed."

The aim of our novel technique

"Since the data is encrypted, retrieving the information from the database is not very meaningful."

"Using an encrypted database seems to be an ideal solution, but one may think of another drawback namely how to search through this database whenever required."

"Our technique allows us to do such a searching by accessing the "index" of the encrypted database."

Searching through an encrypted data—a new approach

"In a symmetric key based scheme, the keys are used to create the encrypted entries and also to search and decrypt the data."

"Boneh et al. [97] introduced a public key version of encrypted data that they call Searchable public key encryption or Public Encryption Key System (PEKS)."

"Whenever an information is required, the entity sends a trapdoor, which is a mathematical function that accepts two parameters: the secret key of the data holder and the information that the entity would like to access, and send this trapdoor to the server (via a secure socket layer interface for example)."

"Receiving this trapdoor, the server can search through the encrypted data and return any match information to the requester."

"We note that to read the encrypted information, the entity (and hence, the requester) is still required to use his secret key to decrypt it."

Discussion and conclusion

"We note that the new approach is very novel since any other mechanism will only secure the access to the information but not the actual protocol itself."

"Using our new approach, a valid secret key is still required and compromising database will not help any attacker to retrieve more information that the attacker has gathered."

"As data can be searched through the encrypted data, patients' confidentiality would not be breached for statistical research data as patients confidentiality can be maintained through encryption and the required data can be retrieved without identifying the individuals."

"This will indeed solve the requirement of maximum access of data and maintaining the confidentiality."

"Using searching through encrypted data will maximise confidentiality of patients anonymity can be maintained and data access will not compromise patients' confidentiality."

4.2.19 Data Privacy Protection in Scientific Publications: Process Implementation at a Pharmaceutical Company [250]

This is a machine-generated summary of:

Maritsch, Friedrich; Cil, Ingeborg; McKinnon, Colin; Potash, Jesse; Baumgartner, Nicole; Philippon, Valérie; Pavlova, Borislava G.: Data privacy protection in scientific publications: process implementation at a pharmaceutical company [250].

Published in: BMC Medical Ethics (2022).

Link to original: https://doi.org/10.1186/s12910-022-00804-w.

If you want to cite the papers, please refer to the original.

For technical reasons we could not place the page where the original quote is coming from.

Abstract-Summary

"Sharing anonymized/de-identified clinical trial data and publishing research outcomes in scientific journals, or presenting them at conferences, is key to data-driven scientific exchange."

"When data from scientific publications are linked to other publicly available personal information, the risk of reidentification of trial participants increases, raising privacy concerns."

"Abstracts, manuscripts, posters, and oral presentations containing study participant information were reviewed and the potential impact on study participant privacy was assessed."

"Risk minimization was sought using a generalized presentation of identifier-relevant information and decision-making on data sharing for further research."

"Additional risk identification was performed based on study participant/ personnel parameters present in materials destined for the public domain."

"The potential for participant/personnel identification was then calculated to facilitate presentation of meaningful but de-identified information."

"We found that in 13% of Takeda clinical trial publications reviewed, either individuals could potentially be reidentified (despite the use of anonymized data sets) or inappropriate data sharing plans could pose a data privacy risk to study participants."

"In 1/110 abstracts, 58/275 manuscripts, 5/87 posters and 3/58 presentations, changes were necessary due to data privacy concerns/rules."

"Risk minimization using de-identification of clinical trial data presented in scientific publications and controlled data sharing conditions improved privacy protection for study participants."

Introduction

"The benefits of open access and data sharing must be weighed against the real risk of reidentification of individuals if data presented in scientific publications can be linked to information available in the public domain, therefore raising legitimate privacy concerns."

"Anonymized clinical trial research outputs available via open access and public information sources linked to study participant data presented in scientific publications would not automatically enable reidentification of study participants."

"The combination of such data from different sources, such as participant data provided by pharmaceutical-industry sponsors, combined with reports from health authorities, real-world data, and information obtained from social media and the study participant data reflected in scientific publications, which is potentially linkable to the original identity of a specific individual, can elevate the risk of reidentification [98]."

Methods

"To check compliance with the protection of privacy of participants in clinical trials, a specification checklist for review of scientific materials before submission for publication was developed based on standards [99, 100] adapted to select identifiers that are applicable to publication materials which, in brief, considered the following rules: elimination of all direct identifiers (participant ID, participant initials) and use of indirect identifiers or quasi-identifiers (e.g. sex, age, race, ethnicity, weight, body mass index [BMI]), avoidance of individual participant listings, use of parametric over non-parametric statistics and exclusion of geographical study locations."

"Clinical trial information in the public domain was searched using pre-defined search terms such as trial identifiers (company internal clinical trial ID numbers, ClinicalTrials.gov identifiers/ NCT number), study-specific terms (e.g. product ID/ name/treatment, disease, patient population, indication) and criteria applicable to the data privacy of study participants (participant indirect identifiers or quasi-identifiers) according to the needs identified for each individual company manuscript reviewed."

Results

"Although the use of a single indirect personal identifier (e.g., sex, age, race, weight, BMI) would not allow study participant reidentification per se, if several of these identifiers are presented together with participant-specific data, there is an increased risk of participant reidentification."

"An Internet search protocol was performed to establish whether a combination of the data and information included in the manuscript (e.g. demographic parameters, age, sex and ethnicity), together with data provided in publicly available regulatory documents (CSRs, Module 2 clinical overview/summaries of the CTD within the scope of EMA transparency regulations [101], public assessment reports within the scope of the Regulation (EC) No 1049/2001)) [102] would increase the risk of study participant identification."

"Our search showed that in several manuscripts and posters, the additional data identified in the public domain were concerning, as targeted mapping of direct and indirect identifiers could enhance the individual participants' profiles; changes were therefore required to these publications to ensure protection of the privacy and anonymity of the clinical trial participants."

Discussion

"We describe a process for analysing and minimizing the risk of potential study participant reidentification in scientific publications that report clinical trial data."

"In several instances, the internet searches described as part of our analyses revealed an increased risk of study participant identification when data and information presented in manuscripts (e.g. demographic parameters, age, gender and ethnicity) were combined with data provided in publicly available clinical regulatory documents (CSRs, Module 2 clinical overviews/summaries of CTD within the scope of EMA transparency regulations [99] and public assessment reports within the scope of EU policy on access to documents [103])."

"Although only anonymized and de-identified clinical trial data were used in both the scientific publications and the publicly available clinical regulatory documents (which were processed up to several levels with data redaction, anonymization, and removal), when these sources were matched to each other there was an increased risk that individuals could be reidentified and/or the data could be used for purposes beyond the scope of scientific research."

Conclusions

"Given the rapid increase in recent years in the amount of publicly accessible clinical trial data, and the introduction of various regulatory policies concerning such data across the globe (for example, EMA transparency policy on the publication of clinical data; EU clinical trial regulations; and Health Canada's regulation on the public release of certain clinical trial information), the need to ensure the privacy of study participants has become an important focus."

"Publications need to be systematically evaluated: clinical trial participant data must be subject to risk minimization and sharing of such data must be carefully controlled."

"Risk-minimization strategies, such as the de-identification of clinical trial data in scientific publications and the use of controlled data sharing conditions, can promote protection of privacy for study participants."

4.2.20 Data Security Crisis in Universities: Identification of Key Factors Affecting Data Breach Incidents [251]

This is a machine-generated summary of:

Li, Jin; Xiao, Wei; Zhang, Chong: Data security crisis in universities: identification of key factors affecting data breach incidents [251].

Published in: Humanities and Social Sciences Communications (2023).

Link to original: https://doi.org/10.1057/s41599-023-01757-0.

If you want to cite the papers, please refer to the original.

For technical reasons we could not place the page where the original quote is coming from.

Abstract-Summary

"The extremely complex and dynamic digital environments of universities make them highly vulnerable to the risk of data breaches."

"After applying Poisson regression for the estimation, increased public disclosure of vulnerabilities was found to escalate the frequency of data breaches, whereas cross-border data flow decreased the number of data breaches."

"The mechanism by which academic strength affects data breaches was examined through the two mediators of cross-border data flow and vulnerability disclosure."

"Cloud adoption also acted as a moderator between the negative impact of vulnerabilities and the positive impact of cross-border data flow on data breaches."

"The estimation and robustness findings revealed the underlying mechanisms that impacted university data security, clarifying the understanding of data breaches and suggesting practical implications for universities and other institutes to improve information security."

Introduction

"These large-scale data interactions and flows among organizations and users inadvertently and continuously expand the digital footprints of universities, potentially leading to information security concerns by increasing the risk of data breaches."

"The number of reported data breaches was relatively high, with a significant upward trend on a monthly basis, reflecting that Chinese universities are also at severe risk of data breaches, which should not be underestimated."

"In this study, which focuses on universities' data breach incidents, we aim to investigate the determinants of data breach risks to better understand the underlying impact mechanisms."

"The aim of this study is to answer the following research questions: (i) What factors impact data breaches in universities and how do these factors interact? (ii) By what mechanism does academic strength impact data breaches? (iii) What is the influence of emerging information technologies, such as cloud storage, on the impact mechanism?"

Literature review

"Prior research has analyzed the motives behind cybersecurity and the influencing factors of data breaches."

"Magura and others [151] highlighted issues affecting database security that could lead to data breaches and data theft, including human factors, work environments, and the technologies used."

"From the listed studies, it can be concluded that when organizations at risk of data breaches have more commercial attributes, the interests involved can be more complex; thus, social perception can significantly affect information security, especially the likelihood of cyberattacks."

"Dolezel and McLeod [152] studied employee behavior, safety culture, training, supplier selection and handling of personal health information, and strong risk management procedures as data breach factors."

"When considering the impact of information security investments on data breaches, companies must consider the impact of institutional factors and balance them."

"Haislip and others [153] found that executives' IT expertize could be an effective factor influencing reported data security breaches."

Theories and hypotheses

"H4(b): The amount of cross-border data flow has a mediating effect on the relationship between research strength and the number of breaches."

"Cloud storage has a moderating effect on the relationship between the number of disclosed vulnerabilities and the number of data breaches."

"H6(a): Adopting cloud storage plays a negative moderating role in the relationship between the number of vulnerabilities and the number of data breaches."

"Cloud storage has a positive moderating effect on the relationship between the number of cross-border data flows and data breaches."

"When cloud storage is adopted, there is greater cross-border data flow and thus fewer data breaches."

"H6(b): Adopting cloud storage plays a positive moderating role in the relationship between the number of cross-border data flows and the number of data breaches."

Data and variables

"By excluding data of research institutes and enterprises, we obtained valid records on the cross-border data of 110 universities, which are included in the list of 681 already collected."

"After matching the data, a valid dataset comprising 110 universities and more than 900 valid data breach records was obtained for further analysis."

"The number of data breaches in universities in 2020 was used as the dependent variable to reflect universities' data security status."

"Economic indicators were found to be positively correlated with the risk of data breaches [154]; thus, we also controlled for the GDP of the city where each university was located."

"Although the data security risks of universities appeared uneven, they generally faced a severe risk of data breaches, with a mean value of 8.5 breaches in 2020."

Empirical analyses

"Poisson regression was employed since the dependent variable Num_Breach was measured using discrete and countable data, representing the number of reported data breaches for university i; X_i is a vector of the independent variables; $Num\_Vulnerability_i$ is the total number of disclosed vulnerabilities of university i; $Num\_Research\_Project_i$ is the number of research projects of university i; $Num\_Data\_Flow_i$ is the number of cross-border data flows of university i in 2020; $Ind\_Cloud\_Storage_i$ indicates the data storage method of university i; Controls denotes a series of control variables, including training commitment, number of doctoral programs and master's programs, number of national key disciplines, years since university establishment, type of the university, attributes of the university, GDP, and number of universities in the city where the university is located."

"In Column (4), the number of research projects is positively related to the number of data breaches, without interference from the mediating variable."

Robustness checks

"Because data breach incidents have different risk levels, we considered the effects of various factors at different levels of risk."

"All p values are less than 0.01, except for the Aroian Test of Num_Data_Flow in Row (2), where $p = 0.01006$, indicating that the mediation effects are highly significant."

"Using Num_Data_Flow as the mediator, there is a significantly negative mediation effect between the independent and dependent variables."

"For low- and medium-risk breach incidents, the cross-border data flow still has a significant negative effect, and the number of publicly disclosed vulnerabilities still has a significant positive effect."

"The results show that cross-border data flow only affects the occurrence of medium-risk breaches and that vulnerabilities tend to mostly increase the occurrence of high-risk breaches, as they have the highest significance in the regression results."

Discussion and conclusion

"We contribute to research on data breaches by discussing the effects of cross-border data flow, which are valued and regulated by numerous countries and regions for their contributions to the digital economy and potential security risks."

"We contribute to the information security literature by identifying the impacts of clouds on data breaches, distinguishing between the effects of different types of cloud adoption on the risk of data breaches."

"Our findings further reveal and strengthen the difference in terms of the impact on data breaches, which has implications for studies focusing on data security in cloud environments."

"Although we focus on a specific industry, and some of the identified factors and key findings are industry specific, they nonetheless provide an impetus for analyzing the causes of data breaches in other contexts, thereby enriching the literature by identifying the factors influencing data security incidents and, in particular, data breaches in the context of universities."

4.2.21 Privacy in Pervasive Environments: Next Generation Labeling Protocols [252]

This is a machine-generated summary of:

Ackerman, Mark S.: Privacy in pervasive environments: next generation labeling protocols [252].

Published in: Personal and Ubiquitous Computing (2004).

Link to original: https://doi.org/10.1007/s00779-004-0305-8.

If you want to cite the papers, please refer to the original.

For technical reasons we could not place the page where the original quote is coming from.

Abstract-Summary

"In pervasive environments, privacy is likely to be a major issue for users, and users will want to be notified of potential data capture."

"Recent experiences with the Platform for Privacy Preferences Project (P3P), an attempt to provide privacy mechanisms for the Web, suggest important lessons for the design of a next generation labeling protocol that will be usable and useful in pervasive environments."

Introduction

"In pervasive environments, privacy will become increasingly important and visible as a problem for users: Because of the envisioned large number of sensors, location-awareness, and fusion systems, the sheer volume of identifiable data will necessarily increase greatly."

"Technical mechanisms to facilitate these informed decisions will be critical in pervasive environments."

"I will refer to a technical mechanism through which users can be informed of data requests and their consequences as a labeling protocol."

"By doing so, the paper will also construct an understanding of what will be required in pervasive environments to suitably facilitate individuals' privacy."

"While P3P is an attempt to find privacy mechanisms for the Web, its lessons will be important for next generation environments."

"The paper concludes by examining these lessons with regards to pervasive environments and their new requirements."

The social side of privacy

"Privacy is about individuals' capabilities in a particular social situation to control what they consider to be personal data."

"Computer users in general are quite concerned with their privacy."

"Culnan [45] argued that privacy concerns were a critical reason why people do not go online and why they provide false information online."

"Culnan and Armstrong [104] make the argument that people have two kinds of privacy concerns."

"Westin [105] found three separate groups: the marginally concerned, the privacy fundamentalists, and the pragmatic majority."

"They are concerned about their privacy, but are willing to trade personal data for some benefit (e.g., customer service)."

"These groupings have been consistent across studies (e.g., [106, 107]). (Spiekermann and others [107] divided the pragmatics into those who were concerned with revealing their identity and those who were more concerned about making their personal profiles available.) In Ackerman and others [106], these groups were 27% marginally concerned, 17% privacy fundamentalists, and 56% pragmatic majority."

"It is critical to note that people differed widely in their types and degrees of concern about their privacy."

Privacy technologies

"There is increasing sentiment that regulation is required for privacy protection (perhaps in combination with technical mechanisms), but regardless, the following technologies are still useful."

"These categories include encryption and security mechanisms, anonymizing mechanisms, infrastructures, and labeling protocols."

"Encryption and other security mechanisms provide some privacy capabilities."

"Even with the tightest security mechanisms, some disclosure will be required (e.g., to provide services to a specific person)."

"Security is necessary for privacy, but security is not sufficient to safeguard against subsequent use, to minimize the risk of sensor-based disclosure, or to reassure users."

"The final privacy mechanism is one that I term here as a labeling protocol."

"In order for there to be informed consent, especially in situations where anonymity is impossible or not desired, some mechanism must exist to both describe and announce the required or preferred personal data, the data collection's scope, and the data collection's intended use and consequences."

P3P: a labeling protocol

"The P3P protocol functions within this architecture to send proposals (i.e., a service's data requests and privacy statements) and to match them with users' privacy preferences: P3P is designed to help users reach informed agreements with services (Web sites and applications that declare privacy practices and make data requests)."

"As the first step towards reaching an agreement, a service sends a machine-readable P3P proposal..., in which the organization responsible for the service declares its identity and privacy practices... Proposals can be automatically parsed by user agents such as Web browsers and compared with privacy preferences set by the user."

"If there is a match between service practices and user preferences, a P3P agreement is reached."

"While the original intention of P3P was to automatically negotiate the release of personal data to services that followed appropriate privacy practices, it was found that users did not want the automatic transfer of most personal data [106]."

P3P proposals

"P3P proposals are XML-based statements that include "slots" for access to identifiable data, dispute resolution mechanisms, desired data, consequence of data release, purpose of data collection, recipients of the data, and data retention policy."

"A P3P statement can have as its data retention disclosure only the values of no-retention, stated-purpose, legal-requirement, business-practices, and indefinitely."

"The site can have a human-readable retention policy as well, but the machine-readable portion is limited to these five values."

"This proposal states that the TheCoolCatalog.com Web site collects data for personalization and their own development."

P3P technologies

"A number of auxiliary technologies have been constructed for P3P. Aside from policy generators and trust engines, prototype user interfaces (e.g., AT&T Privacy Bird [108]) and user agents have been constructed."

"These agents note discrepancies in privacy policies, check privacy statements against third-party databases of problems, and provide other useful assistance to users."

"As Hockheiser [109] noted, P3P is an attempt to construct:... a standardized, machine-readable format for these (privacy) policies, along with a protocol for finding them."

"By making privacy policies easier to find and understand, P3P might help users find the information that they need to make informed decisions regarding sites they visit [109]."

P3P as success

"Even more importantly than adoption, the P3P effort shows what a labeling protocol can accomplish."

"With the Web, a labeling protocol and its resulting architecture are critical: users need to be informed of the data requests and consequences."

"This requires some protocol like P3P. In addition, for users to feel as though they are in control, they must have an adequate understanding of any consequences."

"This requires some protocol like P3P. Only a labeling architecture and protocol can help in this situation, and P3P shows that such an architecture and protocol could be constructed."

"Distilling the tradeoffs, limitations, and open issues from the P3P effort is essential to understanding what needs to be done for next generation labeling protocols."

"These open issues indicate the changes and continued research needed for next generation labeling protocols."

P3P critiques, open issues, and lessons

"In his critique, the vocabulary is structured around the processes and needs of services, rather than the empirically derived preferences of users. (There was, in actuality, some testing of the terms and preferences in [110].) Hockheiser [109] does point out that P3P suffers from an inherent tradeoff between the simplicity needed by users and the complexity required for a concrete and grounded understanding of preference and policy terms."

"This is a tension that must be managed with P3P, labeling protocols in general, and many user-centric technologies, since it involves the conflict between a vocabulary that is succinct and controllable vs. a vocabulary that is completely explanatory."

"The next section argues that privacy will get only worse in pervasive environments, that a labeling protocol will be required, and that the lessons from P3P carry to these next generation environments."

Privacy in pervasive environments

"A next generation labeling protocol, designed for pervasive environments, is necessary."

"New stresses will exist for this next generation labeling protocol: In a context-aware environment with a suite of context-aware applications, an individual will operate within many social and organizational contexts, and surrounding social and organizational environments may make use of many individuals' data."

"This will need to be tackled head-on for large-scale ubiquitous environments. (However, it may be possible to restrict the problem initially in intra-organizational settings.) If only because of the volume of data requests, most users will have an increasingly difficult time controlling their data, given the increased volume, invisibility of the technology, and uncertain regulatory environment."

"If, as argued above, labeling protocols are required for pervasive environments, and as argued above, there are new stresses on even the P3P mechanisms, additional time and efforts are required to find proper solutions."

"This again argues for working out any new labeling protocols while pervasive environments remain research prototypes."

Conclusions

"Best seen as a labeling protocol for describing the consequences of disclosing personal information, P3P's goal is to help users control their privacy on the Web according to their own preferences."

"At times, identification and authentication is required, and some mechanism for signaling the consequences must be created for pervasive environments."

"With the additional time made possible by creating and evaluating numerous research alternatives and prototypes, it is entirely possible that we can help users control their privacy in the upcoming pervasive environments."

4.2.22 Highly Private Blockchain-Based Management System for Digital COVID-19 Certificates [253]

This is a machine-generated summary of:

Pericàs-Gornals, Rosa; Mut-Puigserver, Macià; Payeras-Capellà, M. Magdalena: Highly private blockchain-based management system for digital COVID-19 certificates [253].

Published in: International Journal of Information Security (2022).

Link to original: https://doi.org/10.1007/s10207-022-00598-3.

Copyright of the summarized publication:

If you want to cite the papers, please refer to the original.

For technical reasons we could not place the page where the original quote is coming from.

Abstract-Summary

"Of the declaration of the COVID-19 pandemic, several proposals of blockchain-based solutions for digital COVID-19 certificates have been presented."

"Considering that health data have high privacy requirements, a health data management system must fulfil several strict privacy and security requirements."

"It is required that a digital COVID-19 certificate management protocol provides an easy verification process and also strongly avoid the risk of forgery."

"We present the design and implementation of a protocol to manage digital COVID-19 certificates where individual users decide how to share their private data in a hierarchical system."

"Our protocol introduces an authority to control and regulate the centers that can generate digital COVID-19 certificates and offers two kinds of validation of certificates for registered and non-registered verification entities."

"The paper achieves all the requirements, that is, data sovereignty, high privacy, forgery avoidance, regulation of entities, security and easy verification."

Introduction

"It is therefore proposed to use Digital COVID-19 Certificates [111] where all these cases can be easily checked."

"Thousands of cases of fake Digital COVID-19 Certificates have been already reported, detected in different countries where the use of the certificate is already mandatory."

"Due to certificate falsification, there have already been cases of dead of infected people with COVID-19 who, upon arrival at the hospital, claim to be vaccinated."

"Such digital certificates contain health data, data that today have a high risk."

"Blockchain technology provides the set of features necessary for the implementation of a decentralised application (DApp) to perform the management and certification of healthcare documentation, as well as providing the immutability of the healthcare data [112–114]."

State of the art

"As the authors point out, it does not provide full anonymity of end users, high security against forgery, as well as allowing the blockchain members to examine data without an access permission."

"In [115], Odoom and others propose a protocol solution for Covid-19 test results and vaccination using the blockchain technology and the IPFS system for encrypted results storage."

"Actually there are proposals that follows the World Wide Web Consortium standard "Verifiable Credentials", like in [116], where the blockchain is used in conjunction with the descentralised personal data Solid Pod technology, which involve the user spending part of their device's storage to save their certificates or relying on a cloud provider."

"There are several proposals that obtain solutions for the management of COVID-19 certificates, using innovative and secure technologies, but in general they do not present the technical details of a development and the cryptographic techniques used neither a security analysis that prove the fulfilment of the requirements, so the complete privacy and security of the solutions are not proved."

Contribution

"A good digital COVID-19 certificate proposal must take into account these requirements and achieve an easy verification and a low forgery chance."

"For that, our proposal achieves the following aspects that define our contribution: Supervision of issuers and verifiers The proposed protocol provides a method to control the entities that can issue certificates, avoiding forgery."

"Encryption of all the digital COVID-19 certificates data The encryption and re-encryption procedures used in the protocol make it a high privacy preserving protocol."

"Only the entities authorised by the owners will be able to decrypt the certificate thanks to the re-encryption scheme."

"The paper includes some technical details of the DApp: the proxy re-encryption scheme implemented and the different smart contracts defined."

Background

"To link all data contained in the network (files and directories), IPFS uses Merkle Directed Acyclic Graphs (Merkle DAG) combined with a DHT (Distributed Hash Table) that performs the content rounting."

"IPFS nodes are not required to store all the data contained in the network, generally stores the data like in a cache [117], then, the garbage collector periodically will remove the content of the node."

"The proxy re-encryption (PRE) is a special type of public key encryption that provides the functionality to share encrypted data with third parties, without the need to decrypt the data and re-encrypt again with the cryptographic keys of the recipient."

"The re-encryption functionality is distributed between multiple proxies, achieving the decentralisation of the scheme, also provides a high security."

"The use of PRE is extended for use in protocols that needs to achieve self-sovereignty in the share of confidential data, meaning that have to share the data in a highly securely and privacy form [118, 119]."

System overview

"Small organisation that want to provide secure services to their clients, but they perform the verification phase like a user actor which are determined by the regulatory authority intervention and can only verify a limited number of digital COVID-19 certificates."

"The owner users also can share their digital COVID-19 certificate with different trusted entities or other system users for their digital COVID-19 certificate verification, requesting re-encryption to the PRE service."

"The users can never share a digital COVID-19 certificate directly to another user or non-trusted entity of the system; it must always be verified by the regulatory authority."

"The blue lines represent the request by a trusted entity for access to an external digital COVID-19 certificate for verification, as well as the interaction with the PRE service for the completion of the requested certificate re-encryption key generation and the corresponding decryption."

Protocol

"In order to be able to make a request for a digital COVID-19 certificate from a trusted entity, it is first necessary for the entity to know the Ethereum address of the users to whom it wants to make the request."

"All the necessary information, such as the hash (identifier) of the certificate in the IPFS system and the parameters of the re-encryption key, is sent to the smart contract of the requesting entity (5), so that it has all the necessary information to be able to decrypt the digital COVID-19 certificate."

"In the case where the owner user wishes to share the digital COVID-19 certificate directly without a prior request to give access to a trusted entity, the user must know the Ethereum address of the entity, then enter it together with the hash of the certificate."

Smart contracts

"The rest of the important user management functions refer to the resolution of requests to send digital COVID-19 certificates between users or between a user and a trusted entity, requests which, as already introduced, are first sent to the WHO smart contract."

"For the management of trusted entities, the entity() function is defined, Listing 8, where indicating the owner address of the user's smart contract (_entityAdr), the WHO can activate or remove the trusted entity attribute of the specified smart contract, calling the user activeEntity() function, where the user entity variable simply takes the Boolean value entered as a parameter."

"In the management of external requests, the newSol() function is defined, which is executed when a request received from a trusted entity or user is sent from the smart contract of the regulatory authority, i.e. when the regulatory authority considers a request to be resolved and finally sends it to the recipient user."

Security and privacy analysis

"The parameters for the decryption of the certificates are stored publicly in the user's smart contract, but without the use of the private key of the user with access to it, it is not possible to decrypt the digital COVID-19 certificate, and as it has already been indicated above, this private key must be kept by the user in the most secure way."

"The use of smart contracts for each user of the system provides the service of non-repudiation at source and at destination, by the fact that every transaction is publicly stored on the blockchain together with the sending and receiving Ethereum address information."

"Thanks to the blockchain signed transactions, the system provides the possibility to track the different actions performed to issue or validate a digital COVID-19 certificate."

Performance analysis

"In order to obtain a key point for a results analysis of the decentralised application implemented, a performance analysis has been carried out, studying the waiting time and the cost of executing the different transactions carried out."

"The transactions carried out on the Ethereum blockchain network, where the execution of the smart contracts code is performed on the EVM, have a computational cost, which translates into a certain economic cost."

"As far as the transaction cost of the main functions that make up the implementation is concerned, compared to the deployment of smart contracts, it has a much lower cost."

"Starting with the execution of the removal of a laboratory together with the addition and removal of an entity to the system are the main functions of the implementation that have a lower cost per transaction, due to the fact that they involve the publication of very few parameters in the network."

Conclusions and future work

"A blockchain-based protocol for COVID-19 digital certificate management system using a proxy re-encryption service has been presented, which provides high privacy, authenticity and self-sovereignty of data."

"A useful solution for the management of the pandemic, through a fully decentralised implementation and ensuring the confidentiality of the user's data, a very important feature as it deals with health data."

"The fundamental parts that define the designed protocol are the use of the proxy re-encryption service to achieve data confidentiality and the self-management of the data by its owners."

"The proposed protocol can be used for the management of digital COVID-19 certificates and, following the same criteria, it can be used for the management of other viruses or diseases and also could even be extended to the management of medical data of any kind, such as medical records or test results."

4.2.23 A Survey on Security and Privacy Issues in Contact Tracing Application of Covid-19 [254]

This is a machine-generated summary of:

Sowmiya, B.; Abhijith, V.S.; Sudersan, S.; Sakthi Jaya Sundar, R.; Thangavel, M.; Varalakshmi, P.: A Survey on Security and Privacy Issues in Contact Tracing Application of Covid-19 [254].

Published in: SN Computer Science (2021).

Link to original: https://doi.org/10.1007/s42979-021-00520-z.

Copyright of the summarized publication:

If you want to cite the papers, please refer to the original.

For technical reasons we could not place the page where the original quote is coming from.

Abstract-Summary

"In response to the coronavirus (COVID-19) pandemic, Government and public health authorities around the world are developing contact tracing apps as a way to trace and slow the unfold of the virus."

"Contact tracing apps work by gathering information from people who have tested positive for the virus and so locating and notifying individuals with whom those people are in shut contact, oftentimes by use of GPS, Bluetooth, or wireless technology."

"Contact tracing apps use either a centralized or a decentralized approach to work the user's information."

"The researcher's contributions related to the security and privacy of Contact tracing apps have been discussed and, later research gaps have been identified with proposed solutions."

Introduction

"A contact-tracing application [120] can track every user who has been having been in proximity and then it alerts all affected users when one of them confirms positive for infection."

"Some contact-tracing applications can also inform users when an infected person is nearby and helps in preventing possible infection."

"The major concern of these contact tracing apps is their architecture, data storage, data management, privacy, and security."

"While there are many aspects to consider, we focus on the following ones: (i) which type of privacy-related information (i.e., information that reveals one's identity) has been collected for contact tracing? (ii) Are these apps designed and implemented correctly to avoid privacy leakage? (iii) is the data being transmitted to other parties, e.g., servers or other users, privacy-preserving (e.g., has the data been protected against eavesdropping, tracking, and de-anonymization attacks)?"

COVID Tracing Application

"Contact tracing allows measures such as social distancing, self-quarantining, virus testing to be applied to the infected individuals."

"By estimating the device user's trails or Unique ID, they have collected from those people who were infected with Covid-19, one can identify others who have been near the infected individual."

"Devices exchanges their temporary ID through Bluetooth when they are in proximity to an infected individual."

"Once the individual tested positive for Covid-19, corresponding health workers are mandated to upload the encounter messages to the central server or the user can voluntarily upload all of their details to the server."

"The centralized server maps the temporary ID in the messages to individuals to identify at-risk contacts."

"This approach enhances the user's privacy by generating anonymous identifiers at the user devices and processing the exposure notifications on individual devices instead of the centralized server."

Existing Contact Tracing applications

"GPS-based contact tracing applications create a trail for every user by recording their temporal GPS location."

"Bluetooth-based contact tracing apps create a unique identity for every user which may be static or dynamic in which the application transmits to nearby devices."

"This app uses Bluetooth technology to collect data regarding users who have close to one another."

"This app uses Bluetooth technology to track the users who have been in proximity for more than 15 min."

"France launched a new contact tracing app named "TousAntiCovid" which is the enriched version of its previous application called Stop COVID."

"It uses Bluetooth technology to track user's movements and it is centralized."

"Every business venue was mandated to display QR code posters Users can scan these codes and register their names and mobile numbers, so that if the infected case of coronavirus disease was tracked to the venue and they would be informed by trace workers."

Privacy concerns-contact Tracing

"If such apps record the location history, the user's movement can be traced as well."

"Apart from tracing the location, there are many privacy issues, such as data breach, data collection, obscure data flow [121–131]."

"Considering the concern over data privacy, unnecessary data collection, location tracking, and other issues, users must have free wills to decide by the users who installed the application."

"Users should have manual control over data deletion from the app or the central server."

"The process of data collection, usage, and storing should be transparent to protect user privacy."

"Many applications collect excessive, unnecessary data from their users, for instance, an application named "Aarogya Setu" requires name, mobile number, age, gender, profession, and details of countries visited in the last 30 days Also, Geo-location tracing is unnecessary alongside Bluetooth or other similar wireless technologies."

Security Attacks

"COVID Tracing applications like Trace together, covidSafe, ArogyaSetu uses Bluetooth technology (which generates a random ID for each device) to identify the user and collect information regarding users who have been close to one another."

"Applications that use the centralized architecture, temporary ID, and device model information can be used to identify the device easily."

"The purpose of this attack is to count the number of users who have tested positive for the covid-19 in the centralized architecture."

"In this attack, an attacker can estimate the number of users infected with the Covid-19, who have been volunteered to upload their contact tracing data to the Centralized architecture."

"An address attack is done when the change over time of the Bluetooth MAC address and the temporary identifier have not coincided."

"Complete synchronization of temporary transient identifiers and the Bluetooth random MAC address change are proposed in some Covid tracing applications."

State of the Art

"Whaiduzzaman and others [132] proposed a fog-based IoT healthcare Contact Tracing Application."

"Wenzhe and others [133] proposed a peer-to-peer system of a blockchain protocol, provides users with a unique ID, transparent data storage, location proofing, and zero-knowledge proof-based data ownership authorization."

"Apple and Google have joined to enhance privacy-preserving contact tracing by introducing an exposure notification system [134]."

"Once the individual downloads a contact tracing application that uses the exposure notification API released by Apple and Google and opts in to contact tracing applications."

"Through BLE, those proximity identifiers are periodically shared with nearby devices whose users have also opted into the same contact tracing application."

"Hao et al [135] proposed solutions combined with the use of blockchain, encryption, and anonymization technologies to ensure the user's privacy."

"It also addresses challenges while using some algorithms for storing the user data in the Covid tracing application."

Proposed Solution

"Several Covid-19 tracing apps store user's data using weaker encryption standards."

"User's data can be encrypted using a secure algorithm."

"Advanced Encryption Standard (AES) algorithm is the symmetric block cipher."

"It is the symmetric block cipher that replaces the Data Encryption Standards (DES)."

"The final round of both encryption and decryption consists of three stages namely substitution bytes, MixColumns, and AddRoundKey."

"The encryption consists of 10 rounds for 128-bit keys."

"After block-wise encryption, each block was uploaded to the cloud at different locations with user-encrypted_id and block_id."

"For contact tracing, it is enough to use the User's encrypted id to alert the nearby person."

"The encryption consists of 10 rounds for 128 bit keys."

"The fields in the user's data were split into different blocks and then individual blocks are encrypted separately."

Future Scope

"A mobile application should be user-centric and should protect users from unwanted intervene."

"Future work remains to build an application by adding high security and feasibility in tracing and storing the user's data."

"It is also designed with an automatic generation of reports by self-diagnosis with a high rate of accuracy."

"These types of measures will increase the privacy of the snoopers, adversaries."

Conclusion

"As Coronavirus spreads through close social interaction, contact tracing has become vital for containing its spread."

"There is an increasing concern regarding the collection and use of data and its security."

"We analyzed a huge set of contact-tracing apps implementing different security and privacy measures."

"We analyzed contact-tracing apps for privacy, data storage, and data security."

4.2.24 *Realizing Privacy Aspects in Blockchain Networks [255]*

This is a machine-generated summary of:

Noam, Oded; Rottenstreich, Ori: Realizing privacy aspects in blockchain networks [255].

Published in: Annals of Telecommunications (2021).

Link to original: https://doi.org/10.1007/s12243-021-00861-z.

If you want to cite the papers, please refer to the original.

For technical reasons we could not place the page where the original quote is coming from.

Abstract-Summary

"The decentralized nature of blockchain networks, often involving the publicity of ledger information, can come into conflict with this right."

"We first describe multiple aspects of privacy such as its practical limit towards authorities, its potential vulnerability, and its association with technological systems such as blockchain networks."

"We explain how each can be used to enhance privacy in blockchain networks following their unique nature."

"We further conclude that none of the available techniques can fully answer the complex practical requirements, and that practical solutions should combine both centrally controlled and decentralized systems."

Introduction

"A similar mechanism is used in subsequent blockchain protocols, including general-purpose platforms such as Ethereum [136]."

"In such platforms, the protocol enables operations other than asset transfer, but the data set that programs ("smart contracts") process is duplicated to all validators, similar to the ledger in asset transfer protocols."

"The decentralization of blockchain architectures enables creating systems in which subjecting control to a trusted entity is avoided."

"Thanks to this property, blockchain protocols have the potential to revolutionize processes that are dependent on a trusted intermediary."

"Blockchain's common use cases of settlements, escrows, and asset transfers are discomforting for individuals in societies where income and wealth are considered private."

"The potential of blockchain technology to disrupt other industries that are distorted by powerful monopolies, is unfulfillable without due privacy."

Aspects of privacy

"Norway has gone as far as to make tax payments public and transparent, essentially exposing taxpayers' wealth and income [137]. (ii) Securities regulators require publicly traded companies to reveal significant business information. (iii) Liberal countries choose, by default, the transparency of judicial proceedings over the privacy of those involved. (iv) Many countries expose sex-offender registries, a huge compromise on the privacy of past offenders, to allow caretakers to protect children against the mere possibility of a repeat offence. (v) Countries that cooperate with the Financial Action Task Force (FATF), an intergovernmental organization fighting money laundering, apply banking rules that give enforcement agencies absolute access to account information of their residents."

"While there are doubts of how long Moore's law can persist [138], this does mean encrypted data could be exposed within our lifetimes unless the protocol uses an exceedingly high level of protection against a brute-force attack."

Technical approaches to blockchain privacy

"Off-chain payment channels (such as Lightning [139] or Raiden [140]) use a private ledger where the base layer is a public blockchain."

"These use a combination of several fundamental techniques: (i) Verifiable Computations (VCs) [141, 142] are protocols that enable a party executing a computer program to generate a proof that certain outputs are the result of executing the program on a certain input."

"With ZK-PCPs, one can create proofs that a certain output is the result of executing a program that makes use of secrets without disclosing them. (iv) Non-interactive Zero-Knowledge Proofs [143] are protocols that provide zero-knowledge proofs in a single transmission, without the prover challenging the prover."

"By using strong encryption of the private data and supplying decryption keys only to trusted programs, it is possible to create decentralized networks for processing large pools of private data ("big data")."

Conclusions

"We reviewed several building blocks towards privacy from areas such as cryptography and distributed systems."

"We observe that none of these building blocks provides or will ever provide a general-purpose solution for all privacy needs, all the more so if the requirements are for non-absolute privacy."

"Any practical design of blockchain applications should be a hybrid of privacy-preserving and disclosing parts, by using a tailor-made combination of the building blocks listed in this paper."

4.2.25 *Information Privacy in Institutional and End-User Tracking and Recording Technologies [256]*

This is a machine-generated summary of:

Nguyen, David H.; Hayes, Gillian R.: Information privacy in institutional and end-user tracking and recording technologies [256]

Published in: Personal and Ubiquitous Computing (2009).

Link to original: https://doi.org/10.1007/s00779-009-0229-4.

If you want to cite the papers, please refer to the original.

For technical reasons we could not place the page where the original quote is coming from.

Abstract-Summary

“This work focuses on both institutional and end-user tracking and recording technologies.”

“This paper describes (1) an empirical interview and survey study of everyday institutional tracking and recording technologies and (2) an analysis of these empirical data against a framework originally used to describe tension points for end-user tracking and recording technologies.”

“We provide extensions to the analytic framework for capture and access technologies to address differences, similarities, and interplay between institutional and end-user tracking and recording technologies.”

“The results of this paper contribute to the fields of personal and ubiquitous computing by providing significant insight relevant to the evaluation, design, deployment, and adoption of new tracking and recording technologies.”

Introduction

“Although tracking and recording technologies (TRTs) hold the potential to advance these research areas and address a myriad of domain problems, they may also invoke a variety of privacy-related concerns.”

“Researchers in Ubicomp have long investigated privacy-related issues and concerns surrounding TRTs.”

“The distinction is between institutional and end-user TRTs.”

“Using this framework, we further analyze the potential tension points in the design, use, and policies surrounding the studied institutional TRTs, noting where the framework breaks down and extending it when necessary to consider the particular concerns of capture technologies implemented and used by larger institutions."

"This work identifies and specifically interrogates ways perceptions are constructed around both institutional and end-user TRTs."

"The results of this work contribute to the personal and ubiquitous computing community in understanding and supporting the evaluation, design, deployment, and adoption of both novel and known institutional and end-user TRTs."

Method

"Our approach included the use of a questionnaire to gauge participant attitudes quantitatively, and a follow-up interview focused on their rationales for those attitudes."

"This approach allowed us to poll participants' attitudes surrounding information privacy, data collection, data control and data use around a wide variety of technologies without biasing them towards specific concerns by using potentially loaded terms like privacy and surveillance."

"Participants were then given the opportunity to ask any questions about RFID until they felt comfortable with their understanding of the technology and its uses."

"This section was included to allow for a comparison of the participants in this study with those in Smith and others regarding their attitudes towards information privacy."

"Questions included the number of years the participant had used a particular technology and how concerned the participant was that records were kept when the technologies were used."

Results

"Participants rated their levels of concern with the five studied everyday TRTs: store loyalty cards, credit cards, security cameras, electronic toll collection, and web servers."

"With the exception of web server records ($\mu = 4.43$, $\sigma = 1.90$), participants reported low levels of concern for the records kept by everyday technologies that were studied."

"Technological impossibility, concerns about web tracking technologies were significantly more common than the other everyday TRTs queried, such as pervasive video surveillance, which participants tended to understand more accurately."

"When asked about tracking via RFID, a technology that participants believed to be novel and rare and that has no specific, common usage yet, they replied with similar higher levels of concern as when asked about general information privacy."

"The participants in our study reported not to be concerned with the tracking and recording of store loyalty records."

Comparing end-user and institutional tracking and recording technologies

"We previously proposed a framework through which one can attempt to design adoptable capture technologies and against which one can evaluate their impacts that included seven tension points: ownership of data; choice; visibility and awareness of recording, archival, and deletion; trust; features of rich media; face; and decision point [144]."

"Not knowing the richness of the data collected, another participant said of credit card records: "I can't see anybody would want any of that information anyway."

"When describing use of store loyalty cards, one participant noted "they know what I buy anyway; they scan it on the register" Finally, another interesting finding from examination of these institutional recordings is found in the very masses of data that may be of concern to some."

"Differences include knowing which entity owns the data collection changes the perception of the people being tracked or recorded."

Conclusion and future work

"Some studies have uncovered general privacy concerns; at the same time, other investigations have indicated that people are not concerned with many new Ubicomp technologies."

"Far from claiming that there is a single answer to these potentially conflicting findings, the results of this study demonstrate that people can simultaneously be concerned about data tracking and recording while using these technologies and services on a regular basis."

"A similar argument is that if people are already using these technologies, then they have already consented to the tracking and recording that is a part of these technologies."

"There is a difference in assessing ownership of the recorded data when technology is end-user based or institutional based."

"The first study will gauge people's understanding of and attitudes towards Bluetooth tracking and recording."

"We will treat this study as an example end-user technology."

4.2.26 A Survey of Computational Location Privacy [257]

This is a machine-generated summary of:

Krumm, John: A survey of computational location privacy [257].

Published in: Personal and Ubiquitous Computing (2008).

Link to original: https://doi.org/10.1007/s00779-008-0212-5.

If you want to cite the papers, please refer to the original.

For technical reasons we could not place the page where the original quote is coming from.

Abstract-Summary

"This is a literature survey of computational location privacy, meaning computation-based privacy mechanisms that treat location data as geometric information."

"The survey omits non-computational techniques like manually inspecting geotagged photos, and it omits techniques like encryption or access control that treat location data as general symbols."

Introduction

"This paper surveys research relevant to computational location privacy, i.e., the ways that computation can be used to both protect and compromise location data."

"Location data affords privacy techniques that exploit its geometric nature, and we limit our definition of computational location privacy to largely geometric-based algorithms."

"We concentrate on computational privacy attacks that take advantage of the geometric nature of location data, thus omitting attacks based on manual surveillance or hacking around standard data protection schemes."

"We are most concerned with an attacker gaining access to location data and using it to algorithmically discover a subject's whereabouts and other information."

"These different forms are the subject of many different computational schemes for protecting privacy, such as using a pseudonym instead of an actual name, intentionally adding noise to the data, and reporting location as a region instead of a point."

Why reveal location?

"One category of location measurement technologies use third party infrastructure to compute location, e.g., locations determined by cell phone providers, Google's "My Location" feature using cell tower data [145], the Loki Wi-Fi locator [146], and Quova's IP address locator [147]."

"Some location-sensing systems, such as GPS, are designed to avoid this type of privacy leak by computing location on the local device."

"Such techniques have at least the potential for privacy if they can compute location on the local device as opposed to offloading the computation to a networked server."

"Besides the act of measuring location, another reason for an individual to reveal their location is that institutions are considering the use of location data for variable pricing, pushing their users toward revealing their whereabouts."

"This idea is used in the in-vehicle navigation systems from Dash™ and TomTom®, where users' locations can be transmitted to a server for computing real time traffic speeds."

People do not care about location privacy

"Of university students, Danezis and Lewis [148] asked 74 undergraduates how much they would have to be paid to share a month's worth of their location data."

"The definitive study on peoples' attitudes towards location privacy has yet to be done."

"Attitudes will change depending on the usefulness of the location-based service, the privacy safeguards, the amount of data, and the nature of abuses."

"The Danezis and Lewis study found that the prices set on location data varied depending on who would be using it."

"In terms of sensitivity, the study in [149] found that Greeks demanded a significantly higher price for their location data than the four other EU countries surveyed."

"It may be that people will become much more sensitive about location privacy if there is a major news story about the negative consequences of a location data leak."

Computational threats

"While the above threats come from manual inspection of leaked location data, researchers have explored what can be automatically inferred about a person based on location."

"This includes early work aimed at understanding peoples' movement habits, simulated privacy attacks, and sophisticated algorithms to infer context based on location."

"These subsections are designed to illustrate what an attacker could infer from leaked location data."

"There are four reported studies in which researchers used recorded location data to demonstrate a privacy attack, often referred to as an "inference attack" in the literature."

"By showing how location data can be attacked with inference algorithms, researchers have established a lower bound on what an attacker could do."

"Researchers have used location data to infer other things besides a person's home and work location."

Computational countermeasures

"In their excellent survey, Duckham and Kulik [150] list four general methods for ensuring location privacy: Regulatory strategies: normally government rules on the use of personal information Privacy policies: trust-based agreements between individuals and whomever is receiving their location data Anonymity: use a pseudonym and create ambiguity by grouping with other people Obfuscation: reduce the quality of the location data In this paper we concentrate on computational countermeasures, which encompass the last two points above."

"A privacy policy may specify that only higher levels of the hierarchy may be reported, leaving the user's precise location ambiguous."

"Increasing the time between location reports can also improve privacy."

"Since location can be specified as a single coordinate, one way to measure location privacy is by how much an attacker might know about this coordinate."

"Hoh and others [151] quantify location privacy as the duration over which an attacker could track a subject."

Summary

"This literature survey concentrates on computational location privacy, which means computational algorithms for compromising and protecting location data."

"Studies show that people are generally not concerned about location privacy, although they are sensitive to how their location data could be used, and their sensitivity may rise with their awareness of privacy leaks."

"Researchers have demonstrated many privacy attacks on location data, including algorithms for linking pseudonyms to actual names, linking together completely anonymized location readings, and inferring much about a user based on their location history."

"If the above is possible, what is a good way to inform a user of the privacy threats of revealing location?"

"Given this, how much location data needs to be revealed for each of them to be successful?"

"Is there a single best way to quantify location privacy?"

4.2.27 A Survey of RFID Privacy Approaches [258]

This is a machine-generated summary of:

Langheinrich, Marc: A survey of RFID privacy approaches [258].
Published in: Personal and Ubiquitous Computing (2008).
Link to original: https://doi.org/10.1007/s00779-008-0213-4.

If you want to cite the papers, please refer to the original.

For technical reasons we could not place the page where the original quote is coming from.

Abstract-Summary

"A bewildering number of proposals have offered solutions to the privacy problems inherent in RFID communication."

Introduction

"An April 2008 search for articles on RFID privacy and security in Google Scholar yields over 700 titles, while Gildas Avoine's manually maintained RFID Security and Privacy Bibliography [152] still lists as many as 214 publications on this topic since 2003."

"A June 2007 EU policy document [153] states that "effective action is needed so Europeans can trust that the various applications of RFID and related technologies are as safe, secure and privacy-friendly as they possibly can be.""

"What issues still need to be addressed before "safe, secure and privacy-friendly" RFID tags have become a reality?"

Uses and threats

"A typical concern is thus that "chatty" RFID tags disclose the posession of personal items normally hidden from view, e.g., the brand of underwear one is wearing, the presence of a wig or hip replacement, or even a particular medicine one is carrying [154]."

"This act of sniffing out the data on an RFID tag can only be prevented if tags disclose their identitiy only to authorized readers, i.e., those that are under the control of the item owner or another authorized party."

"As post-sales benefits of tagged items increase (e.g., smart washing machines or RFID-enabled returns), however, permanently disabling tags might force the consumer to choose between privacy and the convenience offered by novel RFID-based services."

Technical approaches to RFID privacy

"Cryptographically controlled access was first proposed by Weis et al. [155] in the form of hash locks: tag data is only released if the correct key is given, which is stored in hashed form directly on the tag."

"This hash value can be read out by any reader, yet only authorized ones would be able to look up the tag's key in a database of key-hash pairs."

"realized that a static hash value would still be traceable, and thus proposed an extension using random values: Instead of simply sending their static hash value, tags choose a random value r and send the pair (r, h(ID||r)) to the reader, prompting the reader to brute-force search its inventory for any ID that matches the given hash if concatenated with r [155]."

Practical issues for deployment

"Of particular interest to any real-world deployment of RFID encryption are approaches that specifically attempt to simplify ownership transfer, i.e., updating the key of an RFID tag in such a way that a prior owner of a tagged item (e.g., the supermarket) cannot read the tag after the item has been given to a new owner."

"Common to this and other approaches is the need for a Trusted Center that holds the actual information about the tag (i.e., its ID or details about the tagged goods), which authenticated readers can query for an encountered tag pseudonym in order to receive the true tag ID."

"If tag ownership changes, the new owner needs to notify the trusted center that previous readers are not authorized to access tag data anymore."

Summary and outlook

"Much work in RFID privacy is concerned with secure and efficient cryptographic algorithms."

"The current activities at the European policy level [156, 157], including the public consultations during March and April 2008, are expected to lead to additional legal instruments, requiring, e.g., operators to conduct privacy impact assessments (PIA) prior to deployment and ensuring the use of up-to-date information security measures in their systems."

"Fabian and others [158] point out that much of the RFID privacy problem might actually lurk in the backend of the envisioned EPC information infrastructure: instead of bothering with localized attacks using deployed readers, smart attackers might simply eavesdrop on the generated (unsecured) traffic in backend systems to track unsuspecting consumers."

"The scope of the envisioned data collections will most certainly require equally large efforts in areas of privacy databases [159] and profile management [160] to be complete."

4.2.28 What it Takes to Boost Internet of Things Privacy beyond Encryption with Unobservable Communication: A Survey And Lessons Learned from the First Implementation of DC-Net [259]

This is a machine-generated summary of:

Staudemeyer, Ralf C.; Pöhls, Henrich C.; Wójcik, Marcin: What it takes to boost Internet of Things privacy beyond encryption with unobservable communication: a survey and lessons learned from the first implementation of DC-net [259].

Published in: Journal of Reliable Intelligent Environments (2019).

Link to original: https://doi.org/10.1007/s40860-019-00075-0.

If you want to cite the papers, please refer to the original.

For technical reasons we could not place the page where the original quote is coming from.

Abstract-Summary

"Privacy requires more than just encryption of data before and during transmission."

"We motivate the need for strong privacy protection in a smart home use case by highlighting the privacy issues that cannot be solved by confidentiality mechanisms like encryption alone."

"We conclude that the computational and network overheads imposed by these techniques do not make them impractical to use in the IoT. To the best knowledge of the authors, this is the first implementation of DC-net on sensors."

"Alongside, we provide a survey of the required strong cryptographic security mechanisms, like encryption of communication, to be in place."

"We describe how the current existing techniques can be facilitated to achieve unobservable communication for the IoT. This includes mechanisms for encrypted IoT communication like DTLS or message authentication like ECDSA signatures on IoT devices."

"For readers unfamiliar with the concepts of MIXing and DC-net, we explain and analyse how those techniques, formerly used to provide private communication in the Internet, can be applied to the IoT. We briefly survey what complementary features from the IoT architecture are helpful in providing strong protection in this particular use case."

Introduction

"Privacy in the Internet of Things (IoT) needs more than encrypted end-to-end communication, as meta-data still leaks."

"We also argue that even though there are still way too many unencrypted channels in the IoT, the challenge of a properly encrypted local IoT network's traffic or the traffic between an IoT device and its cloud back-end is technically already being solved, e.g., DTLS and ECDSA help IoT devices to communicate securely—but as we will highlight in this article: on its own encryption does not enable privacy."

"Even assuming DTLS or the like is enabled, and Hewlett Packard's recent report on IoT security finds just the opposite, i.e., that "70% of devices used unencrypted network service" [161], meta-data still leaks details about the communication."

"This strongly demands an additional layer of privacy protection to prevent the leakage of sensitive information from meta-data, namely unobservable communication (UC)."

Current state of the IoT towards Privacy

"Bandyopadhyay and Sen [162] identify security and privacy as key technologies that will enable IoT. Privacy of humans and confidentiality of business processes are identified as the two major issues in this regard."

"The authors identify data confidentiality, privacy and trust as key challenges in IoT security."

"The authors suggest addressing privacy issues in the system design phase, but admit the lack of a privacy framework tailored for the IoT. An analysis of security challenges in distributed IoT environments is provided by Roman and others [163]."

"The EU-funded project RERUM that developed a framework will allow IoT applications to consider security and privacy mechanisms early in their design phase."

"All EU-funded projects working on security and privacy in the IoT (IERC AC5) reported jointly in January 2015 that there is currently no research project that tackles anonymising the traffic in networks used for IoT applications [, p. 70]."164

Traffic observations and their privacy problems examplified

"To monitor messages within a house or vehicle, the attacker needs to eavesdrop on a user-controlled network of sensors, actuators, and the gateway."

"The event-driven nature of the communication flows in the IoT also lends those messages to be a good basis for attacks that facilitate pattern analysis."

"An active attacker modifies network computations and transmitted messages."

"A timing attack observes the duration of communication between nodes and attempts to correlate patterns of network participation."

"Communication pattern attacks can be launched by simply monitoring communication activity on any network device, i.e., the pattern and timing of message transmission and reception."

"For message frequency attacks, an attacker analyses the traffic flow of messages to fingerprint individual devices and/or communication partners."

"Communication pattern and message frequency attacks are challenging to overcome."

"An attacker may trace every possible path an observed message can take through a network."

Private communications

"Prevention of attacks based on message counting is possible if all network participants send and receive a standard number of messages."

"Anonymity networks offer partial protection against such attacks since neither nodes or nor messages are easy identifiable."

"The DC-net protocol is unconditionally secure and because for the every cryptographer it is equally likely that either his left or right neighbour [165] paid the bill which permits participants to anonymously broadcast one bit messages."

"For one DC-net communication round, the message has to travel twice through the ring: the first time is used to get local outputs from neighbouring nodes (sending) and the second one for broadcasting the final, global message to all nodes (receiving)."

"To compare the costs for the different topologies, the calculations are done twofold: the costs for individual nodes in the DC-net and the overall messages exchanged and their paths in a communication round throughout the whole network."

Building unobservable communication using existing sensors with 6LoWPANs and Contiki

"It features a standard-compliant embedded TCP/IP implementation and supports a number of specifications aiming to optimize the use of TCP/IP networking in embedded devices."

"In order for our use case to be implemented using Contiki and those networking technologies mentioned just now, the following needs to be fulfilled: All embedded devices and sensors must be equipped with an IEEE 802.15.4 low-power radio interface."

"This includes a discussion of physical security, hop-by-hop security between neighbouring devices, security of routing control messages, end-to-end security, ensuring authenticity and integrity with signatures, and privacy-enhancing technologies using overlay networks."

"Once keys are exchanged and parties authenticated, DTLS uses symmetric schemes to provide message integrity and confidentiality."

"The use of per device public keys for confidentiality protection by encryption allows the sender to encrypt data using the intended recipient's public key."

Implementing unobservable communication for the IoT on real sensors

"We now present the security features developed, implemented, and/or ported to the Contiki Operating system running on the Re-Mote."

"To the original protocol, we implemented important optimisations for the message exchange: Instead of sending 1 bit messages in every message round, we exchange messages with an increased payload of 7 bytes."

"A message round is a single round, where each participant commits the combination of its shared secrets and where one single client sends its communication payload."

"In the transmission phase, every node prepares its vectors: 0s and the applied shared secrets in the passive sending slots and n and the applied shared secrets for the active sending slot (if any slot was reserved successfully) with n being the payload encoded as a group number."

"The nodes send their transmission vector and receive all vectors from the others, combine the vectors and are finally able to see the payloads in each slot for each message round."

Complementary privacy-enhancing mechanisms

"Once computational resources are locally available, they can be used to process data before sharing outside the local network and to apply additional privacy preserving techniques."

"The choice of suitable privacy-enhancing processing mechanisms strongly depends on the data the device is collecting (for energy consumption see [166] and for location information see [167, 168])."

"While the gateway cannot make those devices communicate unobservable amongst themselves, it can do the data minimisation and pre-processing before information flows towards outside servers."

"An architecture that channels all data through gateways and then harvests the gateway's smartness, i.e., its local processing power and local knowledge of the local deployment, becomes an enabler for privacy."

"It also allowed the privacy gateway to inform the user of the allowed changes, i.e., alert if the allowed noise level is too low and thus too fine-grained data is requested."

Conclusions

"Our proof-of-concept shows that strong, privacy enhancing technologies can be adapted to use in the IoT. The implementation currently relies on reliable broadcast assumption, which might not hold in real world use cases considering bigger networks like Smart Cities."

"We have further discussed enabling technology in the area of IoT that allows to foster privacy, i.e., flexible networking technology like SDNs could implement privacy preserving routing overlays, local data pre-processing on the IoT gateway could minimize the data being sent to servers, while advanced cryptographic mechanisms of malleable signatures still guarantee the absence of unauthorised subsequent changes to the application."

"All these technological advances brought to the IoT and integrated into each device should allow it to become a safe place—one where we can accept such otherwise too privacy-invasive technology into our daily lives, or into our restricted corporate areas if we consider industrial Internet of Things (IoT) where the threat of privacy is a threat of trade secret protection and corporate espionage."

Recommendations

"We learned from our prototype implementation of DC-net on real hardware that to take the final steps towards a truly private communication between IoT devices, the following recommendations shall be followed: Minimize data collection Data you do not need to fulfil a goal shall not be collected by the IoT. Increase the ROM footprint of devices This is a call to hardware manufacturers, its of course not to store even more data on the devices, but to allow all security mechanisms to fit to make the IoT device a first-class citizen in secure and private communication."

"Optimize network algorithms of lower layers To minimize the negative impact that DC-net's communication overhead generates the lower network protocols could be further optimized to better cope with constant traffic, e.g., adapt solutions from constant-rate transmissions like streaming."

4.2.29 Adaptive Key Management-Based Cryptographic Algorithm for Privacy Preservation in Wireless Mobile Adhoc Networks for IoT Applications [260]

This is a machine-generated summary of:

Pamarthi, Satyanarayana; Narmadha, R.: Adaptive Key Management-Based Cryptographic Algorithm for Privacy Preservation in Wireless Mobile Adhoc Networks for IoT Applications [260].

Published in: Wireless Personal Communications (2021).

Link to original: https://doi.org/10.1007/s11277-021-09360-9.

If you want to cite the papers, please refer to the original.

For technical reasons we could not place the page where the original quote is coming from.

Abstract-Summary

"This paper plans to develop the robust and reliable security protocol in MANET IoT application."

"A chaotic cryptography-based privacy preservation model is proposed in this paper for the purpose of improving the security in the MANET IoT. The key generation process in the chaotic map is optimized by generating optimal key pairs through the newly developed SA-SFO algorithm."

"The key selected from the chaotic map is influenced by selecting the optimal parameters through the proposed Self Adaptive Sail fish Optimization (SA-SFO)."

"The comparative analysis over the conventional models ensures the efficient performance of the proposed model in terms of diverse analysis in MANET and IoT platform."

Introduction

"As many users prefer to use MANET for their services, it is necessary to provide a secured platform such that the unauthorized personnel cannot access the communication network."

"IoT handles a large amount of data at a time, and the data collection and sharing can lead to the security concerns since most users do not wish to reveal entire information to public."

"Providing a secured privacy preservation framework can lead to an increase in the popularity of the IoT based applications."

"Since IoT requires lot of data handling, the cloud computing and the IoT services can be merged to improve the privacy model."

"The contributions of this work towards the privacy preservation in the MANET are presented below: To propose a chaotic cryptography-based privacy preservation model to improve the security in the MANET IoT. To select the key from the chaotic map by the selection of the optimal parameters by the proposed Self Adaptive Sail fish Optimization (SA-SFO)."

Literature Work

"In 2019, Guan and others [169] have proposed a authentication based privacy preservation model namely, Anonymous Privacy-Preserving scheme with Authentication (APPA)."

"The model supported real time security service by improving the device management and update at the network edge."

"The proposed APPA scheme achieved low computational complexity and communication overhead, and the results of the scheme were compared with the several existing models for the demonstration of the security effectiveness."

"For hiding the location information of the user, the privacy preservation model namely, novel dummy location privacy-preserving (DLP) algorithm was formulated in this work."

"The high security performance by the model ensured its usage in the healthcare applications related to IoM. Experimentation of the scheme was carried out in the NS3 simulator, and the results were evaluated based on delivery ratio, end-to-end delay, throughput rates, and routing overhead."

Proposed Privacy Preservation in Wireless Mobile Adhoc Networks for IoT Applications

"The basic processes of the proposed privacy preservation model based on optimized chaotic map are explained as follows: In the MANET, the edge devices communicate with each other through mostly text messages."

"For the encryption process, an optimized key is selected by the proposed chaotic map based on the optimization algorithm to generate the cipher text."

"Key generation After the plain text creation, the optimal key generation is carried out for improving the privacy preservation process."

"The major parameters of the hybrid chaotic map are fine tuned with the proposed SA-SFO and using the optimal parameters optimal key K^{opt}, which is generated for the privacy preservation."

"For the encryption process, both the binary plain text and the binary form of the optimal key undergoes the Binary EX-OR operation, to generate the cipher text."

Optimal Key Generation for Secured Wireless Mobile Adhoc Networks

"Both the parameters are tuned by the newly developed SA-SFO, such that the key generation by the hybrid chaotic map results in maximal confusion and diffusion."

"One of the important steps in the key generation process is optimizing the parameters α, and β of the hybrid chaotic map."

"Update the position of the sailfish and sardines (self adaptive Attack alternation strategy): The proposed SA-SFO algorithm uses the self adaptive attack alteration strategy for improving the solution positions."

"Fit_{y-1} indicates the fitness of the solution at previous iteration $y - 1$ and Fit_y refers to the current fittest solution at iteration y. As the SFO depends on the memory for the updating the solution, using the novel self adaptive parameter A with the current and the previous fittest solution can eventually improve the convergence speed of the proposed SA-SFO."

Enhanced Encryption and Decryption Algorithm for Wireless Mobile Adhoc Networks

"The fitness used for the key generation contains the Jaccard similarity index as its function."

"After initialization of the dimension, for the given optimal key pair $K^{opt}(\alpha, \beta)$, pseudo random values for each dimension are identified through the iteration with the chaotic map."

"The final array of the key element contains the value from the sequence Q_v and its indices are identified through Q_u."

"For the decryption stage, both the optimized key value $K^{opt}(\alpha, \beta)$, and the values initialized for the map during the encryption stage u_0 and v_0 are required."

"The pseudorandom array is generated as same as the procedure mentioned in the encryption phase using the optimized key."

Results and Discussion

"Evaluation of the proposed SA-SFO privacy preservation model is implemented with the different plain texts of varying character length as 20, 40, 60, 80, and 100 respectively."

"The performance of the proposed SA-SFO is analyzed based on the optimal key generated for the varying character length of plain text is analyzed."

"The proposed SA-SFO is compared over the existing heuristics like PSO, GWO, WOA, and SFO for the varying character length of the plain text."

"The analysis proves that the proposed model outperforms other models for security MANET that is applicable for IoT. Here, the overall statistical analysis is done on the proposed SA-SFO and other comparative models for varying length of the plain text."

"For the character length of the plain text as 100, the proposed SA-SFO has achieved best values as 68% higher than PSO, 50% higher than GWO, 49% higher than WOA, and 48% higher than SFO respectively."

Conclusion

"A chaotic cryptography-based privacy preservation model was developed for the purpose of improving the security in the MANET IoT. The proposed SA-SFO was developed by improving the adaptive nature of the existing SFO."

"The proposed cryptography-based privacy preservation algorithm has performed the task of encryption and decryption through the optimal key pair from the chaotic map and the proposed SA-SFO and improved the privacy of the text messages in the communication scenario."

"For the character length of the plain text as 100, the proposed SA-SFO has achieved best values as 68% higher than PSO, 50% higher than GWO, 49% higher than WOA, and 48% higher than SFO respectively."

"For the standard deviation, the proposed SA-SFO has gained values as 90% improved than PSO, 60% improved than GWO, 50% improved than WOA, and 53% improved than SFO respectively."

4.2.30 Enhanced Privacy-Preserving in Student Certificate Management in Blockchain and Interplanetary File System [261]

This is a machine-generated summary of:

Dewangan, Narendra K; Chandrakar, Preeti; Kumari, Saru; Rodrigues, Joel J.P.C.: Enhanced privacy-preserving in student certificate management in blockchain and interplanetary file system [261].

Published in: Multimedia Tools and Applications (2022).

Link to original: https://doi.org/10.1007/s11042-022-13915-8.

If you want to cite the papers, please refer to the original.

For technical reasons we could not place the page where the original quote is coming from.

Abstract-Summary

"This challenge can be solved by using blockchain-based certificate management for students and skill evidence management for employees."

"These types of certificate managements have many issues like security and privacy breaches, fear of cyber-attack, hijacking of the session, identity theft of participants, etc. Because blockchain uses cryptographic hash and digital signature for sending and verifying transactions in blockchain peer-to-peer networks, we can overcome many security and privacy difficulties."

"We proposed a system that generates the identity of students using the tokens and stores them in the interplanetary file system (IPFS)."

"The proposed system makes use of IPFS and EdDSA (Edward-curve Digital Signature Algorithm) for digital signature and verification, as well as SHA-256 for cryptographic hashing."

"We compare our proposed system with the previously developed systems in terms of privacy, transaction cost, large file storage, blockchain implementation and registration cost."

Introduction

"In the traditional record system of student and employee record maintenance, both the studies and skill parts are different and not related."

"In the new-generation system of immutable record, a single person has the same identity for both student and employee life."

"Transferring student data to the employee and verifying the credits earned by the student for employment should be immutable and digitally signed by the institutes."

"Blockchain provides an immutable, secure, non-repudiated distributed ledger, which can serve as a record maintainer for the students and employees during the course and while switching from student life to employee."

"For students and employees, the hard copied records can be stored in the IPFS."

"Implement a public blockchain to store students records and share them with institutes."

"Upgrade the student data on skill change and employment acceptance by adding the employed to the blockchain."

"Maintain privacy and security of student/employee data during all these phases."

Literature review

"Lam and others [170] proposed a hyperledger based proof-of-concept blockchain for the credit management of student in the blockchain."

"Deenmahomed and others [171] proposed design and implementation of a blockchain-based examination, transcript, and certificate system."

"Krejci and others [172–174] proposed ethereum based blockchain to provide security and privacy of participants data with the IPFS and smart contracts."

"Mishra and others [175] proposed an ethereum based blockchain for the storage of students credentials in the blockchain and IPFS."

"For the secure sharing of student credentials, Mishra and others [175] developed a blockchain-based architecture that is tamper-proof, immutable, authentic, non-repudiable, privacy-protected, and simple to distribute."

"Vidal and others [176] proposed customized blockchain for student certificate storage and verification through blockchain and IPFS."

"Liu and others [177] proposed designing a secure authentication protocol for privacy data in blockchains without verifying the third-party participant's encrypted identity signature."

Proposed system

"The registration process is done in the proposed system using the Ed25519 generated key pair based on the student's basic information and unique identification number."

"If the authority wants to share students' certificates with all nodes, they had to store them in the on-chain IPFS."

"Off-chain IPFS used to store only non-sharable information selected by the students, higher education authority and employer."

"Whenever the education institute generated a certificate for the student S, the student can store that certificate into the off-chain IPFS."

"This transaction includes the student's information as in public key form and transaction hash, data as the hash id of IPFS file which is now on-chain before sharing in a transaction."

"The procedure can be expressed as, certificate Ct_i for student S_j is generated by the educational authority He_k, and stored in on-chain IPFS, before storing to IPFS certificate must not have any personally identifiable information within."

System implementation

"Using the data provided in a portal, such as a name, address, gender, and date of birth, a public address and private key are generated for each node."

"When nodes perform transactions, data is sent with the signature of a private key and the public key, and also other data."

"Ed25519, based on 256-bit private and public keys, is generated to register a node as an educational institute/employer."

"The public key of the node is hashed with the registration authority's Ed25519-based generated public key."

"When nodes conduct transactions, data is sent with the signature of the private key and the public key, as well as other data."

"The data from the student is included in the transaction and sent to the requested nodes."

"A newly generated hash id with on-chain IPFS is available for transaction with the employer and other requester nodes."

Security analysis and privacy auditing

"Fake students: If a student's private key is revealed and the attacker knows the public key of the higher education authority, the attacker can generate the student's registration identity."

"Man-in-the-middle attack: In this attack,the adversary has the ability to listen in on student data transactions between students and higher education authorities and disclose it."

"The sharable data of students already exists in on-chain IPFS, and nodes requesting the student's data can obtain the certificates and verify this by verifying the signature in the certificates done by the authority and student."

"Such an attempt is futile because each transaction in a proposed blockchain is compressed by a secure Hash function, such as SHA-256, then signed by the sender using a secure signature algorithm, such as Ed25519, and finally sent to the entire network for verification and approval."

"This UIN is hashed with SHA-256 and used Ed25519 to generate a key pair for the students."

Results and comparison

"Parameters used for the comparison of our proposed system with the previously developed systems are the privacy of the student information, time required for the transactions, security of the shared information, availability of the information, verification of the data and integrity of the shared information."

"The comparison shows that previously developed systems [176–181] and proposed system are equal in terms of transaction bits."

"The proposed system and previously developed systems are taking less time for transactions."

"There are nearly about 17 transactions per second for the proposed system."

"We can see that the proposed system taking 60 ms for a single transaction, but the previously developed systems [177, 182, 183] are taking more time."

"The digital signature is used in the previously developed systems [177, 182, 183] and in the proposed system to authenticate transactions."

Conclusion and future scopes

"As new attributes for user authentication and authorization become more prevalent in modern digital services, traditional identity management systems lose their efficacy and practicality in compiling data privacy regulations and assisting users in creating secure identities."

"We proposed a blockchain-based secure, students privacy-preserving certificate sharing and employment conversion system."

"The use of private blockchains and IPFS enabled the safe and trustworthy management of identities and storage of files in a decentralized manner."

"The experiments illustrated that service providers could request a registry of nodes within the blockchain, and using blockchain transactions, and users can manage and trace their files."

"Our proposed system provides a secure data sharing process with IPFS storage of transactions."

4.2.31 Privacy Enhancing Techniques in the Internet of Things Using Data Anonymisation [262]

This is a machine-generated summary of:

Ren, Wang; Tong, Xin; Du, Jing; Wang, Na; Li, Shancang; Min, Geyong; Zhao, Zhiwei: Privacy Enhancing Techniques in the Internet of Things Using Data Anonymisation [262].

Published in: Information Systems Frontiers (2021).

Link to original: https://doi.org/10.1007/s10796-021-10116-w.

If you want to cite the papers, please refer to the original.

For technical reasons we could not place the page where the original quote is coming from.

Abstract-Summary

"Preserving the privacy in IoT and Industrial 4.0 against re-identification attacks is very challenging."

"We first proposed a stream data anonymisation method based on k-anonymity for data collected by IoT devices; and then privacy enhancing techniques for both continuous data and media data were proposed for different IoT scenarios."

"The experiment results show that the proposed techniques can well preserve privacy without significantly affecting the utility of the data."

Introduction

"In IoT, the data anonymisation techniques are wildly used to protect sensitive information and privacy related to personally identifiable information by erasing or encryption identifiers that connect an individual to stored data."

"In an IoT environment, massive number of smart devices are connected that are used to gather data, this comes with security and privacy risks and the sensitive data breach might can be caused by malicious behaviours or careless."

"Many exciting applications in IoT, such as Facebook, Twitters, Tictok, Skyscanner, Hungryhouse, Trainline, Smart home, etc., collects key information such as user's name, address, credit card, timeline, behaviour patterns of using apps, which may be re-associated with the data at a later time to identify personally identifying information."

"Multiple modal based data anonymisation techniques for continuous dataset and media data (image and video) were proposed in IoT scenarios."

Related Works

"To preserve the privacy in the complicated IoT environments, we need to develop intelligent and automatic data anonymisation techniques that can hide key data attributes and potential patterns in raw data."

"In Pervaiz and others [235], a data stream scheme was proposed for privacy protection in access control, in which k-anonymity or l-diversity is used to generalise stream data."

"Khavkin and others proposed a stream anonymisation scheme based on microaggregation against differential privacy attacks, which includes an algorithm that satisfies k-anonymity and recursive (c, l)-diversity aiming at minimising information lost and reducing data disclosure risks [236, 237]."

"In many IoT scenarios, anonymising personal data is not enough to protect privacy because heavily incomplete dataset will increase the risk to re-identify a specific individual."

"This work will focus on the stream data anonymisation in IoT environments by using existing k-anonymity based techniques."

Data Anonymisation in IoT

"The data generated by sensors in IoT will be aggregated into data stream, which needs to be analysed to identify sensitive information, and then the data will be anonymised using secure anonymistion algorithms before sending to apps or release to public."

"When receives data, the IoT devices needs ζ windows to process."

"Data Streaming DS = {pid,X_t} is an IoT data steam generated by smart sensor, in which pid denotes the identifier, and X_t denotes dataset that include both quasi-identifier attribute set and other attribute set, and t is the time window."

"An IoT system possesses data anonymisation using an anonymise engine to define data anonymisation parameter, which hosts the anonymise algorithms with the parameters to perform the data anonymisation processes."

"Streaming IoT data often involves a large amount of data and a number of devices, which needs to be carefully processed."

Evaluation

"We will evaluate the performance of dynamic k-anonymity algorithms over a context database."

"We use the general k-anonymity, l-closeness anonymisation and the result as follows."

"The noise for different direction were derived based on the walking pattern."

"We first use the VOC and car reg-number dataset trained a model that can be used in the system for recognising vehicle number, pedestrian, etc. The Yolov3 darknet can conduct object detection."

Conclusion and Discussion

"The IoT systems are increasingly gaining importance in both daily life and industrial applications, which create substantial opportunities for users together with huge volume of variety data."

"This work introduced a data anonymisation schemes that can conduct data stream anonymisation before store or share data to other systems or organisation without leaking user privacy or other confidentiality."

"The experimental shows the proposed scheme can effectively anonymise data stream in IoT."

4.2.32 Addressing Privacy Requirements in System Design: The PriS Method [263]

This is a machine-generated summary of:

Kalloniatis, Christos; Kavakli, Evangelia; Gritzalis, Stefanos: Addressing privacy requirements in system design: the PriS method [263].

Published in: Requirements Engineering (2008).

Link to original: https://doi.org/10.1007/s00766-008-0067-3.

If you want to cite the papers, please refer to the original.

For technical reasons we could not place the page where the original quote is coming from.

Abstract-Summary

"Trusting software depends on various elements, one of which is the protection of user privacy."

"Protecting privacy is about complying with user's desires when it comes to handling personal information."

"Users' privacy can also be defined as the right to determine when, how and to what extend information about them is communicated to others."

"This paper describes PriS, a security requirements engineering method, which incorporates privacy requirements early in the system development process."

"PriS considers privacy requirements as organisational goals that need to be satisfied and adopts the use of privacy-process patterns as a way to: (1) describe the effect of privacy requirements on business processes; and (2) facilitate the identification of the system architecture that best supports the privacy-related business processes."

Introduction

"The extended use of various software applications in the context of basic e-services sets additional technology-related requirements for protecting the electronic privacy of individuals."

"Research efforts aiming to the protection of user privacy fall in two main categories: security-oriented requirement engineering methodologies and privacy enhancing technologies."

"The former focus on methods and techniques for considering security issues (including privacy) during the early stages of system development and the latter describe technological solutions for assuring user privacy during system implementation."

"PriS models privacy requirements in terms of organisational goals and uses the concept of privacy-process pattern for describing the impact of privacy goals onto the organisational processes and the associated software systems supporting these processes."

"Formal PriS provides a formal definition of the Pris way-of-working, i.e., it formally defines the processes of: (1) analysing the impact of privacy requirement(s) on organisational goals, subgoals and process and (2) suggesting of appropriate sysstem implementation technique(s) for realising these requirements."

The e-voting case

"PriS method is demonstrated through an e-voting case study, regarding the transformation of an Internet based electronic voting system in order to accommodate the new legal framework regarding privacy protection."

"According to this description, the main objective of the e-voting system is to provide eligible citizens the right to cast a vote over the Internet rather than visiting an election district, aiming to simplify the election processes thus increasing the degree of citizens' participation during elections."

"Generality implies that all citizens above a certain age should have the right to participate in the election process."

"PriS was applied by two teams of postgraduate students of the University of the Aegean that worked in parallel in order to: 1."

"to analyse the impact of privacy issues on the system's goals and processes and propose alternative system implementations (first team), 2."

The PriS method

"If not, a goal modelling method should be used for constructing the goal model prior to PriS's application [238]. (b) Analyse the impact of privacy goals on organisational processes The second step is to analyse the impact of privacy goals on processes and related support systems."

"Repeating this process for every privacy goal and its associated organisational goals leads to the identification of alternative ways for resolving privacy requirements."

"Instead of prescribing a single solution, PriS identifies a number of implementation techniques that best support the realisation of each privacy-related process letting the developer decide which architecture is best for the developing system based on organisation's priorities such as, cost, system efficiency, implementation complexity, etc. In the e-voting case for the realisation of process 'P3: Send Authentication Means to eligible voters' the developer may choose the Tor architecture which satisfies the unlinkability pattern."

Formal PriS

"The link between the operationalised goal G1.1.3.2 and the process P3 will be expressed as follows: Next we must identify which privacy-process patterns need to be applied not only for modelling these processes but also for relating them with the proper implementation techniques."

"Every process is assigned a number of process patterns variables, corresponding to the privacy goals affecting the process."

"Before the selection of the proper process patterns that will be applied on the specific process, PriS identifies the maximum values between every privacy requirement variable of each subgoal and creates a virtual goal G' that contains all seven maximum values."

"The selection of the proper process patterns will be based on the seven values of the G' goal."

Discussion

"The majority of the proposed methodologies (with the exception of GBRAM) focus on the elicitation of security requirements from business goals but neither do address how these requirements are translated into system components, nor do they suggest any relevant implementation techniques."

"While STRAP successfully combines goal-oriented analysis and heuristic-based frameworks for addressing privacy vulnerabilities, it does not take the next step of discovering/suggesting the relevant implementation techniques needed for eliminating these vulnerabilities."

"Alongside the research on requirements engineering methodologies, a number of technological solutions (architectures, tools and protocols) have been designed for protecting user's privacy."

"PriS deals with implementation techniques as solutions that either, satisfy the realisation of a privacy goal, or not."

"Defining a number of context-dependent selection criteria, as well as classifying implementation technologies using fuzzy logic techniques in order to calculate the degree to which each technology contributes to the satisfaction of a particular privacy requirement is a possible solution and is the subject of our current research efforts."

Conclusions

"PriS considers privacy requirements as business goals in the organisation domain and provides a methodological framework for analysing the effect of privacy requirements onto the organisational processes using specific privacy-process patterns."

"Using these patterns, PriS accelerates the modelling of privacy-related business processes indicating where the application of privacy implementation techniques is needed, also suggesting a list of specific implementation techniques that can realise each privacy-related process."

"PriS provides an integrated way-of-working from high-level organisational needs to the IT systems that realise them."

"It provides a set of expressions based on which the whole process, from the goal level to the selection of the appropriate implementation techniques, is accomplished in a more directed way."

"Future work includes the development of a case tool that will automatically identify the impact of privacy goal in the goal-process structure, based on PriS formal definition."

4.2.33 Implications of an Ethic of Privacy for Human-Centred Systems Engineering [264]

This is a machine-generated summary of:

Carew, Peter J.; Stapleton, Larry; Byrne, Gabriel J.: Implications of an ethic of privacy for human-centred systems engineering [264].

Published in: AI & SOCIETY (2007).

Link to original: https://doi.org/10.1007/s00146-007-0149-7.

If you want to cite the papers, please refer to the original.

For technical reasons we could not place the page where the original quote is coming from.

Abstract-Summary

"Given its complicity, there is a moral obligation to redress privacy issues in systems engineering practice itself."

"This paper investigates the role the concept of privacy plays in contemporary systems engineering practice."

"Ontologically a nominalist human concept, privacy is considered from an appropriate engineering perspective: human-centred design."

"Two human-centred design standards are selected as exemplars of best practice, and are analysed using an existing multi-dimensional privacy model."

Introduction

"Given the direct and indubitable role of contemporary technological systems in constituting many intractable privacy issues, there is a moral imperative for the matter to be revisited and redressed at source: the systems engineering process itself."

"This paper considers the current role the concept of privacy plays in modern systems engineering practice."

"The relevance of privacy for artificial intelligence applications is then examined, and a rationale for incorporating the concept in contemporary systems engineering forwarded."

"Recommendations are made to improve the incorporation of privacy in future systems engineering practice."

Research objectives and research questions

"The objective of this paper is to investigate the current role of the concept of privacy in systems engineering best practice as laid out in the human-centred design standards."

"Does best practice for systems engineering adequately incorporate the concept of privacy?"

"How can the incorporation of the concept of privacy be improved in systems engineering practice?"

Antecedents of privacy

"The information systems literature has long recognised the threat posed to individual privacy by technological systems."

"There is a plausible argument therefore, that violations of the concept of privacy occur more frequently in this domain than in pure information systems engineering, where data issues arise."

"Privacy is an important and rational human concept for, and merits a discernable role in, contemporary systems engineering practice."

"In terms of systems engineering, privacy is seemingly most compatible with socio-technical approaches that incorporate human aspects, and "high road" organisational approaches that embrace the human potential in the wider system."

"One well-established tradition that merits particular attention when considering privacy in systems engineering is human-centred design."

"The human-centred design and the philosophy of privacy approaches recognise the inherent value in allowing people to realise intrinsic and personal needs."

"These privacy guidelines remain influential, and offer explicit guidance on data practices for information systems engineering."

Case study: human-centred design standards

"A case study of two ISO human-centred design standards is presented in order to address the question "Does best practice for systems engineering adequately incorporate the concept of privacy?""

"The international human-centred design standards adequately incorporate the concept of privacy."

"This case study refers to two representative standards which have been chosen for systematic and detailed analysis, namely: (A) ISO 13407 (1999)—Human centred design processes for interactive systems (B) ISO TR 18529 (2000)—Human centred lifecycle process descriptions The ISO 13407 standard describes how a human-centred design process can be used to achieve usable systems by providing a supplementary framework to existing lifecycle models."

"The human-centred design processes described by the standards may, therefore, potentially be deemed intrusive and a violation of privacy by those people affected."

"Neither the standards nor human-centred design itself actually preclude the incorporation of privacy issues in systems engineering."

Findings

"The international human-centred design standards do not adequately incorporate the concept of privacy."

"Although the standards engage with many factors of the privacy model, many others factors are overlooked or underrepresented."

"The human-centred standards, therefore, seem to describe systems practice that is most conductive to the concept of privacy."

"It is reasonable and fair to conclude that privacy is not explicitly addressed and is, therefore, currently underrepresented in the human-centred design standards."

Implications

"The analysis presented by this paper demonstrates that the human-centred standards are currently inadequate in addressing issues of privacy in systems engineering and an important question is research question 2, namely: "How can the incorporation of the concept of privacy be improved in systems engineering practice?""

"Because of the individual influence potentially exerted by systems engineers during the systems engineering process, privacy may be most effectively incorporated through a suitable mode of professional ethics."

"By adopting professional ethics, and taking a moral stance on privacy issues, for instance, professional engineers could help curtail the development of intrusive and unethical systems and practices."

"The privacy values held by individual engineers may affect how privacy is incorporated in a given system's design process."

"Since systems engineers use their personal value systems in decision-making [239, 240], it would be worth understanding their attitudes and perspectives on issues of privacy."

Conclusions

"Philosophically, human-centred design demonstrates exceptional resonance with the concept of privacy, and evidently more so than any other systems methodology or approach."

"Taking the international human-centred standards as indicative of best practice for systems engineering, the concept of privacy is currently underrepresented therein."

"In spite of this finding, both the human-centred tradition and the international standards represent practice that is innately conducive to incorporating privacy in systems engineering."

"Given the degree of personal influence individual systems engineers can potentially exert over the process and the consequent system, it may be possible for human issues such as privacy to be effectively promoted and incorporated at this individual level."

"In intelligent systems development, therefore, it is ultimately left to the truly human-centred systems engineer to take personal responsibility and care for the privacy of others."

4.2.34 Privacy-Preserving Process Mining [265]

This is a machine-generated summary of:

Mannhardt, Felix; Koschmider, Agnes; Baracaldo, Nathalie; Weidlich, Matthias; Michael, Judith: Privacy-Preserving Process Mining [265].

Published in: Business & Information Systems Engineering (2019).

Link to original: https://doi.org/10.1007/s12599-019-00613-3.

If you want to cite the papers, please refer to the original.

For technical reasons we could not place the page where the original quote is coming from.

Abstract-Summary

"Since such data, typically available in the form of event log files, contains personalized information on the specific processed entities, it can expose sensitive information that may be traced back to individuals."

"A plethora of methods have been developed to analyse event logs under the umbrella of process mining."

"This paper set out to develop a protection model for event data privacy which applies the well-established notion of differential privacy."

"Starting from common assumptions about the event logs used in process mining, this paper presents potential privacy leakages and means to protect against them."

"The paper also shows at which stages of privacy leakages a protection model for event logs should be used."

"Relying on this understanding, the notion of differential privacy for process discovery methods is instantiated, i.e., algorithms that aim at the construction of a process model from an event log."

Introduction

"The aim of this paper is, therefore, to fill this gap and to provide a privacy-preserving technique for process mining which considers technological challenges."

"We aim to define a protection model for event log privacy with minimum loss of utility for process mining, i.e., process discovery remains useful while the disclosure of sensitive data is reduced For this, however, the following questions must be understood: RQ 1 At which stage of data paths is a protection model for event log privacy required?"

"RQ 2 How can event log privacy be ensured with a minimum loss of utility for process mining?"

"The next section defines the terms used as input to define a privacy-preserving technique for process mining and introduces our use case from a hospital that will be used for illustration throughout the paper."

Foundation

"Once processes conducted by an organization, e.g., the handling and treatment of patients at a hospital, are supported by modern information systems, the conduct of these processes is commonly reflected in event data."

"To formalize the above notion of event logs, we adopt a relational model of events, which is a common model in data stream processing [241]."

"That a notion of a trace has been defined and that an attribute (or an attribute combination) has been selected to signal the activities of interest, i.e., the atomic units of a work, a process model may be discovered from the ordering of the events that represents these activities."

"Event data and particularly healthcare processes have a high demand for privacy preserving process mining."

Privacy Issues for Process Mining of Healthcare Processes

"data storage personal data and events of medical staff, administrative staff and patients are stored in a database or event logs."

"data (re)use At this stage, data from event logs is used for process mining aiming to determine the main paths that are followed by patients or medical staff in the process."

"With regard to RQ 1, we consider the privacy checkpoint 4 and the stage data (re)use as points of privacy leakage for process mining."

"Although event log protection becomes relevant at the data (re)use stage, several privacy concerns must be addressed before."

"Please note that while we use a hospital use case to illustrate our protection model for event logs, there are a many similar situations in which organizations have centralised control over an event log and want to protect the privacy of individuals for which cases are processed (e.g., public administration process as the one in de Leoni and Mannhardt [242])."

Protection Model for Event Logs Based on Differential Privacy

"Differential privacy establishes a theoretical limit on the influence of a single row on a dataset (e.g., individual's data), thus limiting an attacker's ability to infer such a membership."

"Event-driven differential privacy approaches have been proposed for cases where continuous observations are produced [243]."

"Given the fact that in the use case at hand hospitals already have access to the patient's and hospital records, we assume a centralized privacy approach to realize the abstract privacy design strategy to protect the data (re)use of event data for process mining using differential privacy."

"This privacy engine resides in the trusted environment and introduces noise to each query result in order to maintain differential privacy guarantees at all times."

"This enables to safely reuse the collected data for process mining without jeopardizing the privacy of stakeholders."

Differential Privacy for Event Log Queries

"Based on these information requirements, we can identify two queries specific to process mining that the privacy engine needs to support."

"In addition to the queries, the proposed privacy engine should support filtering the event log E based on standard relational algebra operations to provide sub logs, e.g., through the WHERE construct of SQL."

"Different from Query 1, the set of all possible trace variants is infinite and cannot be fixed based on the finite set of activities known for a specific process at hand."

"Glance, it may seem that restricting the access to the event log to the two queries discussed in the previous section already protects privacy of process participants."

"We assume that the set of possible process activities N is publicly known and that we can establish an upper bound for the length of traces of the event log."

Evaluation

"As ground truth we use process models discovered in the original, unprotected event log without any privacy protection."

"As dataset, we use two publicly available event logs from the IEEE task force on process mining repository: Sepsis Cases [244] and Road Traffic Fine Management [242]."

"Whereas the quality of the discovered Sepsis Cases process model decreases, there is still little change for the Road Traffic Fines model."

"The result for the Road Traffic Fines event log is much better and also very stable across repetitions, which indicates that the noise added by our proposed method has only little influence on the quality of the discovered process model."

"To mine process models from such event logs is already a challenge for existing process mining algorithms without added noise."

"The process model starts with Create Fine and may end with either Payment or Send for Credit Collection as in the model discovered on the unprotected log."

Related Work

"In this area, several efforts were made in the last decade to integrate privacy when designing databases by using multi-level security [245] or role-based access control [246] approaches."

"Related to event log data and process mining, a large body of research exists for security-oriented analysis."

"A different event log configuration according to security concerns is suggested in Fazzinga et al. [247] who use security risk as criterion to filter related traces."

"Only the work of Mannhardt et al. [248] discusses privacy challenges for process mining, however, without suggesting any approach for event log protection."

"A privacy-preserving system design for process mining has been suggested in Michael et al. [249], which allows to specify who does what, when, why, where and how with personal data during process mining."

"Our approach could be integrated into the privacy-preserving system design as a privacy engineering technique to protect the event logs."

Conclusion

"Contribution This paper contributes a privacy-preserving technique for process mining and an approach to protect event logs."

"The privacy checkpoint between data storage and data (re)use can be considered as possible privacy leakages while conducting process mining."

"Privacy models could be used to generate an information platform by means of MDA [250] that enables end users to either (a) define privacy policies, in order to determine more precisely who can do what with which data (guided consent process) and (b) monitor compliance with them, in order to allow them to see which privacy mechanisms are provided for which process stages regarding the process mining of their data."

"This paper presents a first technical contribution for privacy-preserving process mining using a differential privacy approach and outlines a roadmap for future research on that field."

4.2.35 *PDGuard: An Architecture for the Control and Secure Processing of Personal Data [266]*

This is a machine-generated summary of:

Mitropoulos, Dimitris; Sotiropoulos, Thodoris; Koutsovasilis, Nikos; Spinellis, Diomidis: PDGuard: an architecture for the control and secure processing of personal data [266].

Published in: International Journal of Information Security (2019).

Link to original: https://doi.org/10.1007/s10207-019-00468-5.

If you want to cite the papers, please refer to the original.

For technical reasons we could not place the page where the original quote is coming from.

Abstract-Summary

"In the context of pdguard, personal data are always stored encrypted as opaque objects."

"Processing them can only be performed through the pdguard application programming interface (api), under data and action-specific authorizations supplied online by third party agents."

"Through these agents, end-users can easily and reliably authorize and audit how organizations use their personal data."

"Following a security by design approach, pdguard changes the problem of personal data management from the, apparently, intractable problem of supervising processes, operations, personnel, and a large software stack to that of auditing the applications that use the framework for compliance."

"We demonstrate the framework's applicability through a reference implementation, by building a pdguard-based e-shop, and by integrating pdguard into the Guardian newspaper's website identity application."

Introduction

"By interacting with escrow agents, data subjects can reliably authorize and audit how data controllers use their personal data."

"For data subjects not interested to setup a relationship with an external escrow agent, pdguard uses a default internal escrow agent, which offers the same functionality and implements the data controller's personal data policy."

"Following the "security by design" [185–189] approach, pdguard improves the effectiveness of personal data protection in the following ways: It decentralizes trust and control by having numerous escrow agents of (hopefully) diverse implementations control the decryption keys."

"This allows data subjects to choose among data controllers based on the concrete protection of their personal data that they offer (adoption of pdguard and actual data use), provides regulators with an easy way to establish whether personal data is effectively protected, and allows market-based mechanisms to spread adoption of competing pdguard implementations, escrow agents, and auditing services."

Background

"Data Controller Applications: The data controller's software systems that process personal data."

"Escrow Agent: A third party software application and service trusted by both the data subject and the data controller to facilitate and monitor the compliance of the data controller's processing activities according to the data subject's instructions."

"In the pdguard framework (qualified) trust is placed on the participating entities, namely the data controller, the data subject, and the escrow agent."

"The data controller's applications must: (1) handle all personal data through the pdguard api, (2) issue decryption requests with correctly specified data type and data use fields, (3) avoid storing personal data locally (in memory or on non-volatile storage) after using them, and (4) never use data or keys obtained for one purpose for a different one (e.g., forward the data subject's contact phone number to a telemarketer)."

Architecture

"The first time that the data subjects provide personal data to the data controller, they should also provide the domain name of the external escrow agent of their choice."

"These rules are provided to the escrow agent the first time that the data controller interacts with a data subject, are marked as such in the escrow's user interface, and cannot be changed thereafter."

"Data subjects can set their own authorization rules regarding the usage of their personal data through the escrow agent they choose."

"All lines represent direct server-to-server api calls that cannot be tampered with by a third party, and the parameters inside square brackets are signed using shared secrets (in the future these will use a public key infrastructure.) The first time that a data controller application requests to process personal data referring to a specific data subject, it obtains from the escrow agent a clientid, and a secret."

Reference implementation and use

"The pdguard escrow agent prototype provides an easy to use web interface for the data subjects to sign up, define or edit rules, and view authorization logs."

"Observe that a key corresponds to a data controller/data subject couple, and an authorization rule for one data type may involve zero or multiple allowable uses and provenances."

"If an application requests to update a specific data type (found on the AUTHORIZATION_RULE table), the agent will check if there is a corresponding record on the ALLOWABLE_PROVENANCE table."

"The code running on the background should first initialize a DataProtection object (e.g., dp) and then invoke the encryptData method in the following manner (assume that surname is a string variable which has Alice's surname as its value): Observe that the data provenance (DATA_SUBJECT_EXPLICIT) indicates that this piece of data comes from the data subject itself."

Evaluation

"The application processes personal data."

"Consider the case where the application attempts to access personal data and display them as part of a webpage (as is done in the EditProfileController class)."

"We measured how much time the application spends to either access or update a piece of data, with and without pdguard."

"It took the application 451 ms to access, and 655 ms to update a piece of data without pdguard."

"When an outsider performs such an attack to retrieve the personal data of multiple users (e.g., addresses), the attack will only return useless encrypted data."

"A privacy breach could occur when an attack targets the meta data stored on the escrow agent's side (e.g., check what kind of data users provide to which sites or when such data is accessed)."

Challenges

"pdguard can help detect and prevent attacks that target personal data."

"Consider a case where an application retrieves credit card data at an unprecedented rate."

"Pdguard does not allow database-layer searches over encrypted data, e.g., obtaining a list of all patients that have been admitted for a particular symptom."

"Each application performed actions directly on the fields of objects, without issuing sql queries with arguments containing personal data."

"To handle this shortcoming, pdguard could be extended to use schemes such as Cryptdb [190], an approach that supports the execution of queries over encrypted data through efficient sql-aware encryption."

Related work

"Diverse approaches have been developed to prevent unauthorized personal data processing, many with different goals and threat models than pdguard."

"In the context of the Ephemerizer, all user data are stored encrypted and the keys are managed centrally by a trusted third party."

"To do so, it stores personal data encrypted on the server, and decrypts them only in users' browsers."

"Their approach provides means for secure computations over encrypted data in a way that does not violate user-privacy."

"By extending the application on the escrow agent's part, we can provide users with means (e.g., data wallets [191]) to decide about releasing different attributes in different ways."

"In pdguard key management differs in the following manner: data types are used as attributes to generate symmetric keys for the encryption or decryption of a specific piece of data."

Concluding remarks

"Adopting a security by design approach, pdguard provides through a standardized, auditable mechanism an effective way for protecting personal data by offering to data subjects direct control and monitoring, and by reducing and strengthening the attack surface of data management systems maintained by data controllers."

"To do so, pdguard combines different concepts including applied cryptography, access control models, an api, authorization protocols, and taint tracking."

"Can pdguard actually make a difference in our everyday lives?"

"Important elements include: the design of performance optimizations at the architectural level; the development of additional auditing tools and methods; the running of large scale trials; the operation of an escrow agent in a production setting; the implementation of pdguard api libraries in more commonly used programming languages; the initial adoption by significant data controllers; the establishment of a community and a governance structure; as well as the education of developers, data controllers, and data subjects."

4.2.36 Using Homomorphic Encryption for Privacy-Preserving Clustering of Intrusion Detection Alerts [267]

This is a machine-generated summary of:

Spathoulas, Georgios; Theodoridis, Georgios; Damiris, Georgios-Paraskevas: Using homomorphic encryption for privacy-preserving clustering of intrusion detection alerts [267].

Published in: International Journal of Information Security (2020).

Link to original: https://doi.org/10.1007/s10207-020-00506-7.

If you want to cite the papers, please refer to the original.

For technical reasons we could not place the page where the original quote is coming from.

Abstract-Summary

"To detect the execution of such attacks, organizations install intrusion detection systems."

"It would be beneficial for such organizations to collaborate, to better assess the severity and the importance of each detected attack."

"A privacy-preserving collaboration system for attack detection is proposed in this paper."

"Results have shown that privacy-preserving clustering of intrusion detection alerts is feasible, with a tolerable performance overhead."

Introduction

"Misuse detection-based systems detect predefined patterns of malicious activity and then produce informative alerts."

"An interesting approach that enables intrusion detection systems to produce better results is collaborative intrusion detection [192, 193], where multiple intrusion detection systems, installed in different organizations, collaborate, to produce more accurate results."

"Independently of the exact information exchanged (this can be raw traffic data or intrusion detection alerts), a significant privacy violation risk is present, which has recently been recognized by the research community [194–196]."

"The standard way to process intrusion detection alerts produced by different organizations is to gather such alerts on a single node and commit the required processing there."

"Intrusion detection alerts reveal vulnerabilities that exist on those networks or attacks that have already happened against them."

"It enables the clustering of alerts produced by different organizations, while at the same time it protects the privacy of each organization's data."

Related work

"The requirement for moving on from single installations of intrusion detection agents to networks of collaborating agents installed in different networks has emerged in different domains such as cloud computing, multiple devices networks or detection of large-scale attacks."

"Morais and Cavalli proposed a distributed intrusion detection systems architecture for wireless mesh networks [197]."

"Large-scale attacks detection has also been proposed as a domain for collaborative intrusion detection application by Zhou [198]."

"The results obtained through experiments show that the proposed system is much more capable of single installations of intrusion detection systems in different ISPs regarding detecting flooding attacks."

"We present one of the first research efforts to develop a complete system, which will enable multiple collaborating intrusion detection systems to unify their results in a privacy-preserving way."

Concept

"The main concept in this paper is the combination of the results got locally at different organizations through their network IDS, which can produce useful information about such attacks and enable the faster and more effective application of mitigation measures."

"By clustering of alerts produced by different organizations, the produced result will be clusters that comprise similar alerts produced by resembling activity in different networks."

"One of the two that is external to the network of the organization (IP of the host potentially committing the attack) is used as the first alert feature."

"Upon this distance metric, the proposed method creates clusters of similar alerts that can reveal information about attacks that affect over one organization at the same time."

"The process comprises an initial phase which is the calculation of the encrypted distance between every pair of alerts and an iterative execution of the K-medoids clustering algorithm."

Prerequisites

"The main difference between the two algorithms is that K-medoids selects existing data points (medoids) as clusters' centers, while k-means calculates an optimal center point for each cluster."

"The encryption function is additionally homomorphic, and the following three properties enable partial processing of encrypted data: Homomorphic addition of encrypted values It is possible to add two encrypted numbers, by calculating the product of the two corresponding ciphertexts."

"Paillier algorithm is partially homomorphic as given two ciphertexts, there is no way to calculate the encryption of the product of the corresponding plaintexts, without knowing the private key."

"We have chosen the Paillier encryption algorithm for the system proposed in the present paper, as it enables for calculations required for alerts clustering to be done while the alerts' data remain in encrypted form."

System design

"The clients and the server commit the clustering for the alerts produced by the local IDSs for a given time window."

"Throughout this procedure, the clients support the server, as some of the required operations are not feasible on the encrypted alert data."

"Clients send to the server the list of encrypted features of alerts, produced locally, along with the two-dimensional matrix (upper triangle) that contains the encrypted distances between these alerts."

"At the end of each time interval, the server waits for all clients to send their input, four encrypted features of their alerts, along with the encrypted distances of all pairs formed between their alerts."

"As soon as the server receives all inputs, it calculates the distances between alerts of different clients."

"The server is required to calculate the distances for all pairs of alerts sent from clients."

Implementation

"An element in this matrix is equal to one, if the alert that corresponds to the row and the alert that corresponds to the column of the element belong to the same cluster."

"An element in this matrix is equal to the distance of the alerts that correspond to the row and the column of the element if these two alerts belong to the same cluster."

"No auxiliary client processing is required at this step: Every row of the resulting DFC matrix corresponds to one alert and holds k values."

"DG matrix For the next step, the server has to calculate DG matrix that holds the distances of alerts from all other alerts that belong to the same cluster."

"For the multiplication of the matrices, the server needs to use the clients' multiplication auxiliary processing: The resulting matrix CCM holds one column for each one of the k clusters."

Experiments

"An intrusion detection-related traffic dataset that would refer to multiple different networks was required to test the proposed system."

"The only thing that the processing client can probably do is recognizing that two different requests (in two different K-medoids rounds) correspond to the same alert or to the same cluster, as in that case many of the elements should remain the same."

"That information might give knowledge to the processing client about what percentage of clusters has changed their centers (along with the set of alerts they comprise) between the two rounds."

"The probability that a client gets a vector for the same alert/cluster over one times is relatively limited as the required number of rounds for K-medoids to converge is 3–5."

"In a setup with six clients, a client may never process vectors for the same alert/cluster for the second time."

Conclusions and future work

"The system proposed enables privacy-preserving alert clustering between multiple organizations."

"Such alerts post-processing may reveal important information about cyber-attacks being conducted and set the organizations able to efficiently protect their networks."

"To comply with the real-world requirements, the proposed system protects the privacy of the participating organizations."

"It has been proven that it is possible to conduct privacy-preserving alerts clustering between different organizations."

"These organizations are willing to invest in the required hardware, according to their needs, it is possible to make this collaboration scheme effective and able to process alerts without inducing any significant overhead."

"A requirement for the system to function properly is that all participants, the server and the clients, operate according to the protocol."

"This would have a positive effect on both performance and privacy protection."

4.2.37 Privacy-Enhanced BPMN: Enabling Data Privacy Analysis in Business Processes Models [268]

This is a machine-generated summary of:

Pullonen, Pille; Tom, Jake; Matulevičius, Raimundas; Toots, Aivo: Privacy-enhanced BPMN: enabling data privacy analysis in business processes models [268].

Published in: Software and Systems Modeling (2019).

Link to original: https://doi.org/10.1007/s10270-01900718-z.

If you want to cite the papers, please refer to the original.

For technical reasons we could not place the page where the original quote is coming from.

Abstract-Summary

"Although business process model and notation (BPMN) is well suited for expressing stakeholder collaboration and business processes support by technical solutions, little is done to depict and analyze the flow of private information and its technical safeguards as it is disclosed to process participants."

"This gap motivates the development of privacy-enhanced BPMN (PE-BPMN)—a BPMN language for capturing PET-related activities in order to study the flow of private information and ease the communication of privacy concerns and requirements among stakeholders."

Introduction

"An alternate approach to privacy analysis using business process modeling through business process model and notation (BPMN) could give us an added means of ensuring minimal privacy leakage while optimizing technological or cost overheads."

"Capturing these privacy characteristics within the visual notation of BPMN can aid communication between process stakeholders and organizational data protection officers."

"We primarily focus on the disclosure of private information in the honest-but-curious adversary cases expressed using business process model and notation (BPMN 2.0) [199, 200]."

"We consider the research question on how BPMN can enable the visualization, analysis and communication of the privacy characteristics of business processes."

"Based on the PET classification in [201], we propose a multi-leveled model of PET abstraction meant to be used with privacy-enhanced BPMN (PE-BPMN)—an extension of the BPMN with privacy-enhancing technologies and discuss information disclosure analysis possibilities."

Background

"BPMN is fairly robust and includes notations to indicate decisional paths (gateways), recurring events, timed events, tasks of different kinds (e.g., user tasks and script tasks), data objects, databases and even processes within processes (subprocesses)."

"We describe our adaptation of the existing PET classifications that we found most suitable for our goal-oriented modeling of privacy in processes."

"The same PET could appear in different categories; for example, encryption is used for data protection and secure communication."

"The computation on protected inputs can be divided to distributed and single party techniques."

"Security means that the protected contents (e.g., using end-to-end encryption, TLS, etc.) can travel without external parties reading or modifying them."

"Computations on confidential inputs allow one to securely process various operations without removing the protection mechanisms."

Extending BPMN to support PETs

"The taxonomy provides us with a link between the concrete technologies we wish to implement in our models and the existing BPMN syntax."

"In this instance, SecureChannel is a generic stereotype of the Security target."

"At the lowest level, we describe actual PET protocols (or their stages) that achieve targets bearing a particular generic stereotype."

"Current standards for secure channel communication include transport layer security (TLS) and end-to-end encryption (E2EE) among others; hence, these are the concrete stereotypes corresponding to the generic SecureChannel stereotype."

"The key reason we separate the taxonomy elements (goals and targets) from the stereotypes (generic and concrete) is that while the taxonomy describes desirable outcomes at a rather high level, the stereotypes offer representation at the level of execution."

"The BPMN Task is extended with a new PET Task to become the parent of the remaining goals and targets in the taxonomy."

Modeling privacy with PE-BPMN

"We go further into each PET that was included in the abstract syntax and illustrate the usage of generic and concrete stereotypes in the metamodel."

"We use the aid of a hypothetical scenario to illustrate how we move in iterations from a standard process model, to one with generic stereotypes and finally its variations in concrete stereotypes."

"With this as a foundation, we are in a position to model and eventually analyze the privacy guarantees and trade-offs that different PETs provide from the perspective of generic and concrete stereotypes."

"To ensure this, Party 1 applies the generic stereotypes related to Data Protection (i.e., ProtectConfidentiality and OpenConfidentiality)."

"Such a generic-stereotyped model provides a basis for PET selection in the next step where the selected PETs are then modeled and their trade-offs analyzed."

Modeling and comparing concrete technologies

"That we only include the inputs and outputs that explicitly appear on the PE-BPMN model in this table, other parameters such as script or access specification are defined as attributes of the concrete stereotype."

"The main difference is the added key input that forces an additional restriction that PKDecryption is valid only if it uses the private key from the key pair with the public key used to encrypt the data."

"To capture that, we define SSComputation tasks to belong to groups where the script of the computations is the same for the whole group and the output shares define one secret shared value."

"SGXProtect protects its input data for the use of one SGX enclave where the access specification defines the enclave."

"To define an enclave we group together the computation tasks with SGXComputation stereotype that are one lane of the process that can use the content protected by concrete SGXProtect task."

Information disclosure analysis

"We propose information disclosure analysis to gain an understanding of communications between pools, their associated data objects and even inter-dependencies between inputs and outputs of activities."

"A Visibility Matrix gives an overview of the data objects that each actor possesses at some point along the process."

"Accessible (A) indicates that a data object is owned or obtained by an actor at some point but it is protected."

"Hidden (H) tells us that a data object is owned or obtained by an actor at some point but its contents are unreadable as it is protected by some PET mechanism (e.g., encryption) and the party does not meet the access specification requirements to recover the protected data."

"For such a data object, we should consider whether it has any other means of protection, e.g., encryption that reduces the disclosure."

Applying PE-BPMN

"The ID of privacy-enhancing activity is prefixed with P and regular activities are prefixed with A. Data objects are assigned IDs in a similar format with the prefix

D. The numbers are not always sequential, but the same ID refers to a conceptually same task in all versions of the processes that we consider."

"The Android OS is responsible for preprocessing and submitting locations (data D2) to the Compute server (task A2)."

"The Compute server processes and updates (task P3) its heatmap information (data D6) to produce an updated heatmap (data D5)."

"Once the request for the movement heatmap is submitted to the Compute server, the updated heatmap (data D5) is used to generate (task P4) the heatmap defined by the query parameters (data D9)."

"The remaining choice about how to proceed from D8 on the Command center side is again context specific: either protected processing or making the data public."

Prototype

"Number of inputs and outputs within stereotype group is as required."

"The latter is required because some of the group based or interdependent stereotype restrictions are hard to check when all stereotypes are not yet added and also to find cases where the model has been modified through other means than the editor for PE-BPMN."

"The roles might be fixed correctly when adding the PKEncryption stereotype but the validation discovers if the key or data object has been deleted from the model afterward."

"Example where encrypted data is secret shared, there would be two sequential tasks with PKEncryption and SSSharing stereotypes, respectively, and one group of SSComputation tasks where computation is specified by PKDecryption stereotype."

"Mainly, we expect that each protection stereotype takes unprotected data as an input."

Related work

"These studies report on the language extensions in the related domain, but do not directly capture security and privacy requirements modeling."

"We will position PE-BPMN with respect to the security requirements modeling, compliance policy management, privacy requirements analysis and data leakage analysis within the extensions of the business process modeling."

"In PE-BPMN we also address similar concerns (i.e., authentication, confidentiality, identification, integrity and secure communication); however, our major emphasis is on the modeling of the privacy requirements and specifically how these requirements could be implemented using PETs."

"In PE-BPMN we do not directly consider compliance or policy management aspects; however, our extensions suggest means (wrt authorization, confidentiality, data sharing, identification, integrity and secure communication) that could be used to define such policies regarding data leakage through the business process [?]."

"Privacy requirements analysis The above studies introduce security extensions to BPMN; however, two studies [202, 203] introduce security related constructs and visual annotation."

Conclusion and future work

"We are continuously expanding the concrete syntax of PE-BPMN with new PETs, especially ones that increase different process modeling and privacy analysis opportunities."

"We have observed that using PE-BPMN requires a different focus from the analyst: (i) analysis requires more details of the data objects that need to be explicit, and (ii) fine-grained separation of stakeholders yields better analysis."

"It is likely that such separation of tasks is clear before adapting sophisticated business processes to this technology and then the corresponding stereotypes can be added to PE-BPMN."

"The core idea of PE-BPMN can still be adapted to allow more levels of precision for the stereotypes to support more technologies or new means of analysis."

"Their analysis can easily be implemented based on PE-BPMN with differential privacy-specific stereotype."

"With sufficient documentation of the used PETs we can have an analysis that lists all the underlying assumptions that need to hold for the technologies to preserve the privacy that they promise."

4.2.38 P-STORE: Extension of STORE Methodology to Elicit Privacy Requirements [269]

This is a machine-generated summary of:

Ansari, Md Tarique Jamal; Baz, Abdullah; Alhakami, Hosam; Alhakami, Wajdi; Kumar, Rajeev; Khan, Raees Ahmad: P-STORE: Extension of STORE Methodology to Elicit Privacy Requirements [269].

Published in: Arabian Journal for Science and Engineering (2021).

Link to original: https://doi.org/10.1007/s13369-021-05476-z.

If you want to cite the papers, please refer to the original.

For technical reasons we could not place the page where the original quote is coming from.

Abstract-Summary

"This research study endeavors to integrate the standards of data protection along with the Security Threat Oriented Requirements Engineering (STORE) methodology in order to recognize the potential threats to privacy requirements."

"We have used the integrated fuzzy AHP with fuzzy TOPSIS technique for the usability assessment of different privacy requirements engineering approaches including the P-STORE methodology."

"The study demonstrates that the P-STORE approach has the capability to elicit more efficient privacy requirements and that it allows the software engineer to arrange privacy requirements efficaciously."

Introduction

"The web applications as well as different software are based upon several Information Technology (IT)-enabled services of computers and networks to store, manage, transfer, and upgrade data across multiple sectors."

"Maintaining software privacy along with preserving data security continues to be a challenge for software professionals as information systems become more complex and dispersed."

"Against this backdrop, this research paper has provided a preliminary extension of the STORE methodology [204] called the P-STORE to address privacy requirements for software system."

"The P-STORE methodology is anticipated to have the capability to elicit the effective and well-organized privacy requirements, which is very important for any security-critical software product."

"A case study of HMS system that has been alluded to demonstrates the provision of the P-STORE methodology for privacy engineering."

"Full procedural processes of P-STORE methodology were performed to elicit complete as well as well-organized privacy requirements toward trustworthy HMS system development."

Literature Review

"Baldassarre and others [205] proposed a system called Privacy Oriented Software Development (POSD) that can technically support application development by integrating privacy protection and security requirements."

"The framework was derived from the existing requirements engineering research for comprehensive examination of privacy potential threats in design as well as recursive preferences adjustments."

"Mai and others [206] implemented an existing methodology to design security and privacy requirements."

"They proposed a new extension to SQUARE's approach and platform to incorporate privacy concerns, and recognize potential possibilities for a more unified security and privacy requirements' development framework."

"He & Antón [207] presented a structured method in the role-engineering procedures by modeling the privacy requirements."

"Meis [208] presented the Problem-based Privacy Analysis (ProPAn) for engineering privacy requirements which is an aspect-oriented requirements engineering methodology."

Overview of STORE Methodology

"The Security Threat Oriented Requirements Engineering (STORE) is a technique focused on security threats that can be used by a software development enterprise to develop effective and efficient security specifications."

"The STORE process is a ten-step operation for software product security specifications engineering."

"With the support of potential stakeholders, the STORE methodology addresses security threats for determining security requirements."

"The STORE methodology is entirely constructed on software security threats since the requirement engineer is able to produce security specifications with the aid of a threat dictionary after recognizing the possible threats."

"The STORE methodology is capable of eliciting only security requirements."

"We have extended the STORE process as P-STORE methodology for incorporating the effective and efficient privacy requirements through a sequential procedure."

STORE for Privacy Requirements

"STORE methodology offers an appropriate and a comprehensive way of obtaining and capturing security requirements from the beginning of the software as well as web-based applications development process."

"It is hard to determine the privacy risks to a software product [209], thus resulting in poor structured strategies for identifying actors with privacy requirements within a system."

"When a software product is to be developed, only a few of the stakeholders seem to be highly concerned about the software product's privacy requirements, whereas most of the stakeholders are apprehensive about the software's security and are also aware of the specific assets in the system which act as safeguards against the attacks."

"The proposed approach discusses the risks to privacy with the aid of potential stakeholders to identify privacy requirements."

"This category of stakeholders facilitates the requirement engineer in software product asset recognition."

P-STORE Methodology

"The only contradiction is that, throughout this case, the software engineering structure of the requirements as well as the stakeholders prepare and prioritize the objectives of privacy instead of the objectives of security."

"For the development, the engineers and stakeholders must adhere to the corresponding software's privacy goals."

"Analyzing any possible privacy infringement for a software application project is an essential step in introducing safe security systems as anticipated privacy requirements would be derived from the identified software privacy attacks."

"A software application may have many dependency points, such as the remote server over which the software system depends, the database server, internet connection, etc. After recognizing the all possible PoA, PoB, PoC, and PoD with the help of the privacy attack analysis in the preceding stage of the P-STORE methodology, it would be easy to determine the privacy risks, the next stage of P-STORE."

Application of P-STORE Methodology for HMS Project

"Prior to the actual production of this software application, the proposed extension of STORE process, P-STORE, was implemented onto the HMS application so that the requirement engineer may obtain complete privacy specifications considered necessary for the development of the software application."

"The case study on an HMS application discussed in this section clearly demonstrates how well the P-STORE methodology elicits efficient privacy requirements."

"After these stakeholders associated with the HMS application project have been recognized, the subsequent stage of the proposed P-STORE privacy requirements engineering approach is to come to an agreement with all the stakeholders upon this planned HMS application's privacy objectives."

"This stage is associated with the verification and validation of the obtained privacy requirements for the HMS application development through the P-STORE methodology."

"The last as well as a crucial phase of the proposed P-STORE methodology is the organized creation of the HMS system's complete and accurate software privacy requirement specification documentation."

Usability Analysis of P-STORE Methodology

"The objective of the survey was to assess the usability impact of different privacy requirements engineering approaches with the functionality of the proposed fuzzy AHP-TOPSIS method."

"Several studies have applied integrated fuzzy AHP with fuzzy TOPSIS approach to assess usability analysis for different situations and problems."

"Sharawat & Dubey [210] implemented the integrated fuzzy AHP-TOPSIS to choose mobile companies based on the findings of usability analysis."

"The researchers adopted a well-established as well as tested decision-making approach, the hybrid fuzzy AHP-TOPSIS."

"This methodology was enlisted to prioritize different privacy requirements engineering techniques that focused on the usability impact evaluation in the current software privacy demand setting."

"Based upon the obtained findings of this research study, P-STORE denoted by A-1 has been established to be preeminent among six qualified alternatives as it provides the highest usability with an Relative Closeness (RCi) of 0.97124 in the different Privacy Requirements Engineering approaches."

Conclusion

"The applicability and the functionality of the P-STORE methodology are pragmatic, easy to comprehend and use."

"Research, we have seen that minor changes need to be made to adjust the STORE process for the privacy requirements engineering."

"The methodology needs to be revised so as to allow the use of STORE for privacy requirements engineering."

"We have successfully implemented P-STORE methodology on a case study of HMS system."

"For an even more convincing validation of the technique's accuracy, P-STORE can be applied on medium-to-large case studies in the field."

"We recommend that P-STORE methodology would establish relatively appropriate privacy requirements as well as enable the software engineer to efficiently organize privacy requirements."

"P-STORE requires a reliable, integrated tool that supports security and privacy, or both."

"We also intend to use the proposed P-STORE methodology for different software as well as web application development projects."

4.2.39 Fuzzy-Based Approach to Assess and Prioritize Privacy Risks [270]

This is a machine-generated summary of:

Hart, Stephen; Ferrara, Anna Lisa; Paci, Federica: Fuzzy-based approach to assess and prioritize privacy risks [270].

Published in: Soft Computing (2019).

Link to original: https://doi.org/10.1007/s00500-019-03986-5.

If you want to cite the papers, please refer to the original.

For technical reasons we could not place the page where the original quote is coming from.

Abstract-Summary

"DPIA allows organizations to identify, assess and prioritize the risks related to the processing of personal information and select suitable mitigations to reduce the severity of the risks."

"The existing DPIA methodologies measure the severity of privacy risks according to analysts' opinions about the likelihood and the impact factors of the threats."

"To reduce subjectivity, we propose a set of well-defined criteria that analysts can use to measure the likelihood and the impact of a privacy risk."

"We adopt the fuzzy multi-criteria decision-making approach to systematically measure the severity of privacy risks while modeling the imprecision and vagueness inherent in linguistic assessment."

Introduction

"Organizations must always conduct a data protection impact assessment (DPIA) when the processing of consumers' personal information could result in a high risk to the consumers' rights and freedoms."

"Existing methodologies to conduct a DPIA [277–280] do not provide a systematic approach to assess the severity of a privacy risk based on likelihood and impact, and they often adopt similar processes to the one used to rate security risks."

"Security risk assessment techniques, rely on security analysts to rate impact and likelihood of threats with respect to simple linguistic scales, e.g., from very low to very high, with no well-specified criteria on how to determine the position on this scale."

"These criteria cannot be applied to evaluate the impact of a privacy risk because the impact of a security risk is rated only from the perspective of an organization rather than from the perspective of the data subject."

Related works

"These methods and standards significantly differ in the way they estimate the value of likelihood and impact and how they combine these values to obtain the risk level."

"Some methodologies like CORAS [281] and the NIST 800-30 [282] standard require analysts to define a qualitative scale for likelihood and impact, assign a value to likelihood and impact and then use a risk evaluation matrix to combine the values of likelihood and impact into a risk level."

"Other methodologies instead define criteria to estimate likelihood and impact, assign a value to these criteria and then use multiplication and/or addition of the values to compute the overall risk level."

"The remaining methodologies provide criteria for likelihood and impact estimation, but they combine these values using approaches that do not reduce the subjectivity of the evaluation."

"Wagner and Boiten [283] propose a set of criteria to assess the likelihood and impact of privacy risks."

Preliminaries

"LINDDUN [284] is a privacy threat modeling technique."

"The first three represent the problem space, while the remaining three correspond to the solution space: Problem space 1. A data flow diagram (DFD) of the system under analysis is created, which represents how information flow into the system, how they are processed and where they are stored; 2. Each element in the DFD is mapped with a number of LINDDUN privacy threat categories; 3. The privacy threat categories are refined in concrete threat scenarios."

"Fuzzy multi-criteria models are used to assess alternatives in situations where crisp data are inadequate."

"In such cases, the ratings of alternatives and weights of the criteria in the problem can be evaluated using linguistic values represented by fuzzy numbers."

"These linguistic variables can be expressed in positive triangular fuzzy numbers."

Proposed methodology

"The relevance of such criteria is assessed with respect to the Cyber Equifax attack, one of the largest recent security breaches carried out against one of the major credit reporting agencies [285]."

"Likelihood The likelihood of a privacy threat estimates the probability that an attacker will discover a vulnerability and will successfully exploit it."

"The Equifax attack did not result in a loss of availability but just in a loss of confidentiality and integrity because cyber criminals gained access to personal information like names, addresses, dates of birth, credit score and social security numbers, and this information could have been included in fraudulent credit rating and loan applications."

"In the rating stage, given m threats and n evaluation criteria, k decision makers express their opinions (or weights) about the importance of each criterion in assessing the likelihood and the impact intensity of privacy threats as well as their ratings about the severity of each threat with respect to each specified criterion."

Case study

"In order to benefit from the pay as you go policy, customers have to install a smart device in their car that collects information about their driving."

"This information is sent via satellite to the car insurance company's central servers where it is stored in a database along with other customer information such as its name and address and the credit card details used to pay the insurance premium."

"For the analysis, we will focus on privacy threats targeting the database storing customer information."

"The two analysts, envisioning a concrete possibility of data breach, associate high weights to the likelihood criteria of capabilities and ease of exploiting with respect to threat T2."

"Since DriveSafe's assets include valuable information for attackers, it is likely that the vulnerability will be exploited if not patched, causing significant financial and reputation losses, as happened with the Equifax attack."

Conclusions

"The existing DPIA methodologies do not provide an effective solution to assess and prioritize privacy risks because they rely upon an analyst to rate the impact and likelihood of the risks."

"We have proposed a set of well-defined criteria that analysts can use to measure the likelihood and the impact of a privacy risk."

"We adopt the fuzzy multi-criteria decision-making approach to systematically measure the severity of privacy risks while modeling the imprecision and vagueness inherent in linguistic assessment."

"We also plan to explore the use of multi-criteria decision-making approaches to select alternative organizational and technical measures to address the highest privacy risks."

4.2.40 Privacy Self Management in Enterprise [271]

This is a machine-generated summary of:

Vidhani, Kumar; Banahatti, Vijayanand; Mirza, Hussain: Privacy self management in enterprise [271].

Published in: CSI Transactions on ICT (2022).

Link to original: https://doi.org/10.1007/s40012-022-00347-w.

If you want to cite the papers, please refer to the original.

For technical reasons we could not place the page where the original quote is coming from.

Abstract-Summary

"Many a times, processing of sensitive data can lead to privacy risks."

"Most laws advocate the privacy self management paradigm based on the notice and choice model to reduce the impact of those risks."

"The literature has documented challenges that should be addressed by an enterprise in order to implement the privacy self management paradigm."

"Two key challenges are: a solution for the management and enforcement of consents obtained from enterprise customers and integrating such solution into diverse set of enterprise applications."

"We propose a consent management and enforcement platform that makes customers an active stakeholder of enterprise application."

"Our platform can enforce consent in real-time at the granularity of data field level."

Introduction

"GDPR has introduced the privacy self management paradigm which equips customers (data subjects) with data subject rights and tasks enterprises with certain responsibilities."

"Notice mechanism helps enterprise inform about data it collects from customers for different purposes and privacy practices it follows."

"There are two types of solutions available to manage data privacy and to some extent the consent management and enforcement."

"Type of solutions are more focused on addressing privacy challenges faced by enterprise."

"It enforces privacy policies collected from customers in form of consent through an identity management solution."

"While the consent manager proposal by goldstein and others do not elaborate their solution in detail and they do not discuss how the privacy self management paradigm can be implemented [211]."

"We propose a consent management solution that can help enterprise meet some of challenges thrown by the privacy self management paradigm."

System

"CMS consists of the following components: Service Integrator and Controller (acts as an orchestrator), Portal (allows enterprise to declare metadata and purposes), Mobile App (enables a customer to express privacy preferences and respond to data processing requests), App Controller (protects sensitive data of a customer according to her privacy preferences), and Utilities (logs all activities and perform analytics over collected logs)."

"Based on the enterprise data policy, customers are provided with an interface where they can choose privacy preferences over their data from available options."

"Once these inputs from enterprise are collected, the portal provides an interface to customers to set privacy preferences over data fields."

"CMS Portal makes it incredibly easy for customers to compare privacy policies simplified in the form of purposes-data field pairs, options offered by different enterprises and state their privacy preferences; which would otherwise be very difficult if they have to read lengthy privacy policies."

Experiments

"The objective of this phase was to demonstrate the feasibility of integrating CMS with existing enterprise information systems, and to let participants experience granular level privacy controls over their data."

"Internal usage purpose was explained to participants as follows 'a person from the bank can access your data for internal operations, for example to provide you with products and services, to facilitate transactions, to service, maintain or collect on accounts etc.' Third party usage purpose was described as 'your data can be shared with third party companies for the marketing product and services of the bank'."

"Out of 139, 125 participants set privacy preferences for their personal and domain sensitive data fields."

"Whenever a DemoBank user opened application in the App controller, the values of data fields which had privacy preferences set by participants were masked, and agents were provided with a pop-up box to raise a data access request."

"DemoBank users randomly chose participants' data where preference was set as either 'Notice' or 'Consent'."

Threats to validity

"There is a huge diversity of participants in terms of age, gender, and ethnicities who participated in the CMS experiment."

"We believe our hypothesis is true as we did not receive even a single query or complaint from any participant during the experiment."

"Threat to our experimental study comes from our lack of inquiry into why 139 people participated in the experiment and not all (300) who were communicated."

"Experiments, we plan to add participants from different countries to reduce the impact of this limitation."

Related work

"Personal Data Stores like MIT's openPDS [212] store and provide privacy to person's meta-data such as digital information about location, phone call logs, web searches etc. In contrast to PDV which shares the data (or a subset of it), openPDS does not share any data with third party services or applications."

"These systems provide granular level data control to customers, these are intended to be used with third party applications such as smartphone applications where customers provide their data voluntarily via smart-phone's sensors or via any other means."

"We argue that these systems are not directly applicable in case of enterprises where data of customer is collected and stored on a mandatory basis."

"Mont and others have proposed an approach to automatically enforce privacy policies over customer's data in enterprise [213]."

"In their approach, privacy preferences from data subjects are first collected, converted into privacy policies and then enforced automatically through identity management systems."

Discussion and conclusion

"CMS users may download the mobile app but not necessarily respond to data processing requests."

"To be fair, such users do not have any motivation to respond to data processing requests."

"CMS needs to create an incentive mechanism to nudge them in giving response to as many data processing requests as possible."

"Mobile app can be augmented with some automation that can respond on behalf of users based on data collected in past interactions."

"Enterprise applications must implement the privacy self management paradigm as recommended by the most of privacy laws to satisfy this legal requirement."

4.2.41 Automated Privacy Negotiations with Preference Uncertainty [272]

This is a machine-generated summary of:

Filipczuk, Dorota; Baarslag, Tim; Gerding, Enrico H.; schraefel, m. c.: Automated privacy negotiations with preference uncertainty [272].

Published in: Autonomous Agents and Multi-Agent Systems (2022).

Link to original: https://doi.org/10.1007/s10458-022-09579-1.

If you want to cite the papers, please refer to the original.

For technical reasons we could not place the page where the original quote is coming from.

Abstract-Summary

"To support users' privacy management, we propose a novel agent-based negotiation framework to negotiate privacy permissions between users and service providers using a new multi-issue alternating-offer protocol based on exchanges of partial and complete offers."

"We compare the take-it-or-leave-it approach, in which users are required to accept all permissions requested by a service, to negotiation, which respects their preferences."

"Negotiation can be less mentally demanding than the take-it-or-leave-it approach and it allows users to align their privacy choices with their preferences."

"Our findings provide insight into users' data sharing strategies to guide the future of automated and negotiable privacy management mechanisms."

Introduction

"To address this problem, Krol and Preibusch envisioned effortless privacy negotiations which provide the opportunity to reach individual agreements about personal data collection between users with heterogeneous privacy preferences and service providers [214]."

"Although privacy issues have received a lot of attention from the multi-agent systems community [215], to date the opportunity of providing an effective solution to enable privacy negotiations between the user and the service provider has not been well-explored."

"There have been very few studies which propose automated negotiation solutions in the privacy domain, and none of these have been evaluated with real users using their real data."

"We argue that privacy negotiations can be automated effectively using autonomous agents that represent users and service providers."

"The results also suggest that agent-supported negotiation might be less mentally demanding compared to setting the permissions manually and, at the same time, it better enables users to align their privacy choices with their actual preferences."

Related work

"In this paper, we explore questions about the viability of an exchange between user and service that goes beyond communicating fine-grained privacy preferences: we investigate the use of automated negotiation as a sustainable, win-win solution that supports a meaningful dialog between the user and the service provider."

"As users' privacy preferences are influenced by the context [216, 217], in our work, we negotiate all the issues together."

"For vertical privacy, several studies based their approach on P3P. Although P3P itself lacks a negotiation mechanism, researchers utilised P3P enhancements to enable privacy policy negotiations between the user and the service provider [218–223]."

"The agent works on behalf of a list of users whose privacy preferences need to be exactly specified by the users themselves prior to the negotiation."

"We build upon previous work on a user agent for an automated negotiation of privacy permissions [224]."

Negotiation framework

"We propose a negotiation framework with a novel protocol that we refer to as the partial-complete offer protocol, which is a variant of the multi-issue alternating-offers protocol."

"We describe the partial-complete offer protocol, where two parties propose offers consisting of values of the negotiable issues."

"In the classic multi-issue alternating offer protocol, offers are exchanged between two parties that specify values for each of the negotiable issues."

"To model this kind of automated negotiation, our premise is that, before the negotiation takes place, the proposer and the responder prescribe which issues they specify values for."

"In the Partial-Complete Offer Protocol, the proposer submits a partial offer specifying the requirements for a subset of issues."

"From here, the proposer can either accept a complete offer and end the negotiation, reject it and submit a new partial one starting the next negotiation round, or break off the negotiation."

Negotiation agent with preference uncertainty

"To address the challenge that preference uncertainty brings into the negotiation, we first discuss how the agent builds the user and the opponent models."

"We propose a negotiation strategy which allows the agent to decide what offer, if any, to send to the opponent and when to terminate the negotiation."

"In our setup, we assume that the proposer agent first negotiates the offer with the responder (its opponent) on behalf of the user, and then proactively interacts with the user to establish whether the negotiated offer aligns with their preferences."

"We can identify two kinds of preference uncertainty in this negotiation: the user's preferences regarding the possible negotiation outcomes, and the opponent's reactions to the offers."

"Using the valuation function, the negotiation strategy needs to determine which of the partial offers, if any, the agent should propose to the opponent."

Negotiation of privacy

"We provide an overview of how the negotiation of privacy permissions is performed with human users in the loop."

"In privacy permission management, the negotiation domain may consist of a set of permissions to access user's device resources such as contacts or text messages, as well as other negotiable issues such as access to certain functionalities of the online service owned by the service provider (e.g. a website, a mobile app or software running on a device)."

"We assume that the user's agent is in control of the values of issues related to the set of permissions and the service provider's agent controls the value of the quote."

"If the user's agent refuses to grant any permissions, the complete offer would be (0, 0, 0, 0, 0, 0)."

"Of negotiating privacy permissions provides two benefits: the user has full control over their privacy and the agent can use the feedback for constructing the user's preference profile."

Experimental evaluation

"Two designs of this interface: the negotiation screen allows the user to request a new negotiation of the permissions and the "Take It Or Leave It" screen that represents the take-it-or-leave-it approach that is still used by many services requiring users to share data."

"When the negotiation between the user's agent and the service provider's agent ends, the agreed permission settings (Share or Don't Share) as well as the agreed quote (a number of points received in exchange) are displayed."

"Following each sharing choice (through either negotiation or take-it-or-leave-it), the app collects three random data points of each shared permission type from the user's device."

"We report on the impact of automated negotiation on data sharing and user's post-sharing regret, aligning users' decisions with their self-reported data sharing sensitivity, the effort required from the users and the accuracy of the proposed agents."

Discussion and future work

"As the number of negotiations increases, the agent can be more effective personalising at offers based on a single user's negotiation data."

"Despite the fact that the human is always in the loop, if a negotiation agent is provided by the service provider, there is a risk that persuasive technologies may be used to influence the behaviour of the user, e.g. based on information mined from other users, or in terms of presenting information to the user that makes clever use of cognitive biases ('nudging')."

"Future work should explore how the interface affects users' decisions and their trust to the offers negotiated by the agent."

"Future work should explore more appropriate valuation models for the negotiated data types and consider how those models impact the user's reactions to the offers."

Conclusions

"In order for an agent to faithfully represent a user, we investigate two approaches to implicit preference elicitation: one approach personalised to each individual user and one personalised depending on the user's privacy profile classification."

"Building on approaches from the field of multi-agent systems, where automated negotiation is widely researched [225], a software agent may eventually negotiate on behalf of the user, based on preferences that are elicited at a time that is convenient to the user."

"An interesting direction for future work is to apply the proposed user preference elicitation to other domains, and to compare it to other approaches, e.g. those developed as part of the Automated Negotiating Agents Competition (ANAC) [226, 227]."

"Unlike our approach, the ANAC competition so far has assumed no noise in the user's own preferences, i.e. the partial preference order is consistent."

4.2.42 Embedding a Hiding Function in a Portable Electronic Health Record for Privacy Preservation [273]

This is a machine-generated summary of:

Huang, Lu-Chou; Chu, Huei-Chung; Lien, Chung-Yueh; Hsiao, Chia-Hung; Kao, Tsair: Embedding a Hiding Function in a Portable Electronic Health Record for Privacy Preservation [273].

Published in: Journal of Medical Systems (2008).

Link to original: https://doi.org/10.1007/s10916-008-9243-8.

If you want to cite the papers, please refer to the original.

For technical reasons we could not place the page where the original quote is coming from.

Abstract-Summary

"In compliance with HIPAA guidelines, a hiding function for personal identifiers in a PEHR was implemented by recognizing and hiding techniques."

"The method emphasizes the feasibility of embedding a hiding function in a PEHR."

"The present study showed that embedding the hiding function in a PEHR is a practical way not only satisfies data confidentiability needs, but also meets patient's personal privacy requirements."

Introduction

"With the increased mobility and accessibility of information, there is also an increased concern over the security of personal data, such as portable electronic health records (PEHR), that people carry with them."

"Given that PEHRs contain information related to disease status, symptoms and related laboratory data, these records include many bits of individually identifiable health information."

"PEHRs contain a lot of people's identifiable information, which should be protected health information."

"Rules regulating privacy limit the ways that health plans, pharmacies, and hospitals can use patients' personal medical information."

"Patients should protect their PEHRs to avoid the disclosure of private material and should safeguard their PEHRs so as not to reveal any personal information to anyone and to prevent other users from reading their personal information."

"We proposed an approach to reduce the disclosure of private information in PEHRs during reading and viewing without the need for special hardware."

Methods

"The function we proposed would allow the user to hide the identifiable information in the PEHR, and this function would be embedded in the PEHR itself."

"The hiding function is meant to make identifiable information invisible, and to prevent the disclosure of any personal and sensitive information while a PEHR being viewed."

"The second step hides personal identifiers and generates the script language code that embedded into the PEHR."

"If the PEHR is in a known and well-defined form such as the Extensible Markup Language (XML) format, PDF with XFA (extensible forms architecture), it is simple to analyze and recognize identifiable information from certain structures such as the XML schema patterns."

"The PEHR with the embedded function that hiding information allows the user to switch between displaying and hiding the identifiers."

Results

"Adding a simple button to execute the hiding function would allow users to obstruct personally identifiable information from their PEHR."

"In the well-defined form, PEHRs with XML-based data would execute this function by using the XSLT transformation to switch between two.XSL files."

"The identifiers in this type will be under control."

"The presentations show that it is possible to embed and implement the hiding function into different XML-based PEHRs such as ASTM CCR, IHE XDM, IHE XPHR, and HL7/ASTM CCD to provide security for patients."

"Even in PEHRs with non-interoperability formats such as PDF-based with free-form type and HTML-based type, we have shown that it is possible to implement the hiding function to reduce the risk of privacy disclosure."

Discussion

"If the protection of identifiers from the HIPAA safe harbor list is effective, then the embedding of the hiding function for PEHRs would help people protect the privacy of personal identifiers in their PEHRs."

"The hiding function could achieve this goal by assuring that others could not view the user's private information while he/she is reading it."

"We addressed some common used file formats to test and verify the possibility for adding the hiding function into PEHRs for privacy preservation."

"With mobile and portable display devices such as notebooks, UMPCs, PDAs and e-paper readers, the capability of these devices would be sufficient to operate the hiding function in the presented file types of the PEHR without any special hardware or software."

"Embedding the function into the PEHR would be protected through cryptography such as digital signatures and encryption [228, 229]."

Conclusion

"People should have a real understanding of privacy and how to protect their PEHR after they copy and view their records from hospitals."

"We presented a method to reduce the disclosure of private information in PEHRs during reading."

"The method works by embedding a hiding function into PEHRs that would give people a way to lessen the risk of leaking private information inappropriately and unintentionally."

"It can also be applied to any PEHR to provide privacy preservation for people during reading."

4.2.43 DeepFake Electrocardiograms Using Generative Adversarial Networks Are the Beginning of the End for Privacy Issues in Medicine [274]

This is a machine-generated summary of:

Thambawita, Vajira; Isaksen, Jonas L.; Hicks, Steven A.; Ghouse, Jonas; Ahlberg, Gustav; Linneberg, Allan; Grarup, Niels; Ellervik, Christina; Olesen, Morten Salling; Hansen, Torben; Graff, Claus; Holstein-Rathlou, Niels-Henrik; Strümke, Inga; Hammer, Hugo L.; Maleckar, Mary M.; Halvorsen, Pål; Riegler, Michael A.; Kanters, Jørgen K.: DeepFake electrocardiograms using generative adversarial networks are the beginning of the end for privacy issues in medicine [274].

Published in: Scientific Reports (2021).

If you want to cite the papers, please refer to the original.

For technical reasons we could not place the page where the original quote is coming from.

Abstract-Summary

"Synthetic data generated to represent real data carrying similar information and distribution may alleviate the privacy issue."

"We present generative adversarial networks (GANs) capable of generating realistic synthetic DeepFake 10-s 12-lead electrocardiograms (ECGs)."

"We trained the GANs with 7,233 real normal ECGs to produce 121,977 DeepFake normal ECGs."

"By verifying the ECGs using a commercial ECG interpretation program (MUSE 12SL, GE Healthcare), we demonstrate that the Pulse2Pulse GAN was superior to the WaveGAN* to produce realistic ECGs."

"The synthetic dataset will be available as open access for researchers at OSF.io and the DeepFake generator available at the Python Package Index (PyPI) for generating synthetic ECGs."

"We were able to generate realistic synthetic ECGs using generative adversarial neural networks on normal ECGs from two population studies, thereby addressing the relevant privacy issues in medical datasets."

Introduction

"Generating realistic synthetic data is an alternative solution to the privacy issue."

"Properly generated synthetic data is a solution to the privacy problem which enables data sharing between research groups."

"We showcase synthetic ECGs as an example of complex medical data."

"Generative adversarial networks (GAN) were introduced in 2014 by Goodfellow and others [230] to generate synthetic data using multi-layer perceptrons."

"A GAN consists of two deep neural networks: a generator network, which creates signals (here ECGs) from random noise, and a discriminator network, which evaluates whether an ECG presented to it is real or fake."

"Since ECGs are time series data, our initial approach was to use a WaveGAN [231] which is capable of generating sound signals."

"We thus present two GANs with the ability to generate an unlimited number of 10-s 12-leads synthetic "DeepFake" ECGs as a solution to overcome the privacy issues related to real ECG data."

Results

"To avoid chimeras between normal and abnormal ECGs, we only trained the neural network with ECGs classified as normal by the MUSE 12SL (version 2.43)."

"Using the Pulse2Pulse model from the optimal number of epochs (2500), we generated 150,000 DeepFake ECGs."

"We found that 81.3% of the 150,000 DeepFake ECGs were classified as "Normal ECG" (vs. 81.6% in the initial training)."

"The real data included all ECGs from GESUS and Inter99 classified as "Normal ECG" which were used for training."

"DeepFake ECGs are presented both as all 150,000 generated ECGs and the subset classified as Normal ECG."

Discussion

"Since we trained our model only on Normal ECGs, the input distribution for the GAN was a truncated asymmetric distribution."

"The left truncation (at low heart rates) will discard more individuals than the right truncation (at high heart rates), and the final distribution of the real ECGs will be close to a truncated normal distribution with asymmetric truncations."

"This could lead to the exclusion of some DeepFake ECGs, simply because the ECG intervals or amplitudes were marginally outside the normal range."

"Although the ECG discarded by the MUSE 12 SL may only have minimal abnormalities (like a heart rate of 59.9 bpm where 60 bpm is normal), the filtering of the DeepFake ECGs ensures that the best epoch is chosen without bias."

"The DeepFake ECGs generated from the Pulse2Pulse model can be used as a replacement to overcome the privacy constraints in real medical datasets."

Methods

"In a GAN, two deep neural networks termed the generator (G) and the discriminator (D) are combined to achieve the generation task."

"The main goal of the generator is to produce a data sample input [ECG(z)] from random noise (z) to present to the discriminator."

"Our Pulse2Pulse implementation is different from the original U-net implementation because the Pulse2Pulse implementation uses 1D convolutional layers for ECG signal generation as opposed to the 2D convolutional layers used for the original image segmentation task."

"The discriminator takes an ECG as input (5000 samples × 8 channels) and outputs a score how close the ECG are to be determined fake or real."

"For evaluation of our two GAN models, we initially generated 10,000 ECGs from every 500 epochs until 3000 epochs from each GAN model."

"Using the best epoch for the best GAN, we generated 150,000 DeepFake ECGs."

4.2.44 Privacy from Promises to Protection [275]

This is a machine-generated summary of:

Maaser, Michael; Langendörfer, Peter: Privacy from Promises to Protection [275]

Published in: Mobile Networks and Applications (2008).

Link to original: https://doi.org/10.1007/s11036-008-0116-7.

If you want to cite the papers, please refer to the original.

For technical reasons we could not place the page where the original quote is coming from.

Abstract-Summary

"The problem is, as soon as a user provides data to a service provider the user looses control over her data."

"The simple solution is not to provide any data but then many useful services, e.g. navigation applications, cannot be used."

"The concept is that the application obtains access to the user data in a specially protected and certified environment, the PGEC."

"PGECs enable applications to access private user data and guarantee that the user data is deleted as soon as the service is quit."

"The PGEC also restricts the communication between the application and the service provider to what is explicitly allowed by the service user."

"The PGEC guarantees a "one time use" of the provided private data."

Introduction

"PGECs securely combine sensitive user data, e.g. current position, with sensitive service provider data, e.g. navigation algorithms and ensure that this data/code cannot be send or stored outside the container as well as that data and code are deleted when the service is finished."

"In order to achieve the claimed "one time use" of user data, the PGEC needs to employ means to protect itself against malicious code executed inside the PGEC as well as against malicious executing hosts."

"To prevent other attacks such as data exchange via shared memory additional means to separate different programs inside the PGEC completely from each other need to be added to the sandbox functionality."

"The PGEC combines ideas from well known sandbox models and ideas for protecting mobile software entities with additional means to prevent data loss via covert channel attacks."

Related work

"P3P and APPEL merely provide a mechanism to describe the intentions of both sides rather than means to protect user data after agreeing to use the service."

"They merely protect media content, which could be considered as the service provider's data."

"Despite the protection of user data is in principle an equivalent problem, these approaches do not provide a solution for protecting service users' data."

"User data is enclosed in an agent and securely transferred into an isolated, closed-door, one-way platform provided by a trusted third party."

"All agents, together with their enclosed data, are deleted after service completion in order to ensure the privacy of the user and to protect the service's data."

"Within the container instances both user and provider data can be brought together and the service can usefully be completed."

"A number of approaches exist to protect code and data of the agent."

Container concept

"In order to provide mutual protection between the service user and the service provider we introduce the concept of privacy guaranteeing execution containers."

"It also needs means to prevent program code provided by the service user or service providers from transmitting data out of the PGEC."

"If property (3) is fulfilled, a service running in a container simultaneously with other services cannot share its knowledge about gained private data with other services that are concurrently executed within the PGEC."

"The communication between the code, which is executed inside the container, and the service provider outside is restricted by a privacy contract."

"We need to differentiate between a container running on the service provider's side suspected to fraudulently obtain private data and a container on the user's side who requires the private data to be protected."

Realization of the container

"In case the executed service in the container must communicate with the outside world at its origin or any other party, this communication must be limited in a way that prevents transmission of privacy relevant data."

"The use of such encrypted files under full control of the container allows further services to persist certain data that is only accessible within the container."

"It provides the necessary infrastructure to services executed inside the container, e.g. access to the processor, volatile memory, etc. It is also responsible for the cleanup operations that must be done when a service is no longer used, i.e. it has to ensure that all data is really deleted."

"The intercommunication channels between the PGEC instances allow the PGEC to provide internal access to private data, permitted by the negotiated privacy contract, transparently to service at each instance."

Conclusions

"This is achieved by providing an execution environment, i.e. the PGEC that is independent of the service user as well as independent of the service provider."

"The PGEC may access user data and also run code from the service provider."

"The PGEC ensures that all data/code is deleted when a services is finished and protects itself against malicious hosts and some covert channel attacks."

"By introducing the concept of distributed PGEC we allow for load balancing between user and service provider side, i.e. large amounts of data such as maps used by location based services can be accessed where they are stored instead of sending them through the Internet to a potentially resource constraint mobile device."

"In order to illustrate the level of security that the PGEC concept provides we have discussed how modified execution environments can be detected as well as how covert channel attacks can be prevented."

References

1. Li WD (2020) The connotation, elements and composition of the Internet of Everything. Acad Front 6:40–45
2. Isiaka F, Adamu Z (2022) Custom emoji based emotion recognition system for dynamic business webpages. Int J Intell Comput 15(4):497–509. https://doi.org/10.1108/IJICC-11-2021-0254
3. Kachur A, Osin E, Davydov D et al. (2020) Assessing the Big Five personality traits using real-life static facial images. Sci Rep 10:8487. https://doi.org/10.1038/s41598-020-65358-6
4. Jain AK, Nandakumar K, Nagar A (2008) Biometric template security. Eurasip J Adv Signal Process 1–17. https://doi.org/10.1155/2008/579416
5. Brous P, Janssen M, Herder P (2020) The dual effects of the Internet of Things (IoT): a systematic review of the benefits and risks of IoT adoption by organizations. Int J Inf Manage 51:101952. https://doi.org/10.1016/j.ijinfomgt.2019.05.008
6. Chin CA, Crosby GV, Ghosh T, Murimi R (2012) Advances and challenges of wireless body area networks for healthcare applications. In: 2012 International Conference on Computing, Networking and Communications (ICNC). IEEE, pp. 99–103
7. Ji S, Gui ZY, Zhou TQ, Yan HY, Shen J (2018) An efficient and certificateless conditional privacy-preserving authentication scheme for wireless body area networks big data services. IEEE Access 6:69603–69611. https://doi.org/10.1109/ACCESS.2018.2880898
8. Li T, Zheng YH, Zhou T (2017) Efficient anonymous authenticated key agreement scheme for wireless body area networks. Secur Commun Netw 8. https://doi.org/10.1155/2017/4167549
9. Kindt EJ (2013) Privacy and data protection issues of biometric applications: a comparative legal analysis. Springer Netherlands, New York. https://doi.org/10.1007/978-94-007-7522-0
10. Kilovaty I (2021) Psychological data breach harms. NCJL & Tech 23:1

11. Jain AK, Ross A (2004) An introduction to biometric recognition. IEEE Trans Circuits Syst Video Technol 14(1):4–20. https://doi.org/10.1109/TCSVT.2003.818349
12. Günther M, El Shafey L, Marcel S (2017) 2D face recognition: an experimental and reproducible research survey. No. REP_WORK. Idiap
13. Sepas-Moghaddam A, Correia P, Nasrollahi K et al. (2019) A double-deep spatio-angular learning framework for light field based face recognition. IEEE Trans Circuits Syst Video Technol 30(12):4496–4512. https://doi.org/10.1109/TCSVT.2019.2916669
14. Ratha NK, Connell JH, Bolle RM (2001) Enhancing security and privacy in biometrics-based authentication systems. IBM Syst J 40(3):614–634. https://doi.org/10.1147/sj.403.0614
15. Tuyls PP, Skoric BB, Kevenaar TT (2007) Security with noisy data: on private biometrics, secure key storage and anti-counterfeiting. Springer, Berlin
16. Kumar S, Singh SK et al. (2018) Privacy preserving security using biometrics in cloud computing. Multimed Tools Appl 77(9):11017–11039
17. Jain AK, Flynn P, Ross AA (2018) Handbook of biometrics. Springer, Berlin
18. Smith M, Mann M, Urbas G (2018) Biometrics, crime and security. Routledge, New York
19. Wang HY, Tang SJ, Ding Y, Wang YJ, Li JH (2020) Survey on biometrics template protection. J Comput Res Dev 05:1003–1021
20. Prasad RS 2019 The Personal Data Protection Bill, 2019
21. Bansod S and Ragha L 2020 Blockchain technology: applications and research challenges. In: International Conference for Emerging Technology (INCET), Belgaum, India, pp. 1–6
22. Eskandari S, Clark J, Barrera D and Stobert E 2015 A first look at the usability of bitcoin key management. In: Proceedings 2015 Workshop on Usable Security
23. Homoliak I, Venugopalan S, Hum Q and Szalachowski P 2019 A security reference architecture for Blockchains. In: IEEE International Conference on Blockchain (Blockchain), Atlanta, GA USA, pp. 390–397
24. Wang H, Wang Y, Cao Z, Li Z and Xiong G 2019 An overview of blockchain security analysis. Cyber Security, Singapore, pp. 55–72
25. Feng T, Chen X, Liu C and Feng X 2019 Research on privacy enhancement scheme of blockchain. Trans. Secur. Priv. 2–6
26. Xiao R, Ren W, Zhu T, Choo K 2019 A mixing scheme using a decentralized signature protocol for privacy protection in bitcoin. Blockchain IEEE Trans. Dependable and Secure Comput., pp. 1–1
27. Hackett R 2019 Zcash discloses vulnerability that could have allowed 'infinite counterfeit'. Cryptocurrency The Ledger—Cybersecurity, Fortune
28. Lu J, Yang B, Liang Z and Lu L 2017 Wanchain White paper Wanchain Foundation Ltd
29. snyktest@3test/wanchain-asset-ledger@1.2.0. https://snyk.io/test/npm/@3test/wanchain-asset-ledger/1.2.0
30. Schwabe P 2017 NewHope Post-quantum key encapsulation [Online]
31. Council E. EU Regulation 2016/679 of the European Parliament and of the Council, "On the protection of natural persons with regard to the processing of personal data and on the free movement of such data", Directive 95/46/EC (General Data Protection Regulation): Official Journal of the European Union; 2016
32. PoSeID-on project. Protection and control of secured information by means of a privacy enhanced dashboard. https://www.poseidon-h2020.eu Accessed Jan 2020
33. Apache Spark. https://spark.apache.org/streaming/. Accessed Jan 2020
34. Omnibus: Hipaa rule in the Federal Register. http://www.gpo.gov/fdsys/pkg/FR-2013-01-25/pdf/2013-01073.pdf (2013)
35. Wang, W., Zhang, Q.: Towards long-term privacy preservation: a context aware perspective. IEEE Wirel. Commun. 22, 34–40 (2015)
36. Wang, W., Chen, L., Zhang, Q.: Outsourcing high-dimensional healthcare data to cloud with personalized privacy preservation. Comput. Netw. 88, 136–148 (2015). ISSN 1389–1286. https://doi.org/10.1016/j.comnet.2015.06.014
37. Zhang, K., et al.: Sedic: privacy-aware data intensive computing on hybrid clouds. In: Proceedings of the 18th ACM Conference on Computer and Communications Security. ACM (2011)

38. Zhou, Z., et al.: Prometheus: privacy-aware data retrieval on hybrid cloud. In: INFOCOM, 2013 Proceedings IEEE. IEEE (2013)
39. Sun, X., Sun, L., Wang, H.: Extended k-anonymity models against sensitive attribute disclosure. Comput. Commun. 34(4), 526–535 (2011)
40. Terrovitis, M., Mamoulis, N., Kalnis, P.: Privacy-preserving anonymization of set-valued data. Proc. VLDB Endow. 1(1), 115–125 (2008)
41. Liu, J.: Optimal anonymization for transaction publishing. Chin. J. Electron. 20(2), 238–242 (2011)
42. Pandilakshmi, K.R., Rashitha Banu, G.: An advanced bottom up generalization approach for big data on cloud. Int. J. Commun. Netw. Syst. 3, 1054–1059 (2014)
43. Shrivastva, K.M.P.D., Rizvi, M.A., Singh, S.: Big data privacy based on differential privacy a hope for big data. In: 2014 International Conference on Computational Intelligence and Communication Networks (CICN). IEEE (2014)
44. Zhang, X., et al.: A hybrid approach for scalable sub-tree anonymization over big data using MapReduce on cloud. J. Comput. Syst. Sci. 80(5), 1008–1020 (2014)
45. EU Commission Website. https://ec.europa.eu/commission/priorities/justice-and-fundamental-rights/data-protection/2018-reform-eu-data-protection-rules_en. Accessed 28 May 2019
46. Mewada S, Gautam SS, Sharma P (2020) Artificial bee colony-based approach for privacy preservation of medical data. Int J Inform Syst Mod Des (IJISMD) 11(3):22–39
47. Wu X, Xu X, Dai F, Gao J, Ji G, and Qi L (2020) An Ensemble of Random Decision Trees with Personalized Privacy Preservation in Edge-Cloud Computing. In: 2020 International Conferences on Internet of Things (iThings) and IEEE Green Computing and Communications (GreenCom) and IEEE Cyber, Physical and Social Computing (CPSCom) and IEEE Smart Data (SmartData) and IEEE Congress on Cybermatics (Cybermatics), pp. 779–786. IEEE
48. Abdo MA, Abdel-Hamid AA. and Elzouka HA (2020) A Cloud-based Mobile Healthcare Monitoring Framework with Location Privacy Preservation. In: 2020 International Conference on Innovation and Intelligence for Informatics, Computing and Technologies (3ICT) (pp. 1–8). IEEE
49. Yang X, Wang M, Wang X, Chen G, Wang C (2020) Stateless cloud auditing scheme for non-manager dynamic group data with privacy preservation. IEEE Access 8:212888–212903
50. Jayaram R, Prabakaran S (2020) Onboard disease prediction and rehabilitation monitoring on secure edge-cloud integrated privacy preserving healthcare system. Egypt Inform J. https://doi.org/10.1016/j.eij.2020.12.003
51. Mansour HO, Siraj MM, Ghaleb FA, Saeed F, Alkhammash EH, Maarof MA (2021) Quasi-Identifier recognition algorithm for privacy preservation of cloud data based on risk reidentification. Wirel Commun Mob Comput 2021:7154705. https://doi.org/10.1155/2021/7154705
52. Sathya A, Raja SKS (2021) Privacy preservation-based access control intelligence for cloud data storage in smart healthcare infrastructure. Wirel Pers Commun 118(4):3595–3614
53. Xie Q, Sundararaj V, Mr R (2021) Analyzing the factors affecting the attitude of public toward lockdown, institutional trust, and civic engagement activities. J Community Psychol. https://doi.org/10.1002/jcop.22681
54. Tu NA, Wong KS, Demirci MF, Lee YK (2021) Toward efficient and intelligent video analytics with visual privacy protection for large-scale surveillance. J Supercomput. https://doi.org/10.1007/s11227-021-03865-7
55. Zhu Y, Xiong L, Verdery C (2010) Anonymizing user profiles for personalized web search. In: Proceedings of the 19th international conference on World Wide Web, ACM, New York, NY, USA, WWW '10, pp 1225–1226. https://doi.org/10.1145/1772690.1772886
56. Liu X, Liu A, Zhang X, Li Z, Liu G, Zhao L, Zhou X (2017b) When differential privacy meets randomized perturbation: a hybrid approach for privacy-preserving recommender system. In: International conference on database systems for advanced applications. Springer, pp 576–591
57. Shou L, Bai H, Chen K, Chen G (2014) Supporting privacy protection in personalized web search. IEEE Trans Knowl Data Eng 26(2):453–467. https://doi.org/10.1109/TKDE.2012.201

58. Wu Z, Li R, Zhou Z, Guo J, Jiang J, Su X (2020) A user sensitive subject protection approach for book search service. J Assoc Inf Sci Technol 71(2):183–195
59. SEAL (2020) Microsoft SEAL (release 3.5). https://github.com/Microsoft/SEAL, microsoft Research, Redmond, WA
60. Makkaoui KE, Beni-Hssane A, Ezzati A, El-Ansari A (2017) Fast cloud-rsa scheme for promoting data confidentiality in the cloud computing. Procedia Comput Sci 113:33–40, https://doi.org/10.1016/j.procs.2017.08.282. (the 8th International Conference on Emerging Ubiquitous Systems and Pervasive Networks (EUSPN 2017)/The 7th International Conference on Current and Future Trends of Information and Communication Technologies in Healthcare (ICTH-2017)/Affiliated Workshops)
61. O'Neil, C. (2016). Weapons of math destruction: How big data increases inequality and threatens democracy. Crown
62. Wachter, S., & Mittelstadt, B. (2018). A right to reasonable inferences: Re-thinking data protection law in the age of Big Data and AI. Preprint. LawArXiv. Retrieved December 20, 2019, from https://osf.io/mu2kf
63. Mühlhoff, R. (2020a). Automatisierte Ungleichheit: Ethik der Künstlichen Intelligenz in der biopolitischen Wende des Digitalen Kapitalismus. Deutsche Zeitschrift für Philosophie, 68(6), 867–890. https://doi.org/10.1515/dzph-2020-0059
64. Mittelstadt, B. D., Allo, P., Taddeo, M., et al. (2016). The ethics of algorithms: mapping the debate. Big Data & Society, 3(2), 1–21
65. Amoore, L. (2020). Cloud ethics: Algorithms and the attributes of ourselves and others. Duke University Press
66. Coeckelbergh, M. (2020a). AI ethics. The MIT press essential knowledge series. The MIT Press
67. Basu, R. (2019). What we epistemically owe to each other. Philosophical Studies, 176(4), 915–931. https://doi.org/10.1007/s11098-018-1219-z
68. Trinckes JJ. The definitive guide to complying with the HIPAA/HITECH privacy and security rules. Boca Raton: CRC Press; 2013
69. Sharma S. Data privacy and GDPR handbook. Hoboken: Wiley; 2020
70. Christen P, Ranbaduge T, Schnell R. Linking sensitive data: methods and techniques for practical privacy-preserving information sharing. Cham: Springer; 2020
71. Gutmann MP, Witkowski K, Colyer C, O'Rourke JM, McNally J. Providing spatial data for secondary analysis: issues and current practices relating to confidentiality. Popul Res Policy Rev. 2008;27(6):639–65
72. Armstrong MP, Rushton G, Zimmerman DL. Geographically masking health data to preserve confidentiality. Stat Med. 1999;18:497–525
73. Kroll M, Schnell R. Anonymisation of geographical distance matrices via lipschitz embedding. Intern J Health Geogr. 2016;15(1):1–14
74. Kroll M. A graph theoretic linkage attack on microdata in a metric space. Trans Data Privacy. 2015;8(3):217–43
75. Vatsalan D, Christen P, Verykios VS. A taxonomy of privacy preserving record linkage techniques. Inform Syst. 2013;38:946–69
76. Lwin, M. O., & Williams, J. D. (2003). A model interpreting the multidimensional developmental theory of privacy and theory of planned behavior to examine fabrication of information online. Marketing Letters, 14(4), 257–272
77. Culnan, M. J., & Bies, R. J. (2003). Consumer privacy: Balancing economic and justice considerations. Journal of Social Issues, 59(2), 323–342
78. Culnan, M. J., & Milne, G. R. (2001). The Culnan–Milne survey on consumers and online privacy notices. (accessed August 8, 2006). [available at: http://www.ftc.gov/bcp/workshops/glb/supporting/culnan-milne.pdf]
79. Gauzente, C., & Ranchhod, A. 2001. Ethical marketing for competitive advantage on the internet. Academy of Marketing Science Review (Online). 01 (10)
80. Westin, A. F. (2003). Social and political dimensions of privacy. Journal of Social Issues, 59(2), 431–453

81. Bloom, P. N., Milne, G. R., & Adler, R. (1994). Avoiding misuse of new information technologies: Legal and societal consideration. Journal of Marketing, 58(January), 98–110
82. Caudill, E. M., & Murphy, P. E. (2000). Consumer Online Privacy: Legal and Ethical Issues. Journal of Public Policy and Marketing, 19(Spring), 7–19
83. Culnan, M. J. (2000). Protecting privacy online: Is self-regulation working? Journal of Public Policy & Marketing, 19(Spring), 20–26
84. Karson, E. J. (2002). Exploring a valid and reliable scale of consumer privacy and security concerns on the internet and implications for e-commerce. Paper presented at Academy of Marketing Science Conference, May 29–June 2, Sanibel, FL
85. Bennett, C. J. (1992). Regulating Privacy. Ithaca, New York: Cornell University Press
86. Milberg, S. J., Burke, S. J., Smith, H. J., & Kallman, E. A. (1995). Values, personal information, privacy and regulatory approaches. Communications of the ACM, 38(12), 65–74
87. Cranor, L. F., Reagle, J., & Ackerman, M. S. (1999). Beyond concern: Understanding net users' attitudes about online privacy. AT&T Labs–Research Technical Report TR 99.4.3
88. Milne, G. R., & Boza, M. E. (1999). Consumers' trust and concern about organizations use of personal information in direct marketing. Journal of Interactive Marketing, 13(Winter), 7–24
89. Sheehan, K. B., & Hoy, M. G. (2000). Dimensions of privacy concern among online consumers. Journal of Public Policy & Marketing, 19(Spring), 62–73
90. Georgia Tech Research Corporation (1998). 10th WWW user survey, (accessed January 8, 2001), [available at: http://www.cc.gatech.edu/gvu/user_surveys]
91. Fox, S., Lee, R., Horrigan, J., Lenhart, A., Tom, S., & Carter, C. (2000). Trust and privacy online: Why Americans want to rewrite the rules
92. Milne, G. R., & Culnan, M. J. (2004). Strategies for Reducing Online Privacy Risks: Why Consumers Read [or Don't Read] Online Privacy Notices. Journal of Interactive Marketing, 18(3)
93. Peixoto M, Silva C, Lima R, Ara´ujo J, Gorschek T, Silva J (2019) PCM Tool: Privacy Requirements Specification in Agile Software Development. In: 10th Brazilian Software Conference: Theory and Practice (CBSoft'19), Extended Annals of the, SBC, Porto Alegre, RS, Brasil, pp 108–113. https://doi.org/10.5753/cbsoft_estendido.2019.7666
94. Peixoto M, Ferreira D, Cavalcanti M, Silva C, Vilela J, Araújo J, Gorschek T (2020) On undertanding how developers perceive and interpret privacy requirements research preview. In: Requirements Engineering: Foundation for Software Quality: 26th International Working Conference, REFSQ 2020, Pisa, Italy, March 24–27, 2020, Proceedings, Springer-Verlag, Berlin, Heidelberg, pp 116–123. https://doi.org/10.1007/978-3-030-44429-7_8
95. Medeiros J, Vasconcelos A, Silva C, Goulão M (2018) Quality of software requirements specification in agile projects: A cross-case analysis of six companies. J Syst Softw 142:171–194. https://doi.org/10.1016/j.jss.2018.04.064
96. Canedo ED, Bandeira IN, Calazans ATS, Costa PHT, Cançado ECR, Bonifácio R (2022) Privacy requirements elicitation: a systematic literature review and perception analysis of it practitioners. Requir Eng. https://doi.org/10.1007/s00766-022-00382-8
97. Martin, Y.S., Kung, A.: Methods and tools for GDPR compliance through privacy and data protection engineering. In: 2018 IEEE European Symposium on Security and Privacy Workshops (EuroS&PW), pp. 108–111. IEEE (2018)
98. Ramezanifarkhani, T., Owe, O., Tokas, S.: A secrecy-preserving language for distributed and object-oriented systems. J. Logic. Algebr. Methods Program. 99, 1–25 (2018)
99. Diver, L., Schafer, B.: Opening the black box: Petri nets and privacy by design. Int. Rev. Law Comput. Technol. 31(1), 68–90 (2017)
100. Hjerppe, K., Ruohonen, J., Leppänen, V.: Extracting layered privacy language purposes from web services. In: 2020 IEEE European Symposium on Security and Privacy Workshops (EuroS PW), pp. 318–325 (2020)
101. Bhavnani, S. P., Parakh, K., Atreja, A., Druz, R., Graham, G. N., Hayek, S. S., Krumholz, H. M., Maddox, T. M., Majmudar, M. D., Rumsfeld, J. S., and Shah, B. R., 2017 Roadmap for Innovation—ACC Health Policy Statement on Healthcare Transformation in the Era of Digital Health, Big Data, and Precision Health: A Report of the American College of Cardiology Task

Force on Health Policy Statements and Systems of Care. Journal of the American College of Cardiology 70(21):2696–2718, 2017. https://doi.org/10.1016/j.jacc.2017.10.018

102. Verizon, Protected health information data breach report, 2018
103. Zawoad, S., Dutta, A. K., and Hasan R., SecLaaS: secure logging-as-a-service for cloud forensics. In: Proceedings of the 8th ACM SIGSAC symposium on Information, computer and communications security. ACM, pp 219–230, 2013
104. Nanda, S., Hansen, R. A., Forensics as a Service: Three-tier Architecture for Cloud based Forensic Analysis. In: Parallel and Distributed Computing (ISPDC), 2016 15th International Symposium on, 2016. IEEE, pp 178–183
105. Zawoad, S., and Hasan, R., Faiot: Towards building a forensics aware eco system for the internet of things. In: Services Computing (SCC), 2015 IEEE International Conference on. IEEE, pp 279–284, 2015
106. Raju, B. K., Moharil, B., Geethakumari G FaaSeC: Enabling Forensics-as-a-Service for Cloud Computing Systems. In: 2016 IEEE/ACM 9th International Conference on Utility and Cloud Computing (UCC). pp 220–227, 2016
107. Liu X, Wang Z, Jin C, Li F, Li G (2019) A blockchain-based medical data sharing and protection scheme. IEEE Access 7:118943–118953
108. Shen B, Guo J, Yang Y (2019) MedChain: Efficient healthcare data sharing via blockchain. Appl Sci 9(6):1207
109. Chen Y, Ding S, Xu Z, Zheng H, Yang S (2019) Blockchain-based medical records secure storage and medical service framework. J Med Syst 43(1):1–9
110. Shi S, He D, Li L, Kumar N, Khan MK, Choo KK (2020) Applications of blockchain in ensuring the security and privacy of electronic health record systems: A survey. Comput Secur 97:101966
111. Lu X, Cheng X (2019) A secure and lightweight data sharing scheme for Internet of medical things. IEEE Access 8:5022–5030
112. Pushpa B (2020) Hybrid Data Encryption Algorithm for Secure Medical Data Transmission in Cloud Environment. In: 2020 Fourth International Conference on Computing Methodologies and Communication (ICCMC) IEEE 329–334
113. Cheng X, Chen F, Xie D, Sun H, Huang C (2020) Design of a secure medical data sharing scheme based on blockchain. J Med Syst 44(2):1–1
114. Qiu H, Qiu M, Liu M, Memmi G (2020) Secure health data sharing for medical cyber-physical systems for the healthcare 4.0. IEEE J Biomed Health Inform 24(9):2499–505
115. Fang L, Yin C, Zhu J, Ge C, Tanveer M, Jolfaei A, Cao Z (2020) Privacy protection for medical data sharing in smart healthcare. ACM Transactions on Multimedia Computing, Communications, and Applications (TOMM 16(3s):1–8
116. Akash SS, Ferdous MS (2022) A Blockchain Based System for Healthcare Digital Twin. IEEE Access
117. Xu Z, Wu L, Khan MK, Choo KK, He D (2017) A secure and efficient public auditing scheme using RSA algorithm for cloud storage. J Supercomput 73(12):5285–5309
118. Gangireddy VK, Kannan S, Subburathinam K (2021) Implementation of enhanced blowfish algorithm in cloud environment. J Ambient Intell Humaniz Comput 12(3):3999–4005
119. Akhil KM, Kumar MP, Pushpa BR (2017) Enhanced cloud data security using AES algorithm. In: 2017 International Conference on Intelligent Computing and Control (I2C2). IEEE, pp 1–5
120. Ganorkar SS, Vishwakarma SU, Pande SD (2014) An information security scheme for cloud based environment using 3DES encryption algorithm. International Journal of Recent Development in Engineering and Technology 2(4)
121. Basri M, Mawengkang H, Zamzami EM (2018) Cloud Computing Security Model with Combination of Data Encryption Standard Algorithm (DES) and Least Significant Bit (LSB). InJournal of Physics: Conference Series, IOP Publishing 970(1):012027
122. Deepa N, Pandiaraja P (2021) E health care data privacy preserving efficient file retrieval from the cloud service provider using attribute based file encryption. J Ambient Intell Humaniz Comput 12(5):4877–4887

123. Barati M, Aujla GS, Llanos JT, Duodu KA, Rana OF, Carr M, Rajan R (2021) Privacy-Aware cloud auditing for gdpr compliance verification in online healthcare. IEEE Transactions on Industrial Informatics
124. Zulkifl Z, Khan F, Tahir S, Afzal M, Iqbal W, Rehman A, Saeed S, Almuhaideb AM (2022) FBASHI: Fuzzy and Blockchain-Based Adaptive Security for Healthcare IoTs. IEEE Access 10:15644–15656
125. Namasudra S, Sharma P, Crespo RG (2022) Shanmuganathan V. Blockchain-based medical certificate generation and verification for IoT-based healthcare systems. IEEE Consumer Electronics Magazine
126. Pang Z, Yao Y, Li Q, Zhang X, Zhang J (2022) Electronic Health Records Sharing Model based on Blockchain with Checkable State PBFT Consensus Algorithm. IEEE Access 10:87803–87815
127. Carter AB. Considerations for genomic data privacy and security when working in the cloud. J Mol Diagnost. 2019;21(4):542–52
128. Frizzo-Barker J, Chow-White PA, Charters A, Ha D. Genomic big data and privacy: challenges and opportunities for precision medicine. Comput Support Coop Work. 2016;25(2):115–36
129. Schwab AP, Luu HS, Wang J, Park JY. Genomic privacy. Clin Chem. 2018;64(12):1696–703
130. Shen H, Ma J. Privacy challenges of genomic big data. Adv Exp Med Biol. 2017;1028:139–48
131. Ayday E, De Cristofaro, Hubaux J-P, Tsudik G. The chills and thrills of whole genome sequencing. Computer. 2013a. https://doi.org/10.1109/mc.2013.333
132. Ayday E, Raisaro JL, Hubaux J-P, Rougemont J. Protecting and evaluating genomic privacy in medical tests and personalized medicine. In: Proceedings of the 12th ACM workshop on workshop on privacy in the electronic society, 95–106: ACM; 2013b
133. Erlich Y, Narayanan A. Routes for breaching and protecting genetic privacy. Nat Rev Genet. 2014;15(6):409–21
134. Jagadeesh, Karthik A., David J. Wu, Johannes A. Birgmeier, Dan Boneh, and Gill Bejerano. 2017. Deriving genomic diagnoses without revealing patient genomes. Science 357 (6352): 692–695. https://doi.org/10.1126/science.aam9710
135. Lauter K, López-Alt A, Naehrig M. Private computation on encrypted genomic data. In: Progress in cryptology—LATINCRYPT 2014, 3–27: Springer International Publishing; 2015
136. Sousa JS, Lefebvre C, Huang Z, Raisaro JL, Aguilar-Melchor C, Killijian M-O, et al. Efficient and secure outsourcing of genomic data storage. BMC Med Genom. 2017;10(Suppl 2):46
137. Kubiritova Z, Gyuraszova M, Nagyova E, Hyblova M, Harsanyova M, Budis J, et al. On the critical evaluation and confirmation of germline sequence variants identified using massively parallel sequencing. J Biotechnol. 2019;298(June):64–75
138. Cogo VV, Bessani A, Couto FM, Verissimo P. A high-throughput method to detect privacy-sensitive human genomic data. In: Proceedings of the 14th ACM workshop on privacy in the electronic society, 101–10: ACM; 2015
139. Decouchant J, Fernandes M, Voelp M, Couto FM, Esteves-Verissimo P. Accurate filtering of privacy-sensitive information in raw genomic data; 2018. https://doi.org/10.1101/292185
140. Lowrance, W., Learning from experience: privacy and the secondary use of data in health research. J. Health Serv. Res. Policy 8(supp 1):S1:2–7, 2003
141. Willison, D. J., Keshavjee, K., Nair, K., Goldsmith, C., and Holbrook, A. M., Patients' consent preferences for research uses of information in electronic medical records: Interview and survey data. BMJ 326:373-376, 2003
142. Dreiseitl, S., Vinterbo, S., and Ohno-Machado, L., Disambiguation data: Extracting information from anonymized sources. J. Am. Med. Inform. Assoc. 9(6):110–114, 2002
143. Muralidhar, K., and Sarathy, R., Can you have your cake and eat it too? Sharing healthcare data without compromising privacy and confidentiality. In: Proceedings of the 11th Americas Conference on Information Systems. Omaha, NE, 2005
144. Boneh, D., Crescenzo, G. D., Ostrovsky, R., and Persiano, G., Searchable Public Key Encryption. http://eprint.iacr.org/2003/195, 2003
145. Opinion 05/2014 on Anonymisation Techniques, (2014)

146. European Medicines Agency. External guidance on the implementation of the European Medicines Agency policy on the publication of clinical data for medicinal products for human use. 2018. https://www.ema.europa.eu/en/documents/regulatory-procedural-guideline/external-guidance-implementation-european-medicines-agency-policy-publication-clinical-data_en-3.pdf . Accessed 27 Jan 2022
147. PhUSE De-Identification Working Group. Providing de-identification standards to CDISC data models. 2015. https://www.pharmasug.org/proceedings/2015/DS/PharmaSUG-2015-DS10.pdf . Accessed 07 Feb 2022
148. European Medicines Agency. European Medicines Agency policy on publication of clinical data for medicinal products for human use. 2019. https://www.ema.europa.eu/en/documents/other/european-medicines-agency-policy-publication-clinical-data-medicinal-products-human-use_en.pdf . Accessed 27 Jan 2022
149. Cohen IG, Mello MM. Big data, big tech, and protecting patient privacy. JAMA. 2019;322(12):1141–2. https://doi.org/10.1001/jama.2019.11365
150. European Parliament. Regulation (EC) No 1049/2001 of the European Parliament and of the Council of 30 May 2001 regarding public access to European Parliament, Council and Commission documents. 2001. https://eur-lex.europa.eu/legal-content/EN/TXT/?uri=CELEX:32001R1049. Accessed 27 Jan 2022
151. Magura Z, Zhou TG, Musungwini S (2021) A guiding framework for enhancing database security in state-owned universities in Zimbabwe. Afr J Sci Technol Innov Dev 14(7):1761–1775. https://doi.org/10.1080/20421338.2021.1984010
152. Dolezel D, McLeod A (2019) Managing security risk: modeling the root causes of data breaches. Health Care Manag 38(4):322–330. https://doi.org/10.1097/HCM.0000000000000282
153. Haislip J, Lim JH, Pinsker R (2021) The impact of executives' IT expertise on reported data security breaches. Inf Syst Res 32(2):318–334. https://doi.org/10.1287/isre.2020.0986
154. Sen R, Borle S (2015) Estimating the contextual risk of data breach: an empirical approach. J Manag Inf Syst 32(2):314–341. https://doi.org/10.1080/07421222.2015.1063315
155. Culnan MJ, Armstrong PK (1999) Information privacy concerns, procedural fairness, and impersonal trust: an empirical investigation. Organ Sci 10(1):104–115
156. Westin AF (1991) Harris-Equifax consumer privacy survey 1991. Equifax Inc., Atlanta, Georgia
157. Ackerman MS, Cranor L, Reagle J (1999) Privacy in e-commerce: examining user scenarios and privacy preferences. In: Proceedings of ACM conference on electronic commerce (EC'99), Denver, Colorado, November 1999, pp 1–8
158. Spiekermann S, Grossklags J, Berendt B (2001) E-privacy in 2nd generation e-commerce: privacy preferences versus actual behavior. In: Proceedings of ACM conference on electronic commerce (EC 2001), Tampa, Florida, October 2001, pp 38–46
159. Cranor LF (2003) Web privacy with P3P. O'Reilly, Cambridge, Massachusetts
160. Hockheiser H (2002) The platform for privacy preference as a social protocol: an examination within the US policy context. ACM Trans Internet Technol 2(4):276–306
161. Ackerman MS (2000) The intellectual challenge of CSCW: the gap between social requirements and technical feasibility. Hum Comput Interact 15(2–3):179–204
162. EU Digital COVID Certificate (Online). https://ec.europa.eu/info/live-work-travel-eu/coronavirus-response/safe-covid-19-vaccines-europeans/eu-digital-covid-certificate
163. Yaqoob, I., Salah, K., Jayaraman, R., et al.: Blockchain for healthcare data management: opportunities, challenges, and future recommendations. Neural Comput. Appl. (2021). https://doi.org/10.1007/s00521-020-05519-w
164. Ahmad, R.W., Salah, K., Jayaraman, R., Yaqoob, I., Ellahham, S., Omar, M.: The role of blockchain technology in telehealth and telemedicine. Int. J. Med. Inform. 148, 104399 (2021)
165. Griggs, K.N., Ossipova, O., Kohlios, C.P., et al.: Healthcare blockchain system using smart contracts for secure automated remote patient monitoring. J. Med. Syst. 42, 130 (2018). https://doi.org/10.1007/s10916-018-0982-x

166. Odoom, J., Soglo, R.S., Danso, S.A., Xiaofang, H.: A privacy-preserving Covid-19 updatable test result and vaccination provenance based on blockchain and smart contract. In: 2019 International Conference on Mechatronics, Remote Sensing, Information Systems and Industrial Information Technologies (ICMRSISIIT), pp. 1–6 (2019). https://doi.org/10.1109/ICMRSISIIT46373.2020.9405872
167. Eisenstadt, M., Ramachandran, M., Chowdhury, N., Third, A., Domingue, J.: Covid-19 antibody test/vaccination certification: there's an app for that. IEEE Open J. Eng. Med. Biol. 1, 148–155 (2020)
168. Guidi, B., Michienzi, A., Ricci, L.: Data persistence in decentralized social applications: the IPFS approach. In: 2021 IEEE 18th Annual Consumer Communications & Networking Conference (CCNC), pp. 1–4 (2021). https://doi.org/10.1109/CCNC49032.2021.9369473
169. Manzoor, A., Braeken, A., Kanhere, S.S., Ylianttila, M., Liyanage, M.: Proxy re-encryption enabled secure and anonymous IoT data sharing platform based on blockchain. J. Netw. Comput. Appl. 176, 102917 (2021). https://doi.org/10.1016/j.jnca.2020.102917
170. Meena, D.K., Dwivedi, R., Shukla, S.: Preserving patient's privacy using proxy re-encryption in permissioned blockchain. In: 2019 Sixth International Conference on Internet of Things: Systems, Management and Security (IOTSMS), pp. 450–457 (2019). https://doi.org/10.1109/IOTSMS48152.2019.8939226
171. AM Arshad, J Akmal, A Abdullah, S Ahmad, F Imran, M Riaz, F (2020) A first look at contact tracing apps. Url https://arxiv.org/pdf/2006.13354.pdf
172. D. Zeinalipour-Yazti and C. Claramunt (2020) COVID-19 mobile contact tracing apps (MCTA): a digital vaccine or a privacy demolition?, 2020 21st IEEE International Conference on Mobile Data Management (MDM), Versailles, France https://doi.org/10.1109/MDM48529.2020.00020
173. Jaiswal R, Agarwal A, Negi R. Smart solution for reducing the COVID-19 risk using smart city technology. IET Smart Cities. 2020;2(2):82–8
174. Ahmed N, Michelin R, Xue W, Ruj S, Malaney R, Kanhere S, Seneviratne A, Hu W, Janicke H, Jha S. A Survey of COVID-19 contact tracing apps. IEEE Access. 2020;8:134577–601. https://doi.org/10.1109/ACCESS.2020.3010226
175. Michael K, Abbas R. Getting behind COVID-19 contact trace apps: the google-apple partnership. IEEE Cons Electr Mag. 2020. https://doi.org/10.1109/MCE.2020.3002492
176. Abbas R, Michael K. COVID-19 contact trace app deployments: learnings from Australia and Singapore. IEEE Cons Electr Mag. 2020;9(5):65–70. https://doi.org/10.1109/MCE.2020.3002490
177. Hernández-Orallo E, Manzoni P, Calafate CT, Cano J. Evaluating how smartphone contact tracing technology can reduce the spread of infectious diseases: the case of COVID-19. IEEE Access. 2020;8:99083–97. https://doi.org/10.1109/ACCESS.2020.2998042
178. Dubov A, Shoptawb S. The value and ethics of using technology to contain the COVID-19 epidemic. Am J Bioethic. 2020;20(7):W7–11. https://doi.org/10.1080/15265161.2020.1764136
179. Basu S. Effective contact tracing for COVID-19 using mobile phones: an ethical analysis of the mandatory use of the aarogya setu application in India. Camb Quarter Healthcare Ethics. 2020. https://doi.org/10.1017/S0963180120000821
180. J Bay, J Kek, A Tan, CS Hau, L Yongquan, J Tan, and TA Quy (2020) BlueTrace: a privacy-preserving protocol for community-driven contact tracing across borders. Govt Technol Agency-Singapore Tech Rep. https://bluetrace.io/static/bluetrace_whitepaper-938063656596c104632def383eb33b3c.pd . Accessed 31 Jan 2021
181. Bassi S, Chaudhary A. Cloud computing data security-background and benefits. Int J Comp Sci Commun. 2015;6(1):34–40
182. A Ako. (2017) Advanced encryption standard (AES) algorithm to encrypt and decrypt data. Url: https://www.researchgate.net/publication/317615794. Accessed 31 Jan 2021
183. Whaiduzzaman M, Hossain MR, Shovon AR, Roy S, Laszka A, Buyya R, Barros A. A privacy-preserving mobile and fog computing framework to trace and prevent COVID-19 community transmission. IEEE J Biomed Health Informat. 2020;24(12):3564–75. https://doi.org/10.1109/JBHI.2020.3026060

184. W Lv, S Wu, C Jiang, Y Cui, X Qiu, Y Zhang (2020) Decentralized blockchain for privacy-preserving large-scale contact tracing Url: https://arxiv.org/abs/2007.00894
185. DJ Leith, S Farrell (2020) Measurement-based evaluation of google/apple exposure notification API for proximity detection in a commuter bus. Url: https://arxiv.org/abs/2006.08543
186. Xu H, Zhang L, Onireti O, Fang Y, Buchanan WJ, Imran MA. BeepTrace: blockchain-enabled privacy-preserving contact tracing for COVID-19 pandemic and beyond. IEEE Int Things J. 2020. https://doi.org/10.1109/JIOT.2020.3025953
187. Wood G (2014) Ethereum: A secure decentralised generalised transaction ledger. [Online]. Available: http://gavwood.com/Paper.pdf
188. Bø EE, Slemrod J, Thoresen TO (2015) Taxes on the internet: Deterrence effects of public disclosure. American Economic Journal: Economic Policy 7(1):36–62
189. Theis T N, Wong H P (2017) The end of Moore's Law: A new beginning for information technology. Comput Sci Eng 19(2):41–50
190. Poon J, Dryja T (2016) The Bitcoin lightning network: scalable off-chain instant payments
191. The Raiden network. [Online]. Available: https://raiden.network/
192. Goldwasser S, Micali S, Rackoff C (1989) The knowledge complexity of interactive proof systems. SIAM J Comput 18(1):186–208
193. Micali S (2000) Computationally sound proofs. SIAM J Comput 30(4):1253–1298
194. Groth J (2010) Short non-interactive zero-knowledge proofs. In: International conference on the theory and application of cryptology and information security. Springer, Berlin, pp 341–358
195. Hayes GR (2007) Documenting and understanding everyday activities through the selective archiving of live experiences. Doctoral thesis, Georgia Institute of Technology, Atlanta
196. Google.com. http://google.com/gmm/mylocation.html
197. Loki. http://www.loki.com/
198. Quova. http://www.quova.com/
199. Danezis G, Lewis S, Anderson R (2005) How much is location privacy worth? In: Fourth workshop on the economics of information security. Harvard University, Cambridge
200. Cvrček D et al (2006) A study on the value of location privacy. In: Fifth ACM workshop on privacy in the electronic society. ACM, Alexandria, VA, USA, pp 109–118
201. Duckham M, Kulik L (2006) Location privacy and location-aware computing. In: Drummond J (ed) Dynamic & mobile GIS: investigating change in space and time. Boca Raton, CRC Press, pp 34–51
202. Hoh B et al (2007) Preserving privacy in GPS traces via uncertainty-aware path cloaking. In: 14th ACM conference on computer and communication security (ACM CCS 2007), Alexandria, VA, USA
203. Avoine G (2006) Bibliography on security and privacy in RFID systems. http://www.epfl.ch/~gavoine/rfid/
204. European Union (2007) European policy outlook RFID (draft version). Working document, German Federal Ministry of Economics and Technology, June 2007. http://www.nextgenerationmedia.de/Nextgenerationmedia/Navigation/en/rfid-conference.html
205. Langheinrich M (2007) RFID and privacy. In: Petkovic M, Jonker W (eds) Security, privacy, and trust in modern data management. Springer, Heidelberg, pp 433–450
206. Weis SA, Sarma SE, Rivest RL, Engels DW (2003) Security and privacy aspects of low-cost radio frequency identification systems. In: Hutter D, Müller G, Stephan W, Ullmann M (eds) Security in pervasive computing—first international conference, Boppard, Germany, 12–14 March 2003, Revised Papers, Lecture Notes in Computer Science, vol 2802. Springer, Heidelberg, pp 201–212. http://www.springerlink.com/openurl.asp?genre=issue&issn=0302-9743&volume=2802
207. European Commission (EC) (2007) Communication from the Commission to the European Parliament, the Council, the European Economic and Social Committee and the Committee of the Regions on radio frequency identification RFID in Europe: Steps towards a policy framework. COM/2007/0096 final, March 2007. http://eur-lex.europa.eu/LexUriServ/LexUriServ.do?uri=CELEX:52007DC0096:EN:NOT

208. European Data Protection Supervisor (EDPS) (2007) Opinion of the European Data Protection Supervisor on the communication from the Commission to the European Parliament, the Council, the European Economic and Social Committee and the Committee of the Regions on radio frequency identification (RFID) in Europe: steps towards a policy framework COM(2007)96, December 2007. http://www.edps.europa.eu/EDPSWEB/webdav/site/mySite/shared/Documents/Consultation/Opinions/2007/07-12-20_RFID_EN.pdf
209. Fabian B, Günther O, Spiekermann S (2005) Security analysis of the object name service for RFID. In: Proceedings of the first international workshop on security, privacy and trust in pervasive and ubiquitous computing, SecPerU 2005, in conjunction with IEEE ICPS'05, Santorini, Greece, July 14, 2005. http://cgi.di.uoa.gr/~spu2005/
210. Agrawal R, Kiernan J, Srikant R, Xu Y (2002) Hippocratic databases. In: Proceedings of the 28th international conference on very large databases (VLDB 2002). Morgan Kaufmann, Hong Kong, pp 143–154. http://www.vldb.org/conf/2002/S05P02.pdf
211. Kobsa A, Schreck J (2003) Privacy through pseudonymity in user-adaptive systems. ACM Trans Internet Technol 3(2):149–183
212. Hewlett Packard Enterprise (2015) Internet of Things research study. Tech. Rep. jul, HP
213. Bandyopadhyay D, Sen J (2011) Internet of things: applications and challenges in technology and standardization. Wirel Pers Commun 58(1):49–69
214. Roman R, Zhou J, Lopez J (2013) On the features and challenges of security and privacy in distributed internet of things. Comput Netw 57(10):2266–2279
215. Baldini G, Peirce T, Botterman M, Talacchini MC, Pereira A, Handte M, Rotondi D, Pöhls HC, Vermesan O, Baddii A, Copigneaux B, Schreckling D, Vigano L, Steri G, Piccione S, Vlacheas P, Stavroulaki V, Kelaidonis, D., Neisse, R., Tragos E, Smadja P, Hennebert C, Serrano M, Severi S, Abreu G, Kirstein PT, Varakliotis S, Skarmeta A (2015) Internet of Things: IoT governance, privacy and security issues. In: Position paper activity chain 05, IERC-European research cluster on the Internet of Things
216. Chaum DL (1988) The dining cryptographers problem: unconditional sender and recipient untraceability. J Cryptol 1(1):65–75
217. Jawurek M (2013) Privacy in smart grids. Ph.D. thesis, Friedrich-Alexander-University Erlangen-Nuernberg
218. Cuellar J, Ochoa M, Rios R (2012) Indistinguishable regions in geographic privacy. In: Proceedings of the 27th annual ACM symposium on applied computing, SAC '12. ACM, New York. pp 1463–1469. https://doi.org/10.1145/2245276.2232010
219. Rios R, Lopez J, Cuellar J (2016) Location privacy in wireless sensor networks, 1st edn. CRC Press Inc, Boca Raton
220. Guan, Z., Zhang, Y., Wu, L., Wu, J., Li, J., Ma, Y., & Hu, J. (2019). APPA: An anonymous and privacy preserving data aggregation scheme for fogenhanced IoT. Journal of Network and Computer Applications, 125, 82–92
221. Lam TY, Dongol B (2022) A blockchain-enabled e-learning platform. Interact Learn Environ 30 (7):1229–1251. https://doi.org/10.1080/10494820.2020.1716022
222. Deenmahomed HA, Didier MM, Sungkur RK (2021) The future of university education: Examination, transcript, and certificate system using blockchain. Comput Appl Eng Educ 29(5):1234–1256
223. Bellini E, Ceravolo P, Damiani E (2019) Blockchain-based e-vote-as-a-service. In: 2019 IEEE 12th International Conference on Cloud Computing (CLOUD), pp 484–486. https://doi.org/10.1109/CLOUD.2019.00085
224. Krejci S, Sigwart M, Schulte S (2020) Blockchain- and ipfs-based data distribution for the internet of things. In: Brogi A, Zimmermann W, Kritikos K (eds) service-oriented and cloud computing, pp 177–191. Springer, Cham
225. Mamun MAA, Alam SMM, Hossain MS, Samiruzzaman M (2020) A novel approach to blockchain-based digital identity system. In: Arai K, Kapoor S, Bhatia R (eds) Advances in information and communication, pp 93–112. Springer, Cham
226. Mishra RA, Kalla A, Singh NA, Liyanage M (2020) Implementation and analysis of blockchain based dapp for secure sharing of students' credentials. In: 2020 IEEE 17th Annual

Consumer Communications Networking Conference (CCNC), pp 1–2. https://doi.org/10.1109/CCNC46108.2020.9045196
227. Vidal FR, Gouveia F, Soares C (2020) Revocation mechanisms for academic certificates stored on a blockchain. In: 2020 15th Iberian Conference on Information Systems and Technologies (CISTI), pp 1–6. https://doi.org/10.23919/CISTI49556.2020.9141088
228. Liu B, Xiao L, Long J, Tang M, Hosam O (2020) Secure digital certificate-based data access control scheme in blockchain. IEEE Access 8:91751–91760. https://doi.org/10.1109/ACCESS.2020.2993921
229. Cheng H, Lu J, Xiang Z, Song B (2020) A permissioned blockchain-based platform for education certificate verification. In: Zheng Z, Dai H-N, Fu X, Chen, B (eds) blockchain and trustworthy systems, pp 456–471. Springer, Singapore
230. Curmi A, Inguanez F (2019) Blockchain based certificate verification platform. In: Abramowicz W, Paschke A (eds) business information systems workshops, pp 211–216. Springer, Cham
231. Gresch J, Rodrigues B, Scheid E, Kanhere SS, Stiller B (2019) The proposal of a blockchain-based architecture for transparent certificate handling. In: Abramowicz, W, Paschke A (eds) workshops, business information systems, pp 185–196. Springer, Cham
232. Turkanović M, Hölbl M, Košič K, Heričko M, Kamišalić A (2018) Eductx: a blockchain-based higher education credit platform. IEEE Access 6:5112–5127. https://doi.org/10.1109/ACCESS.2018.2789929
233. Liu L, Han M, Zhou Y, Parizi RM, Korayem M (2020) Blockchain-based certification for education, employment, and skill with incentive mechanism, pp 269–290. Springer, Cham. In: Choo K-KR, Dehghantanha A, Parizi RM (eds). https://doi.org/10.1007/978-3-030-38181-3_14
234. Nguyen BM, Dao T-C, Do B-L (2020) Towards a blockchain-based certificate authentication system in vietnam. Peer J Computer Science 6:266
235. Pervaiz, Z., Ghafoor, A., Aref, W.G. (2015). Precision-bounded access control using sliding-window query views for privacy-preserving data streams. IEEE Transactions on Knowledge and Data Engineering, 27 (7), 1992.
236. Khavkin, M., & Last, M. (2018). Preserving differential privacy and utility of non-stationary data streams. In 2018 IEEE International Conference on Data Mining Workshops (ICDMW) (pp. 29–34)
237. Soria-Comas J, Domingo-Ferrer J, Sánchez D, Megías D (2017) Individual differential privacy: a utility-preserving formulation of differential privacy guarantees. IEEE Trans Inf Forensics Secur 12(6):1418–1429
238. Kavakli V (2002) Goal oriented requirements engineering: a unifying framework. Req Eng J 6(4):237–251. Springer, London
239. Kreie J, Cronan TP (2000) Making ethical decisions. Commun ACM 43(12):66–71
240. Kumar K, Bjorn-Andersen N (1990) A cross-cultural comparison of IS designer values. Commun ACM 33(5):528–538
241. Arasu A, Babcock B, Babu S, Cieslewicz J, Datar M, Ito K, Motwani R, Srivastava U, Widom J (2016) STREAM: the Stanford data stream management system. In: Garofalakis MN, Gehrke J, Rastogi R (eds) Data stream management: processing high-speed data streams, data-centric systems and applications. Springer, Berlin, pp 317–336
242. de Leoni M, Mannhardt F (2015) Road traffic fine management process. Eindhoven University of Technology, Eindhoven (Dataset)
243. Dwork C, Naor M, Pitassi T, Rothblum GN (2010) Differential privacy under continual observation. In: Proceedings of the 42nd ACM symposium on theory of computing-STOC '10. ACM Press, New York
244. Mannhardt F (2016) Sepsis cases-event log. Eindhoven University of Technology, Eindhoven (Dataset)
245. Macedo R, Paulo J, Pontes R, Portela B, Oliveira T, Matos M, Oliveira R (2017) A practical framework for privacy-preserving NoSQL databases. In: SRDS. IEEE Computer Society, pp 11–20

246. Colombo P, Ferrari E (2015) Privacy aware access control for big data: a research roadmap. Big Data Res 2:145–154
247. Fazzinga B, Flesca S, Furfaro F, Pontieri L (2018) Online and offline classification of traces of event logs on the basis of security risks. J Intell Inf Syst 50(1):195–230
248. Mannhardt F, Petersen S, de Oliveira MFD (2018) Privacy challenges for process mining in human-centered industrial environments. In: 14th international conference on intelligent environments (IE). IEEE Xplore, pp 64–71
249. Michael J, Koschmider A, Mannhardt F, Baracaldo N, Rumpe B (2019) User-centered and privacy-driven process mining system design for IoT. In: information systems engineering in responsible information systems-CAiSE forum 2019, Rome, Proceedings, pp 194–206
250. Adam K, Netz L, Varga S, Michael J, Rumpe B, Heuser P, Letmathe P (2018) Model-based generation of enterprise information systems. In: Fellmann M, Sandkuhl K (eds) Enterprise modeling and information systems architectures (EMISA'18), volume 2097 of CEUR workshop proceedings, pp 75–79. CEUR-WS.org
251. Barnum, S., Gegick, M.: Design principles. https://buildsecurityin.us-cert.gov/articles/knowledge/principles/design-principles 19 Sept (2005)
252. Howard, M., LeBlanc, D.: Writing Secure Code, 2nd edn. Microsoft Press, Redmond, WA (2003)
253. McGraw, G.: Software Security: Building Security. Addison-Wesley Professional, Boston (2006)
254. Spiekermann S (2012) The challenges of privacy by design, vol 55. ACM, New York. https://doi.org/10.1145/2209249.2209263
255. Viega, J., McGraw, G.: Building Secure Software: How to Avoid Security Problems the Right Way. Addison-Wesley, Boston, MA (2001)
256. Popa, R.A., Redfield, C., Zeldovich, N., Balakrishnan, H.: CryptDB: protecting confidentiality with encrypted query processing. In: Proceedings of the Twenty-Third ACM Symposium on Operating Systems Principles, pp. 85–100. ACM, New York, NY, USA (2011)
257. Karegar, F., Lindegren, D., Pettersson, J.S., Fischer-Hübner, S.: Assessments of a cloud-based data wallet for personal identity management. In: Information Systems Development: Advances in Methods, Tools and Management—Proceedings of the 26th International Conference on Information Systems Development, ISD 2017, Larnaca, Cyprus, University of Central Lancashire Cyprus, September 6–8 2017 (2017)
258. Vasilomanolakis, E., Karuppayah, S., Mühlhäuser, M., Fischer, M.: Taxonomy and survey of collaborative intrusion detection. ACM Comput. Surv. (CSUR) 47(4), 55 (2015)
259. Zhou, C.V., Leckie, C., Karunasekera, S.: A survey of coordinated attacks and collaborative intrusion detection. Comput. Secur. 29(1), 124–140 (2010)
260. Benali, F., Bennani, N., Gianini, G., Cimato, S.: A distributed and privacy-preserving method for network intrusion detection. In: OTM Confederated International Conferences On the Move to Meaningful Internet Systems, pp. 861–875. Springer (2010)
261. Dara, S., Muralidhara, V.: Privacy preserving architectures for collaborative intrusion detection. arXiv preprint arXiv:1602.02452 (2016)
262. Zhang, P., Huang, X., Sun, X., Wang, H., Ma, Y.: Privacy-preserving anomaly detection across multi-domain networks. In: 2012 9th International Conference on Fuzzy Systems and Knowledge Discovery (FSKD), pp. 1066–1070. IEEE (2012)
263. Morais, A., Cavalli, A.: A distributed and collaborative intrusion detection architecture for wireless mesh networks. Mobile Netw. Appl. 19(1), 101–120 (2014). https://doi.org/10.1007/s11036-013-0457-8
264. Zhou, C.V., Karunasekera, S., Leckie, C.: Evaluation of a decentralized architecture for large scale collaborative intrusion detection. In: 2007 10th IFIP/IEEE International Symposium on Integrated Network Management, pp. 80–89 (2007)
265. Dumas, M., La Rosa, M., Mendling, J., Reijers, H.: Fundamentals of Business Process Management. Springer, Berlin (2013)
266. OMG. Business Process Model and Notation (BPMN). http://www.omg.org/spec/BPMN/2.0/

267. Danezis, G., Domingo-Ferrer, J., Hansen, M., Hoepman, J.-H., Metayer, D.L., Tirtea, R., Schiffner, S.: Privacy and data protection by design-from policy to engineering. Technical report, European Union Agency for Network and Information Security (2015)
268. Rodriguez, A., Fernandez-Medina, E., Piattini, M.: A BPMN extension for the modeling of security requirements in business processes. IEICE Trans. Inf. Syst. 90(4), 745–752 (2007)
269. Salnitri, M., Dalpiaz, F., Giorgini, P.: Modelling and verifying security policies in business processes. Lect. Notes Bus. Inf. Process. LNBIP 175, 200–214 (2014)
270. Ansari, M.T.J., Pandey, D., & Alenezi, M.: Store: security threat oriented requirements engineering methodology. J. King Saud Univ. Comput. Inf. Sci. (2018)
271. Baldassarre, M. T., Santa Barletta, V., Caivano, D., & Scalera, M.: Integrating security and privacy in software development. Softw. Qual. J. pp. 1–32 (2020)
272. Mai PX, Goknil A, Shar LK, Pastore F, Briand LC, Shaame S (2018) Modeling security and privacy requirements: a use case-driven approach. Inf Softw Technol 100:165–182. https://doi.org/10.1016/j.infsof.2018.04.007
273. He, Q., & Antón, A. I.: A framework for modeling privacy requirements in role engineering. In: Proceediings of REFSQ, Vol. 3, pp. 137–146 (2003)
274. Meis, R.: Problem-Based Privacy Analysis (ProPAn): A Computer-aided Privacy Requirements Engineering Method. Universitaet Duisburg-Essen, Germany (2018)
275. Beckers, K., Fabender, S., Heisel, M., & Meis, R. A problem-based approach for computer-aided privacy threat identification. In: Annual Privacy Forum, pp. 1–16. Springer, Berlin, Heidelberg (2012)
276. Sharawat, K., & Dubey, S. K.: An approach to vendor selection on usability basis by AHP and fuzzy topsis method. In: Soft Computing: Theories and Applications, pp. 595–604. Springer, Singapore (2018)
277. CNIL (2018) Privacy impact assessment methodology. https://www.cnil.fr/sites/default/files/atoms/files/cnil-pia-1-en-methodology.pdf. Accessed 15 Nov 2018
278. ICO (2017) Data protection impact assessment (DPIA). https://ico.org.uk/for-organisations/guide-to-the-general-data-protection-regulation-gdpr/data-protection-impact-assessments-dpias/. Accessed 13 Nov 2018
279. AEPD (2017) Guia Practica de Anlisis de riesgos en los tratamientos de datos personales sujetos al RGPD). https://www.aepd.es/media/guias/guia-analisis-de-riesgos-rgpd.pdf. Accessed 10 Nov 2018
280. SDM (2017) North Rhine-Westphalia State Commissioner for Data Protection and Freedom of Information: The Standard Data Protection Model (SDM) (2017)
281. Lund MS, Solhaug B, Stlen K (2010) Model-driven risk analysis: the CORAS approach. Springer, Berlin
282. Stoneburner G, Goguen AY, Feringa A (2002) Sp 800-30. Risk management guide for information technology systems. Technical report, Gaithersburg, MD, USA
283. Wagner I, Boiten E (2018) Privacy risk assessment: from art to science, by metrics, pp 225–241
284. Wuyts K, Scandariato R, Joosen W (2014) Empirical evaluation of a privacy-focused threat modeling methodology. J Syst Softw 96:122–138
285. Moore T (2017) On the harms arising from the equifax data breach of 2017. Int J Crit Infrastruct Prot 19(C):47–48. https://doi.org/10.1016/j.ijcip.2017.10.004
286. Goldsteen A, Garion S, Nadler S, Razinkov N, Moatti Y, Ta-Shma P (2017) Brief announcement: a consent management solution for enterprises. In: International conference on cyber security cryptography and machine learning. Springer, pp 189–192 (2017)
287. De Montjoye YA, Shmueli E, Wang SS, Pentland AS (2014) openpds: Protecting the privacy of metadata through safeanswers. PLoS ONE 9(7):e98790
288. Mont MC, Thyne R (2006) A systemic approach to automate privacy policy enforcement in enterprises. In: International workshop on privacy enhancing technologies. Springer, pp 118–134
289. Krol, K., Preibusch, S.: Effortless privacy negotiations. IEEE Security & Privacy 13(3), 88–91 (2015)

290. Such, J.M., Espinosa, A., García-Fornes, A.: A survey of privacy in multi-agent systems. The Knowledge Engineering Review 29(03), 314–344 (2014)
291. Nissenbaum, H.: Privacy in context: Technology, policy, and the integrity of social life. Stanford University Press (2009)
292. Nissenbaum, H. (2011). A contextual approach to privacy online. Daedalus, 140(4), 32–48
293. Bennicke, M., & Langendörfer, P. (2003). Towards automatic negotiation of privacy contracts for internet services. In The 11th IEEE International Conference on Networks, pp. 319–324. IEEE
294. Cheng, V.S., Hung, P.C., & Chiu, D.K. (2007). Enabling web services policy negotiation with privacy preserved using XACML. In Proceedings of the 40th Annual Hawaii International Conference on System Sciences, pp. 33–33. IEEE
295. Kalyani, Y., & Adams, C. (2006). Privacy negotiation using a mobile agent. In Electrical and Computer Engineering, 2006. CCECE'06. Canadian Conference on
296. Maaser, M., & Langendoerfer, P. (2005). Automated negotiation of privacy contracts. In Computer Software and Applications Conference, 2005. COMPSAC 2005. 29th Annual International, vol. 1, pp. 505–510. IEEE
297. Maaser, M., Ortmann, S., & Langendörfer, P. (2006). NEPP: Negotiation enhancements for privacy policies. In W3C Workshop on Languages for Privacy Policy Negotiation and Semantics-Driven Enforcement
298. Preibusch, S. (2006). Implementing privacy negotiations in e-commerce. In Asia-Pacific Web Conference, pp. 604–615. Springer
299. Baarslag, T., Alan, A.T., Gomer, R., Alam, M., Perera, C., & Gerding, E.H., et al. (2017). An automated negotiation agent for permission management. In Proceedings of the 16th Conference on Autonomous Agents and Multi-Agent Systems, pp. 380–390. International Foundation for Autonomous Agents and Multiagent Systems
300. Fatima, S., Kraus, S., & Wooldridge, M. (2014). Principles of automated negotiation. Cambridge University Press
301. Aydoğan, R., Baarslag, T., Fujita, K., Mell, J., Gratch, J., Jonge, D.d., Mohammad, Y., Nakadai, S., Morinaga, S., & Osawa, H., et al. (2020). Challenges and main results of the automated negotiating agents competition (ANAC) 2019. In Multi-agent systems and agreement technologies, pp. 366–381. Springer
302. Jonker, C., Aydogan, R., Baarslag, T., Fujita, K., Ito, T., & Hindriks, K. (2017). Automated negotiating agents competition (ANAC). In Proceedings of the AAAI Conference on Artificial Intelligence, vol. 31
303. Seitz, L., Pierson, J. M., and Brunie, L., Encrypted storage of medical data on a grid. Methods Inf. Med. 44:2198–201, 2005
304. Brandner, R., Van Der Haak, M., Hartmann, M., Haux, R., and Schmucker, P., Electronic signature for medical documents—integration and evaluation of a public key infrastructure in hospitals. Methods Inf. Med. 41:4321–330, 2002
305. Goodfellow, I. et al. Generative adversarial nets. Adv. Neural Inf. Process. Syst. 27, 9 (2020)
306. Donahue, C., McAuley, J. & Puckette, M. Adversarial audio synthesis. in International Conference on Learning Representations (2019)
307. Wang, Meng; Qin, Yalin; Liu, Jiaojiao; Li, Weidong Identifying personal physiological data risks to the Internet of Everything: the case of facial data breach risks. Humanities and Social Sciences Communications (2023). https://doi.org/10.1057/s41599-023-01673-3
308. Bansod, Smita; Ragha, Lata Challenges in making blockchain privacy compliant for the digital world: some measures. Sādhanā (2022). https://doi.org/10.1007/s12046-022-01931-1
309. Silva, Paulo; Casaleiro, Rui; Simões, Paulo; Antunes, Nuno; Curado, Marilia; Monteiro, Edmundo Risk Management and Privacy Violation Detection in the PoSeID-on Data Privacy Platform. SN Computer Science (2020). https://doi.org/10.1007/s42979-020-00198-9
310. Holub, Petr; Müller, Heimo; Bíl, Tomáš; Pireddu, Luca; Plass, Markus; Prasser, Fabian; Schlünder, Irene; Zatloukal, Kurt; Nenutil, Rudolf; Brázdil, Tomáš Privacy risks of whole-slide image sharing in digital pathology. Nature Communications (2023). https://doi.org/10.1038/s41467-023-37991-y

311. Kanwal, Tehsin; Anjum, Adeel; Khan, Abid Privacy preservation in e-health cloud: taxonomy, privacy requirements, feasibility analysis, and opportunities. Cluster Computing (2020). https://doi.org/10.1007/s10586-020-03106-1
312. Tavani, Herman T. KDD, data mining, and the challenge for normative privacy. Ethics and Information Technology (1999). https://doi.org/10.1023/a:1010051717305
313. Anand, K.; Vijayaraj, A.; Vijay Anand, M. Privacy preserving framework using Gaussian mutation based firebug optimization in cloud computing. The Journal of Supercomputing (2022). https://doi.org/10.1007/s11227-021-04173-w
314. El-Ansari, Anas; Beni-Hssane, Abderrahim; Saadi, Mostafa; El Fissaoui, Mohamed PAPIR: privacy-aware personalized information retrieval. Journal of Ambient Intelligence and Humanized Computing (2021). https://doi.org/10.1007/s12652-020-02736-y
315. Mühlhoff, Rainer Predictive privacy: towards an applied ethics of data analytics. Ethics and Information Technology (2021). https://doi.org/10.1007/s10676-021-09606-x
316. Agahari, Wirawan; Ofe, Hosea; de Reuver, Mark It is not (only) about privacy: How multi-party computation redefines control, trust, and risk in data sharing. Electronic Markets (2022). https://doi.org/10.1007/s12525-022-00572-w
317. Schnell, Rainer; Klingwort, Jonas; Farrow, James M. Locational privacy-preserving distance computations with intersecting sets of randomly labeled grid points. International Journal of Health Geographics (2021). https://doi.org/10.1186/s12942-021-00268-y
318. Lwin, May; Wirtz, Jochen; Williams, Jerome D. Consumer online privacy concerns and responses: a power–responsibility equilibrium perspective. Journal of the Academy of Marketing Science (2007). https://doi.org/10.1007/s11747-006-0003-3
319. Peixoto, Mariana; Silva, Carla; Araújo, João; Gorschek, Tony; Vasconcelos, Alexandre; Vilela, Jéssyka Evaluating a privacy requirements specification method by using a mixed-method approach: results and lessons learned. Requirements Engineering (2022). https://doi.org/10.1007/s00766-022-00388-2
320. Hjerppe, Kalle; Ruohonen, Jukka; Leppänen, Ville Extracting LPL privacy policy purposes from annotated web service source code. Software and Systems Modeling (2022). https://doi.org/10.1007/s10270-022-00998-y
321. Chernyshev, Maxim; Zeadally, Sherali; Baig, Zubair Healthcare Data Breaches: Implications for Digital Forensic Readiness. Journal of Medical Systems (2018). https://doi.org/10.1007/s10916-018-1123-2
322. Vidhya, S.; Kalaivani, V. A blockchain based secure and privacy aware medical data sharing using smart contract and encryption scheme. Peer-to-Peer Networking and Applications (2023). https://doi.org/10.1007/s12083-023-01449-1
323. Hekel, Rastislav; Budis, Jaroslav; Kucharik, Marcel; Radvanszky, Jan; Pös, Zuzana; Szemes, Tomas Privacy-preserving storage of sequenced genomic data. BMC Genomics (2021). https://doi.org/10.1186/s12864-021-07996-2
324. Susilo, Willy; Win, Khin Than Security and Access of Health Research Data. Journal of Medical Systems (2007). https://doi.org/10.1007/s10916-006-9035-y
325. Maritsch, Friedrich; Cil, Ingeborg; McKinnon, Colin; Potash, Jesse; Baumgartner, Nicole; Philippon, Valérie; Pavlova, Borislava G. Data privacy protection in scientific publications: process implementation at a pharmaceutical company. BMC Medical Ethics (2022). https://doi.org/10.1186/s12910-022-00804-w
326. Li, Jin; Xiao, Wei; Zhang, Chong Data security crisis in universities: identification of key factors affecting data breach incidents. Humanities and Social Sciences Communications (2023). https://doi.org/10.1057/s41599-023-01757-0
327. Ackerman, Mark S. Privacy in pervasive environments: next generation labeling protocols. Personal and Ubiquitous Computing (2004). https://doi.org/10.1007/s00779-004-0305-8
328. Pericàs-Gornals, Rosa; Mut-Puigserver, Macià; Payeras-Capellà, M. Magdalena Highly private blockchain-based management system for digital COVID-19 certificates. International Journal of Information Security (2022). https://doi.org/10.1007/s10207-022-00598-3
329. Sowmiya, B.; Abhijith, V.S.; Sudersan, S.; Sakthi Jaya Sundar, R.; Thangavel, M.; Varalakshmi, P. A Survey on Security and Privacy Issues in Contact Tracing Application of Covid-19. SN Computer Science (2021). https://doi.org/10.1007/s42979-021-00520-z

330. Noam, Oded; Rottenstreich, Ori Realizing privacy aspects in blockchain networks. Annals of Telecommunications (2021). https://doi.org/10.1007/s12243-021-00861-z
331. Nguyen, David H.; Hayes, Gillian R. Information privacy in institutional and end-user tracking and recording technologies. Personal and Ubiquitous Computing (2009). https://doi.org/10.1007/s00779-009-0229-4
332. Krumm, John A survey of computational location privacy. Personal and Ubiquitous Computing (2008). https://doi.org/10.1007/s00779-008-0212-5
333. Langheinrich, Marc A survey of RFID privacy approaches. Personal and Ubiquitous Computing (2008). https://doi.org/10.1007/s00779-008-0213-4
334. Staudemeyer, Ralf C.; Pöhls, Henrich C.; Wójcik, Marcin What it takes to boost Internet of Things privacy beyond encryption with unobservable communication: a survey and lessons learned from the first implementation of DC-net. Journal of Reliable Intelligent Environments (2019). https://doi.org/10.1007/s40860-019-00075-0
335. Pamarthi, Satyanarayana; Narmadha, R. Adaptive Key Management-Based Cryptographic Algorithm for Privacy Preservation in Wireless Mobile Adhoc Networks for IoT Applications. Wireless Personal Communications (2021). https://doi.org/10.1007/s11277-021-09360-9
336. Dewangan, Narendra K; Chandrakar, Preeti; Kumari, Saru; Rodrigues, Joel J.P.C. Enhanced privacy-preserving in student certificate management in blockchain and interplanetary file system. Multimedia Tools and Applications (2022). https://doi.org/10.1007/s11042-022-13915-8
337. Ren, Wang; Tong, Xin; Du, Jing; Wang, Na; Li, Shancang; Min, Geyong; Zhao, Zhiwei Privacy Enhancing Techniques in the Internet of Things Using Data Anonymisation. Information Systems Frontiers (2021). https://doi.org/10.1007/s10796-021-10116-w
338. Kalloniatis, Christos; Kavakli, Evangelia; Gritzalis, Stefanos Addressing privacy requirements in system design: the PriS method. Requirements Engineering (2008). https://doi.org/10.1007/s00766-008-0067-3
339. Carew, Peter J.; Stapleton, Larry; Byrne, Gabriel J. Implications of an ethic of privacy for human-centred systems engineering. AI & SOCIETY (2007). https://doi.org/10.1007/s00146-007-0149-7
340. Mannhardt, Felix; Koschmider, Agnes; Baracaldo, Nathalie; Weidlich, Matthias; Michael, Judith Privacy-Preserving Process Mining. Business & Information Systems Engineering (2019). https://doi.org/10.1007/s12599-019-00613-3
341. Mitropoulos, Dimitris; Sotiropoulos, Thodoris; Koutsovasilis, Nikos; Spinellis, Diomidis PDGuard: an architecture for the control and secure processing of personal data. International Journal of Information Security (2019). https://doi.org/10.1007/s10207-019-00468-5
342. Spathoulas, Georgios; Theodoridis, Georgios; Damiris, Georgios-Paraskevas Using homomorphic encryption for privacy-preserving clustering of intrusion detection alerts. International Journal of Information Security (2020). https://doi.org/10.1007/s10207-020-00506-7
343. Pullonen, Pille; Tom, Jake; Matulevičius, Raimundas; Toots, Aivo Privacy-enhanced BPMN: enabling data privacy analysis in business processes models. Software and Systems Modeling (2019). https://doi.org/10.1007/s10270-019-00718-z
344. Ansari, Md Tarique Jamal; Baz, Abdullah; Alhakami, Hosam; Alhakami, Wajdi; Kumar, Rajeev; Khan, Raees Ahmad P-STORE: Extension of STORE Methodology to Elicit Privacy Requirements. Arabian Journal for Science and Engineering (2021). https://doi.org/10.1007/s13369-021-05476-z
345. Hart, Stephen; Ferrara, Anna Lisa; Paci, Federica Fuzzy-based approach to assess and prioritize privacy risks. Soft Computing (2019). https://doi.org/10.1007/s00500-019-03986-5
346. Vidhani, Kumar; Banahatti, Vijayanand; Mirza, Hussain Privacy self management in enterprise. CSI Transactions on ICT (2022). https://doi.org/10.1007/s40012-022-00347-w
347. Filipczuk, Dorota; Baarslag, Tim; Gerding, Enrico H.; schraefel, m. c. Automated privacy negotiations with preference uncertainty. Autonomous Agents and Multi-Agent Systems (2022). https://doi.org/10.1007/s10458-022-09579-1

348. Huang, Lu-Chou; Chu, Huei-Chung; Lien, Chung-Yueh; Hsiao, Chia-Hung; Kao, Tsair Embedding a Hiding Function in a Portable Electronic Health Record for Privacy Preservation. Journal of Medical Systems (2008). https://doi.org/10.1007/s10916-008-9243-8
349. Thambawita, Vajira; Isaksen, Jonas L.; Hicks, Steven A.; Ghouse, Jonas; Ahlberg, Gustav; Linneberg, Allan; Grarup, Niels; Ellervik, Christina; Olesen, Morten Salling; Hansen, Torben; Graff, Claus; Holstein-Rathlou, Niels-Henrik; Strümke, Inga; Hammer, Hugo L.; Maleckar, Mary M.; Halvorsen, Pål; Riegler, Michael A.; Kanters, Jørgen K. DeepFake electrocardiograms using generative adversarial networks are the beginning of the end for privacy issues in medicine. Scientific Reports (2021). https://doi.org/10.1038/s41598-021-01295-2
350. Maaser, Michael; Langendörfer, Peter Privacy from Promises to Protection. Mobile Networks and Applications (2008). https://doi.org/10.1007/s11036-008-0116-7

Chapter 5
Technical Growth vis-à-vis Privacy and Data Protection Issues

5.1 Introduction by the Editor

This chapter starts on a note that the use of technology and its transformative power must balance with the privacy concerns that individuals may have. There is no stopping the development but it comes with additional responsibility on the part of a data controller.

Recital 6 of the GDPR states that:

"Rapid technological developments and globalisation have brought new challenges for the protection of personal data. The scale of the collection and sharing of personal data has increased significantly. Technology allows both private companies and public authorities to make use of personal data on an unprecedented scale in order to pursue their activities. Natural persons increasingly make personal information available publicly and globally. Technology has transformed both the economy and social life, and should further facilitate the free flow of personal data within the Union and the transfer to third countries and international organisations, while ensuring a high level of the protection of personal data."[1]

The Recital underlines the requirement to balance several of the situational changes that advancement of technology brings. The scale of data collection changes amongst various stakeholders because technological growth is changing the social and economic construct in an appreciable way. Data subjects are in environments that encourage data sharing, thereby placing the legal requirement at the confluence of the free flow of data and the process of finding an optimal way to protect personal

[1] Regulation (EU) 2016/679 of the European Parliament and of the Council of 27 April 2016 on the protection of natural persons with regard to the processing of personal data and on the free movement of such data, and repealing Directive 95/46/EC (General Data Protection Regulation) (Text with EEA relevance) *OJ L 119, 4.5.2016, pp. 1–88.*

I. Gupta (ed.), *Operationalizing Expectations and Mapping Challenges of Information Privacy and Data Protection Measures*, https://doi.org/10.1007/978-981-99-7371-2_5

data.[2] Every change in how a technology functions would also need a corresponding mapping of the legal requirements. It can become rather challenging to map a technological advancement to the present legal needs owing to the complex landscape of the legal provisions.[3] Therefore, there is a need to follow an optimal approach to evaluate the legal compliance of a technology.[4] Privacy Oriented Software Development has been one such approach that helps in deciphering the software vulnerabilities.[5] It looks to integrate privacy and security practices and provides adequate guidelines to developers and designers.[6] Often it can be a difficult proposition to put in such a requirement in the operational practice of a development process.[7]

There is a need to explore security and privacy requirements in the area of critical research that requires sharing of large-scale data amongst institutions.[8] The privacy-preserving infrastructures aim to keep such data anonymous; however, there could be operational challenges.[9] The idea of privacy must be preserved alongside harnessing the benefits of research in critical areas.[10] For instance, research concerning rare disease is critical. Equally, patients' willingness to share data depends on the degree of trust that they repose on data controllers.[11] The questions are about the efforts that are required to protect privacy alongside the legal basis of processing personal

---

[2] See generally, *Devriendt, Thijs; Shabani, Mahsa; Lekadir, Karim; Borry, Pascal Data sharing platforms: instruments to inform and shape science policy on data sharing?. Scientometrics (2022)* https://doi.org/10.1007/s11192-022-04361-2.

[3] *Massey, Aaron K.; Otto, Paul N.; Hayward, Lauren J.; Antón, Annie I. Evaluating existing security and privacy requirements for legal compliance. Requirements Engineering (2009)* https://doi.org/10.1007/s00766-009-0089-5.

[4] Ibid.

[5] *Baldassarre, Maria Teresa; Barletta, Vita Santa; Caivano, Danilo; Scalera, Michele Integrating security and privacy in software development. Software Quality Journal (2020)* https://doi.org/10.1007/s11219-020-09501-6.

[6] Ibid.

[7] Ibid; See, *Compagna, Luca; El Khoury, Paul; Krausová, Alžběta; Massacci, Fabio; Zannone, Nicola How to integrate legal requirements into a requirements engineering methodology for the development of security and privacy patterns. Artificial Intelligence and Law (2008)* https://doi.org/10.1007/s10506-008-9067-3.

[8] *Manion, Frank J; Robbins, Robert J; Weems, William A; Crowley, Rebecca S Security and privacy requirements for a multi-institutional cancer research data grid: an interview-based study. BMC Medical Informatics and Decision Making (2009)* https://doi.org/10.1186/1472-6947-9-31.

[9] *Wirth, Felix Nikolaus; Meurers, Thierry; Johns, Marco; Prasser, Fabian Privacy-preserving data sharing infrastructures for medical research: systematization and comparison. BMC Medical Informatics and Decision Making (2021)* https://doi.org/10.1186/s12911-021-01602-x.

[10] *DeCew, Judith Wagner Alternatives for protecting privacy while respecting patient care and public health needs. Ethics and Information Technology (1999)* https://doi.org/10.1023/a:1010012018213.

[11] *Courbier, Sandra; Dimond, Rebecca; Bros-Facer, Virginie Share and protect our health data: an evidence based approach to rare disease patients' perspectives on data sharing and data protection—quantitative survey and recommendations. Orphanet Journal of Rare Diseases (2019)* https://doi.org/10.1186/s13023-019-1123-4; See, *Reischl, J; Schröder, M; Luttenberger, N; Petrov, D; Schümann, B; Ternes, R; Stürzebecher, S Pharmacogenetic research and data protection—challenges and solutions. The Pharmacogenomics Journal (2006)* https://doi.org/10.1038/sj.tpj.6500372.

sensitive data.[12] Developing a right kind of privacy perception amongst developers is a critical task. Using a vocabulary of security to address privacy concerns may not be a way forward.[13] An improved organisational privacy climate[14] improves developers' approach towards privacy and their interpretation. Research suggests that they need to move to privacy-by-architecture solution, and it is attainable by bringing about changes at multiple levels.[15] Developers often receive contrary indications by the management and other stakeholders leading to lowering the priority that privacy concerns should receive.[16] Therefore, organisational factors positively or negatively influence "developers' work environment, which has a critical role in shaping their mindsets"[17] about prioritising privacy and security. Similar research suggests how organisations assimilate privacy in their strategies in the backdrop of possible external influence.[18] Findings suggest that "... senior management support is a key factor that mediates the impact of external forces on privacy related business strategy within organizations".[19]

Advancements in the healthcare and use of IoT and industrial internet of things (IIoT)[20] enabled processes bring along questions of risking personal data. The approach has been to follow a privacy preserving data collection practices that reduce the risks associated with the disclosure of sensitive data.[21] Continual use of AI robots and collection of sensitive personal data for further growth has to be optimised with the design elements of privacy.[22] With the need for digitisation of

---

[12] *Austin, Melissa A; Harding, Sarah E; McElroy, Courtney E Monitoring ethical, legal, and social issues in developing population genetic databases. Genetics in Medicine (2003)* https://doi.org/10.1097/01.gim.0000093976.08649.1b.

[13] *Hadar, Irit; Hasson, Tomer; Ayalon, Oshrat; Toch, Eran; Birnhack, Michael; Sherman, Sofia; Balissa, Arod Privacy by designers: software developers' privacy mindset. Empirical Software Engineering (2017)* https://doi.org/10.1007/s10664-017-9517-1.

[14] Ibid.

[15] Ibid.

[16] *Arizon-Peretz, Renana; Hadar, Irit; Luria, Gil; Sherman, Sofia Understanding developers' privacy and security mindsets* via*climate theory. Empirical Software Engineering (2021)* https://doi.org/10.1007/s10664-021-09995-z.

[17] Ibid.

[18] *Attili, V. S. Prakash; Mathew, Saji K.; Sugumaran, Vijayan Information Privacy Assimilation in IT Organizations. Information Systems Frontiers (2021)* https://doi.org/10.1007/s10796-021-10158-0.

[19] Ibid.

[20] *Abdullah, Sidrah; Arshad, Junaid; Khan, Muhammad Mubashir; Alazab, Mamoun; Salah, Khaled PRISED tangle: a privacy-aware framework for smart healthcare data sharing using IOTA tangle. Complex & Intelligent Systems (2022)* https://doi.org/10.1007/s40747-021-00610-8.

[21] *Onesimu, J. Andrew; Karthikeyan, J.; Sei, Yuichi An efficient clustering-based anonymization scheme for privacy-preserving data collection in IoT based healthcare services. Peer-to-Peer Networking and Applications (2021)* https://doi.org/10.1007/s12083-021-01077-7.

[22] *Ishii, Kaori Comparative legal study on privacy and personal data protection for robots equipped with artificial intelligence: looking at functional and technological aspects. AI & SOCIETY (2017)* https://doi.org/10.1007/s00146-017-0758-8.

health records, companies handling such personal sensitive data must work towards improving trustworthiness of adopted processes.[23]

There are instances when technology adoption is an absolute necessity. For instance, use of assistive technology for supporting the elderly and vulnerable individuals.[24] Neglecting privacy issues would lead to non-adoption of these technologies. The devices using these technologies often process personal sensitive data. Therefore, organisations involved in the development of these devices must adopt privacy impact assessment measures to reduce data privacy risks.[25] These devices will further provide effective healthcare support alongside promoting "evidence-driven care practice and research activities".[26] Informed consent becomes a potent legal basis that ensures data subjects' trust towards such data processing and helps moving critical developments.[27] Providing e-Health services relying upon connected health care systems can improve upon the quality of services that patients receive. The degree of user control over their personal data becomes pivotal to the success of these services.[28]

User authentication systems should follow the broad parameters of anonymous credentials and anonymous communication channels. They should follow the data minimisation principle and the requirement of maximising privacy concerns at the design stage of a particular system.[29] (Privacy Enhancing Technologies) PETs

---

[23] *Fox, Grace; James, Tabitha L. Toward an Understanding of the Antecedents to Health Information Privacy Concern: A Mixed Methods Study. Information Systems Frontiers (2020)* https://doi.org/10.1007/s10796-020-10053-0.

[24] *Wilkowska, Wiktoria; Offermann, Julia; Colonna, Liane; Florez-Revuelta, Francisco; Climent-Pérez, Pau; Mihailidis, Alex; Poli, Angelica; Spinsante, Susanna; Ziefle, Martina Interdisciplinary perspectives on privacy awareness in lifelogging technology development. Journal of Ambient Intelligence and Humanized Computing (2022)* https://doi.org/10.1007/s12652-022-04486-5.

[25] Ibid.

[26] *Jusob, Farad Rafique; George, Carlisle; Mapp, Glenford Exploring the need for a suitable privacy framework for mHealth when managing chronic diseases. Journal of Reliable Intelligent Environments (2017)* https://doi.org/10.1007/s40860-017-0049-7.

[27] *Wicks, Paul; Wahlstrom-Edwards, Lindsey; Fillingham, Sam; Downing, Andrea; Davies, Elin Haf So You Want to Build Your Disease's First Online Patient Registry: An Educational Guide for Patient Organizations Based on US and European Experience. The Patient—Patient-Centered Outcomes Research (2023)* https://doi.org/10.1007/s40271-023-00619-w; See, *Ambrosini, Anna; Calabrese, Daniela; Avato, Francesco Maria; Catania, Felice; Cavaletti, Guido; Pera, Maria Carmela; Toscano, Antonio; Vita, Giuseppe; Monaco, Lucia; Pareyson, Davide The Italian neuromuscular registry: a coordinated platform where patient organizations and clinicians collaborate for data collection and multiple usage. Orphanet Journal of Rare Diseases (2018)* https://doi.org/10.1186/s13023-018-0918-z; See, *Scheibner, James; Ienca, Marcello; Vayena, Effy Health data privacy through homomorphic encryption and distributed ledger computing: an ethical-legal qualitative expert assessment study. BMC Medical Ethics (2022)* https://doi.org/10.1186/s12910-022-00852-2.

[28] *Abdullah, Sidrah; Arshad, Junaid; Khan, Muhammad Mubashir; Alazab, Mamoun; Salah, Khaled PRISED tangle: a privacy-aware framework for smart healthcare data sharing using IOTA tangle. Complex & Intelligent Systems (2022)* https://doi.org/10.1007/s40747-021-00610-8.

[29] *Diaz, Claudia; Kosta, Eleni; Dekeyser, Hannelore; Kohlweiss, Markulf; Nigusse, Girma Privacy preserving electronic petitions. Identity in the Information Society (2008)* https://doi.org/10.1007/s12394-009-0012-8.

provide technological interventions to the cause of protecting privacy; however, research suggests that PETS are most helpful when they help capturing data leakages.[30]

General concerns have been raised about harnessing the utility of generated data vis-à-vis associated privacy risks.[31] Use of synthetic data has emerged as an alternative to balance "…loss of information and privacy protection" compared to other approaches that increase risk of reidentification of data subjects.[32]

There are privacy concerns about the possible use of sensitive visual data[33] focussing on image and video data and deep learning of such visual data as both are vulnerable to external attacks.[34] Large scale datasets like data available on social media platforms extends innumerable opportunities to researchers. However, it is difficult to perceive data subjects' sentiments about the acceptable use of social media data compared to their understanding of acceptable use of personal sensitive data in other environments.[35]

With an advancement of technology, those collecting personal data must ensure that the control over personal data remains with the data subjects. Privacy policies should enable the data subjects to understand their rights and not become a deterrent to the understanding of the privacy risks.[36] Transparency should act as a bridge towards building trust between the data subjects and the data controllers with an overall improvement of all round accountability of data controllers.[37]

This chapter ends on a note that technology and its related growth continue to push the boundaries of data privacy. Every change in technology leads to a complex situation that forces us to revisit the optimal solution to a previous problem, and its suitability in a newer environment.

---

[30] *Dumas, Marlon; García-Bañuelos, Luciano; Jääger, Joosep; Laud, Peeter; Matulevičius, Raimundas; Pankova, Alisa; Pettai, Martin; Pullonen-Raudvere, Pille; Toots, Aivo; Tuuling, Reedik; Yerokhin, Maksym Multi-level privacy analysis of business processes: the Pleak toolset. International Journal on Software Tools for Technology Transfer (2021)* https://doi.org/10.1007/s10009-021-00636-w.

[31] *Koebe, Till; Arias-Salazar, Alejandra; Schmid, Timo Releasing survey microdata with exact cluster locations and additional privacy safeguards. Humanities and Social Sciences Communications (2023)* https://doi.org/10.1057/s41599-023-01694-y.

[32] Ibid.

[33] *Climent-Pérez, Pau; Florez-Revuelta, Francisco Protection of visual privacy in videos acquired with RGB cameras for active and assisted living applications. Multimedia Tools and Applications (2021)* https://doi.org/10.1007/s11042-020-10249-1.

[34] *Zhang, Guangsheng; Liu, Bo; Zhu, Tianqing; Zhou, Andi; Zhou, Wanlei Visual privacy attacks and defenses in deep learning: a survey. Artificial Intelligence Review (2022)* https://doi.org/10.1007/s10462-021-10123-y.

[35] *Hemphill, Libby; Schöpke-Gonzalez, Angela; Panda, Anmol Comparative sensitivity of social media data and their acceptable use in research. Scientific Data (2022)* https://doi.org/10.1038/s41597-022-01773-w.

[36] *Vidhani, Kumar; Banahatti, Vijayanand; Lodha, Sachin Challenges in enabling privacy self management. CSI Transactions on ICT (2021)* https://doi.org/10.1007/s40012-021-00335-6.

[37] Ibid.

5.2 Machine Generated Summaries

Disclaimer: The summaries in this chapter were generated from Springer Nature publications using extractive AI auto-summarization: An extraction-based summarizer aims to identify the most important sentences of a text using an algorithm and uses those original sentences to create the auto-summary (unlike generative AI). As the constituted sentences are machine selected, they may not fully reflect the body of the work, so we strongly advise that the original content is read and cited. The auto generated summaries were curated by the editor to meet Springer Nature publication standards. To cite this content, please refer to the original publications.

Machine generated keywords: patient, sharing, disease, software, genetic, population, datum share, developer, share, registry, security privacy, organizational, healthcare, datum sharing, survey.

5.2.1 Evaluating Existing Security and Privacy Requirements for Legal Compliance [1]

This is a machine-generated summary of:

Massey, Aaron K.; Otto, Paul N.; Hayward, Lauren J.; Antón, Annie I.: Evaluating existing security and privacy requirements for legal compliance [1]

Published in: Requirements Engineering (2009)

Link to original: https://doi.org/10.1007/s00766-009-0089-5

If you want to cite the papers, please refer to the original.

For technical reasons we could not place the page where the original quote is coming from.

Abstract-Summary

"In response, requirements engineers must specify compliant system requirements to satisfy applicable legal security and privacy obligations."

"We discuss our evaluation of the requirements for iTrust, an open-source Electronic Health Records system, for compliance with legal requirements governing security and privacy in the healthcare domain."

"We illustrate some of the challenges that practitioners face when specifying requirements for a system that must comply with law and close with a discussion of needed future research focusing on security and privacy requirements."

Introduction

"This paper describes shortcomings and improvements with respect to HIPAA compliance for the security and privacy software requirements in the iTrust medical records system."

"Requirements engineers evaluating systems like iTrust for legal compliance must be able to establish and maintain traceability from software artifacts to the relevant governing legal texts to demonstrate due diligence in satisfying security and privacy requirements [2]."

"This paper demonstrates how requirements engineers can evaluate security and privacy software requirements for compliance with relevant law in the form of a specific legal text."

"We discuss our study of the iTrust security and privacy requirements with respect to the HIPAA regulation."

"Our study entailed three primary activities: (1) mapping iTrust system terminology to the corresponding HIPAA terminology; (2) clarifying software security and privacy requirements with respect to legal requirements; and (3) improving traceability between the HIPAA regulations and the software requirements."

Related Work

"The methodology presented herein entails evaluating and improving a set of existing requirements that must comply with relevant privacy and security law This section discusses relevant work in legal compliance relating to software requirements as well as requirements engineering techniques employed in our methodology."

"Legal texts, however, present special problems for managing compliance: ambiguities, cross-references, domain-specific definitions, acronyms, and the evolution in law have all been identified as confounding factors facing requirements engineers working with such texts [2]."

"The regulations' broad and, at times, ambiguous nature complicates the legal compliance landscape for requirements engineers attempting to implement healthcare software systems."

"Because our focus is on security and privacy requirements, our legal compliance efforts are focused on Part 164 of the HIPAA regulations—titled "Security and Privacy"—and §160.103, which contains definitions that are a part of the "General Provisions" governing all HIPAA regulations."

Methodology to Evaluate Legal Compliance of Security and Privacy Requirements

"The three terminology categories that must be mapped are Actors, Data Objects, and Actions, which we define as follows: Actors/Stakeholders are individuals or organizations explicitly mentioned in either the software documentation under analysis or the legal text with which compliance must be achieved."

"Even if the legal text with which the software must comply is actor-focused, it still may be necessary to create a data object hierarchy for the legal text and map data objects from the requirements document to the legal text."

"Data object hierarchies may still be used, even with HIPAA, if requirements engineers and legal experts believe the complexity of either the requirements document or the legal text calls for such a mapping."

"Requirements engineers must identify all actions described by the software requirements and map them to those described in the relevant legal texts."

Evaluating Software Requirements for Compliance with Legal Texts

"Identifying the iTrust actors proved much easier than identifying the data objects or actions because the original iTrust requirements document was comprised of use cases, which describe user interactions with the resulting system."

"Aligning the different actions of iTrust and HIPAA enables the requirements engineer to specify the security and privacy requirements so that all actions are legally compliant."

"Requirement 15 has an engineering issue as described below: Requirement 15 Description: iTrust shall allow a user, using their authorized account, to request their current password be emailed to their recorded email address and shall require the user's user ID be provided to satisfy this request."

"Requirement 21, which has a legal issue, handles the management of personal representative lists as described below: Requirement 21 Description: iTrust shall allow a patient, using their authorized account, to read or update their demographic information, the demographic information for the patients they represent, their list of personal representatives, the list of personal representatives for the patients they represent, their list of designated physicians, and the list of designated physicians for the patients they represent."

Lessons Learned

"We discuss lessons regarding the use of actor (or stakeholder) hierarchies, complications resulting from unresolved ambiguity, requirements prioritization, and tool support."

"Building actor hierarchies for both the requirements document and the legal text allow the requirements engineer to clearly and unambiguously state the legal rights and obligations for actors in the resulting software."

"The single most frustrating aspect of extracting requirements from use cases was the lack of tool support."

"Support for searching particular elements of the document while ignoring others (e.g., searching just the issues and not the requirements or the glossary terms)."

"These features address several key elements identified by Otto and Antón as supporting legal compliance within requirements engineering [2]: automatic ordering of both requirements and glossary terms supports prioritizing requirements."

"The use of hyperlinks supports tracing from requirements to legal texts."

Summary and Future Work

"The regular evolution of laws and regulations creates a pressing need for a methodology for evaluating the legal compliance of existing software."

"We presented a basic approach for evaluating and improving the legal compliance of an existing software system."

"Systems built prior to the HIPAA regulations coming into effect or without HIPAA domain knowledge face legal compliance issues similar to those found in iTrust."

"Many of these software systems have insufficient requirements documentation and poor traceability to both software artifacts and legal texts."

"Testing techniques that verify functionality alone are not enough to ensure security, privacy, or legal compliance; the software requirements and design must be validated against the relevant legal texts."

"Although we attempted to evaluate and improve legal compliance as a part of our methodology, we have by no means provided a comprehensive overview of the extensive legal compliance issues faced by an EHR system [2–6]."

5.2.2 Integrating Security and Privacy in Software Development [7]

This is a machine-generated summary of:

Baldassarre, Maria Teresa; Barletta, Vita Santa; Caivano, Danilo; Scalera, Michele: Integrating security and privacy in software development [7]

Published in: Software Quality Journal (2020)

Link to original: https://doi.org/10.1007/s11219-020-09501-6

If you want to cite the papers, please refer to the original.

For technical reasons we could not place the page where the original quote is coming from.

Abstract-Summary

"In such a context, this paper proposes a software development approach, Privacy Oriented Software Development (POSD), that complements traditional development processes by integrating the activities needed for addressing security and privacy management in software systems."

"The approach is based on 5 key elements (Privacy by Design, Privacy Design Strategies, Privacy Pattern, Vulnerabilities, Context)."

"The approach can be applied in two directions forward and backward, for developing new software systems or re-engineering an existing one."

"Results show that POSD is able to discover software vulnerabilities, identify the remediation patterns needed for addressing them in the source code, and design the target architecture to be used for guiding privacy-oriented system re-engineering."

Introduction

"The lack of system security can compromise privacy and for this reason privacy emerges as a proactive, integrative, and creative approach to strengthen security requirements, starting from the design of new systems and protection of information assets and data in case of existing ones."

"The challenges that companies and developer communities need to face within this context are many, but to start implementing defenses operatively, three major issues have to be addressed: (i) translate best practices for both secure application development and data privacy into operational guidelines that can be traced back to code structures and software architectures; (ii) share security and privacy competences within the development team."

"It is necessary to share and transfer knowledge effectively (Baldassarre et al. [8]; Ardimento et al. [9]); (iii) integrate new methodologies for data privacy and secure software development into existing business processes."

Related Work

"Hilbrich and others (Hilbrich and Frank [10]) propose a strategy for a clear and engineer-like decision-making process and to include security and privacy requirements in software development processes."

"With respect to privacy, Privacy by Design (PbD) (Cavoukian [11]) is an approach to address data protection during software development and to integrate privacy throughout the system development life cycle."

"In today's environment, privacy needs to be integrated into software development to protect sensitive data in growing systems and to enhance software quality."

"The approach presented in this work (POSD) strives to overcome these weaknesses: (i) POSD integrates privacy and security practices by providing guidelines to developers that can be translated into operational practices, software architecture, and software code. (ii) POSD can be used both when developing a software system and when re-engineering an existing one. (iii) POSD can be used jointly with the legacy development processes adopted."

Privacy Knowledge Base

"The proposed approach, named Privacy Oriented Software Development (POSD), provides operational guidelines for integrating security and privacy management into any development process."

"PKB has been developed following a Model-View-Controller architecture: the model provides all the methods to access the elements of the PKB; the view visualizes the relations between the 5 elements and manages interaction with the developer; the controller receives user requests and fulfills them changing the status of the two components."

"The PKB engine integrates the results of static code analysis so that each vulnerability identified in the legacy system is associated to a privacy pattern and the architectural defects are fixed."

"Privacy by Design (PbD) integrates the principles of privacy and data protection within all stages of the software development process."

"Privacy as the default setting: PbD seeks to deliver the maximum degree of privacy by ensuring that personal data are automatically protected in any given IT system or business practice."

"Respect for User Privacy: It requires architects and operators to treat the interests of the individual by offering specific solutions for a strong privacy default, appropriate notice and empowering user-friendly options."

"Data-Oriented Strategies focus on the privacy-friendly processing of the data: Minimize: Limit the processing of personal data as much as possible."

"Process-Oriented Strategies focus on the processes surrounding the responsible handling of personal data: Inform: Inform data subjects about the processing of their personal data in a timely and adequate manner."

"Control: Provide data subjects an adequate control over the processing of their personal data."

"Enforce: Commit to processing personal data in a privacy-friendly way, and adequately enforce this."

"Demonstrate: Demonstrate personal data is being processed in a privacy-friendly manner."

"Privacy patterns support documenting common solutions to privacy problems can improve the re-engineering of existing systems, describing classes, collaborations between objects, and their purposes, and help designers identify and address privacy concerns."

"In this context, a privacy pattern represents an answer to the following questions: (i) Which Privacy Design Strategies must be implemented? (ii) Which vulnerabilities are mitigated/eliminated with the privacy solution?"

"An example of a Privacy Pattern is the following: Name: Anonymous reputation-based blacklisting."

"Context: A service provider provides a service to users who access anonymously, and who may make bad use of the service."

"A service provider can assign a reputation score to its users, based on their interactions with the service."

"Solution: First, the service provider provides their users with credentials for anonymous authentication."

"Every time an authenticated user holds a session at the service, the service provider assigns and records a reputation value for that session, depending on the user behavior during the session."

"When the user returns and starts a new session at the service, the service provider challenges the user to prove in zero-knowledge that he is not linked to any of the offending sessions (those that have a negative reputation associated)."

"Zero-knowledge proofs allow the user to prove this, without revealing their identity to the service provider."

"OWASP Top 10 is based primarily on data and information provided by firms specialized in application security or collected by using industry surveys."

"A short description of the macro categories of vulnerabilities included in the PKB follows: A1-Injection: Untrusted data is sent to an interpreter as part of a command or query."

"In SQL injection vulnerability, the root cause is the ability of an attacker to change context in the SQL query, causing a value that the programmer intended to be interpreted as data to be interpreted as a command instead."

"The query that this code intends to execute follows: However, since the query is constructed dynamically by concatenating a constant base query string and a user input string, the query only behaves correctly if itemName does not contain a single-quote character."

"If an attacker with the user name Wiley enters the string 'name' OR 'a' = 'a' for itemName, then the query becomes the following: The addition of the OR 'a' = 'a' condition causes the where clause to always evaluate to true, so the query becomes logically equivalent to the much simpler query: This simplification of the query allows the attacker to bypass the requirement that the query only return items owned by the authenticated user; the query now returns all entries stored in the items table, regardless of their specified owner."

"A2-Broken Authentication: Authentication and session management not properly implemented allow attackers to compromise data and application."

"It is crucial to use less complex data formats and avoid serialization of sensitive data; patch or upgrade all XML processors and libraries in use by the application or on the underlying operating system; disable XML external entity and DTD processing in all XML parsers in the application; implement positive server-side input validation, filtering, or sanitization to prevent hostile data within XML documents, headers, or nodes; etc. Consequently, to identify and mitigate this type of vulnerability, it is essential to train developers."

"This allows to access unauthorized functionality and/or data, modify or destroy data, or perform a business function outside of the user's limits."

"Database access control errors occur when data enters a program from an untrusted source; the data is used to specify the value of a primary key in a SQL query."

"The following example implements the same functionality but imposes an additional constraint requiring that the current authenticated user have specific access to the invoice."

"The underlying flaws are that applications include untrusted data in a new web page without proper validation, and updates an existing web page with user-supplied data using a browser API that can create HTML or JavaScript."

"If the value of name originates from user-supplied data, then the database can be a conduit for malicious content."

"Without proper input validation on all data stored in the database, an attacker may execute malicious commands in the user's web browser."

"There is also an insecure call to unserialize()with user-supplied data."

"The application may be expecting a serialized User object but an attacker may actually provide a serialized version of SomeAvailableClass with a predefined value for its command property: The destructor method will be called as soon as there are no other references to the $user object and then it will execute the command provided by the attacker."

"Components, such as frameworks, libraries, and other software modules, run with the same privileges as the application and if a vulnerable component is exploited, an attack can facilitate serious data loss or server takeover."

"In software development, it becomes necessary to identify all the components or libraries that applications use and the versions involved; establish security policies and best practices for component use; monitor known security vulnerabilities in any published databases, etc., and keep components upgraded to the latest available versions."

"The following code uses a vulnerable method: An attacker could send requests where the parameter d is a value in the range $[2^{(-1022)} - 2(^{-1075}):2^{(-1022)} - 2^{(-1076)}]$, such as "0.0222507385850720119e-00306," to cause the program to hang while processing the request."

"To ensure these objectives, a key element in the PKB is the context that integrates the necessary requirements to preserve the security of the data and the application."

"It consists of (i) architectural requirements, (ii) use cases and scenario, and (iii) privacy enhancing technologies described below: (i) Architectural requirements determine the flow of data within the system, components, roles, and responsibilities."

"The operational flow is described below: The user interacts with the system via the Graphical User Interface and entrusts the personal data management to the system."

"For principle of minimization, the system has three fundamental modules to reduce the personal data that can be shared on the network: Metadata Manager module acts when a user shares documents or web resources with third party services; it allows users who are not fully aware of the attached metadata to manage them and then delete them; Location Granularity module to share user geographic location with a third party service."

"After minimizing the data, the system separates them so that the user's profile cannot be reconstructed: User Data Confinement module allows a user to manage personal data directly from his device."

"The system then allows data to be aggregated by means of a module: Trustworthy Plug-in module creates a set of aggregated records; they are the set of all personal data grouped so that the server cannot correlate individual information with the user profile ensuring privacy."

"Before sending the data to the server, the system hides it: Encryption with user-managed keys module enables encryption with keys managed directly by the user (asymmetric encryption)."

"The identity is generated before the client establishes a connection to the server so that no private information is known; Onion Routing module allows the identity to be hidden by encapsulating the data in different levels of encryption, limiting the knowledge of each node along the delivery path."

"The system provides three modules that are fundamental for the respect of the Inform principle: Multi-Factor Authentication module is integrated in the GUI and allows multi-factor authentication using CPU ID, UUID device, and machine IP addresses; Unusual Account Activity module identifies unusual activities or unauthorized access by means of cookies, metadata, browser, type of architecture, etc.; Proxy module (P3P) defines a proxy capable of understanding the policies produced by the server and translating them into a format that the user can understand."

"The system allows the user to control the data by means of two modules: Privacy Selection module avoids overly general privacy practices and provides the possibility of defining an adaptive privacy level for the contents shared with the controller; Web Local Server module defines a local web server that, by interfacing with a local database, allows the management of personal tokens to be shared over the network."

"The system demonstrates the Control Authority the system's security (Confidentiality, Integrity, Availability), using a fundamental module: Federated Privacy Assessment provides an impact assessment on the privacy of the user applying strategies oriented to processes and data on the system."

"They allow to better specify the use of personal data and their processing; Orchestrator acts on behalf of the user and is controlled only by the latter, it ensures that the Identity Broker cannot correlate the original request from the service provider with the assertions that are returned by the Identity Provider."

"Countermeasures: define a process for granting and revoking an account that at least includes the use of individual User ID so that users can be made responsible for their own actions."

"The use of the group ID should be allowed only for: business or operational needs after approval and production of supporting documentation; verifying that the level of access requested is in line with the principle of "need to know"; immediately disabling or removing the User ID of users who no longer use the system (for example, the end of the employment relationship); periodically (at least quarterly) verifying the absence of inconsistent, redundant or obsolete accounts and their elimination."

"This means that the password must be composed of an increasing number of characters (at least 8) depending on the criticality of the information to be defended (e.g., 15 characters for administrative users); must contain characters in at least three of the following four categories: (i) uppercase letters of the Latin alphabet (from A to Z), (ii) lowercase letters of the Latin alphabet (from a to z), (iii) numbers to base 10 (from 0 to 9), (iv) special non-alphanumeric characters; should not refer to something that someone else can easily guess or obtain using information about the person (e.g., names, phone numbers)."

"The following rules must be observed: prohibit predefined account names and rename standard accounts such as, for example, the administrator account; do not show the password on the screen when it is entered and do not give indications on its length; the temporary password must be changed at the first log-on; it must be forced for all users, and particularly for administrators, to change the password periodically; the password change procedure must prevent the reuse of all passwords previously used and include an effective procedure that takes into account input errors; limit the number of attempts allowed in a given time period or, alternatively, block the account for end-user access after a given number of attempts; do not allow access until the log-on process has been successfully completed; validate the log-on information only upon completion of all input data; limit the time within which the log-on procedure must be completed; consider viewing the following information successfully after log-on: (i) date and time of the previous successful log-on, (ii) detail any failed log-on attempts since the last successful log-on; the persistence and transmission of passwords must be protected; contrast the possibility provided by the browser cache to allow access, implement a policy that allows the user to choose not to save credentials or to force this policy as default; track both successful accesses and failed access attempts; perform the Audit of unsuccessful accesses to detect attempts to hack passwords; always check and validate the source IP address of the client used by the user. (iii) Privacy Enhancing Technologies (PETs) (Van Blarkom et al. [12]) are a set of tools and technologies that help to protect the personal information handled by the applications: Anonymity System is an anonymization technique through which the User's sensitive data is made inaccessible and encrypted."

"Privacy Preserving Authentication is a useful technology to hide sensitive data of the user."

"Privacy preserving cryptographic protocols are widely used for secure application-level data transport."

"Data Anonymization consists in making user data anonymous and impossible to recompose in order to go back to personal information."

"Encryption techniques describe the set of techniques that apply to ensure that personal data belonging to a user is stored and made secure against possible external attacks."

"Access control requires to define measures and procedures to prevent unauthorized access to locations and information systems where personal data is held."

"Policy and feedback tools is to inform the user of the privacy policies and how the data will be protected while using the service."

Privacy Oriented Software Development

"Without proper input validation on all data stored in the database, an attacker may execute malicious commands in the user's web browser."

"Since XSS vulnerabilities occur when an application includes malicious data in its output, one logical approach is to validate data immediately before it leaves the application."

"The most secure approach to validation for XSS is to create a whitelist of safe characters that are allowed to appear in HTTP content and accept input composed exclusively of characters in the approved set."

"This solution is often infeasible in web applications because many characters that have special meaning to the browser should still be considered valid input once they are encoded, such as a web design bulletin board that must accept HTML fragments from its users."

Application to a Real Case

"The people involved in the project were 4 developers and 1 system administrator with more than 5 years of experience in the field but with no specific knowledge and competences on software security and privacy."

"The list of vulnerabilities produced by the Security Assessment activity were imported and analyzed in the PKB in order to identify the violated principles of PbD. The result showed that all 7 principles were violated, and thus the need to implement both data and process-oriented strategies in the system by using the set of Privacy Patterns identified in the Privacy Assessment phase."

"For what concerns the Coding phase that includes Security Fix and Privacy Coding, the team had fixed 1200 vulnerabilities by applying the remediation patterns."

"The remaining 78 vulnerabilities refer to architectural flaws that affect privacy of the legacy system."

"The results obtained show that all the security and privacy issues that affected the original legacy system have been overcome."

Discussion and Limitations

"The existing application park was for the most part designed and developed when issues related to cyber security and privacy were not perceived in such a way."

"POSD and PKB represent a useful cookbook that can be used to address security and privacy problems on behalf of both developers and systems engineers, even in the absence of specific security and privacy skills."

"The results obtained in the real case application of the POSD show that it was able to address both security and privacy issues."

"It allowed to fix the vulnerabilities identified during the static code analysis through the application of remediation patterns provided in the Security Report and furthermore to addresses the privacy flaws listed in the Privacy Report."

"PKB supported the team in the privacy-oriented re-engineering of the legacy system, in assuming design decisions even though the team members had no specific skills and knowledge about security and privacy."

Conclusions

"This research work proposes an approach named Privacy Oriented Software Development (POSD) for addressing privacy and security problems in software systems."

"It uses a Privacy Knowledge Base (PKB) that developers can use whenever necessary during software development or re-engineering."

"The results obtained show that it was able to provide best practices for both secure application development and data privacy, operational guidelines, software architectures, and code structures to use."

"Despite the limitations of the validation carried out, this work has allowed us to start a discussion about this research idea and lay the foundations for future work such as validate the POSD approach in forward mode on systems to develop; improve the

PKB effectiveness by adding further knowledge (Remediation Patterns and Privacy Patterns); and increase the number of supported programming languages."

5.2.3 An Efficient Clustering-Based Anonymization Scheme for Privacy-Preserving Data Collection in IoT Based Healthcare Services [13]

This is a machine-generated summary of:

Onesimu, J. Andrew; Karthikeyan, J.; Sei, Yuichi: An efficient clustering-based anonymization scheme for privacy-preserving data collection in IoT based healthcare services [13]

Published in: Peer-to-Peer Networking and Applications (2021)

Link to original: https://doi.org/10.1007/s12083-021-01077-7

If you want to cite the papers, please refer to the original.

For technical reasons we could not place the page where the original quote is coming from.

Abstract-Summary

"Privacy preserving data collection (PPDC) is on-demand to ensure the privacy of patient data."

"We propose a novel privacy-preserving data collection scheme for IoT based healthcare services systems."

"A clustering-based anonymity model is utilized to develop an efficient privacy-preserving scheme to meet privacy requirements and to prevent healthcare IoT from various privacy attacks."

"We formulated the threat model as client-server-to-user to ensure privacy on both ends."

"On the client-side, a modified clustering-based k-anonymity model with α-deassociation is used to anonymize the data generated from the IoT nodes."

"The base-level privacy is then ensured through a bottom-up clustering method which generates clusters of records as per the privacy requirements."

"On the server-side, the cluster-combination method-UPGMA is utilized to reduce communication costs and to achieve a better level of privacy."

"The proposed scheme is efficient in tackling privacy attacks such as attribute disclosure, identity disclosure, membership disclosure, sensitivity attacks, similarity attacks, and skewness attacks."

Introduction

"Privacy-preserving data collection of personal and health data has become vital and essential, and has made privacy in healthcare IoT a major research focus [14, 14–20]."

"Research studies have proposed various privacy-preserving data collection methods, out of those, encryption-based privacy protection [21, 22] and anonymization-based methods are yielding promising results."

"Traditional anonymity models such as k-anonymity [23], l-diversity [24] and t-closeness [25] are prone to various privacy attacks such as identity disclosure, membership disclosure, attribute disclosure, background knowledge attacks, homogeneity attacks, sensitive attribute attacks, similarity attacks, skewness attacks [25, 26, 26–28] etc. Privacy-preserving data collection scheme for healthcare IoT service systems is different from regular approaches."

"It is essential to develop an efficient privacy-preserving scheme for healthcare IoT. The clustering-based k-anonymity model [29] is efficient in handling healthcare IoT data."

"An efficient privacy-preserving data collection scheme based on the clustering-based anonymity model is proposed for IoT based healthcare services."

"The proposed scheme addressed the privacy violations of data collection through a client-server-to-user model."

"The clustering-based anonymity model for privacy-preserving data collection scheme is proposed for the IoT based healthcare services."

System Architecture and Threat Model

"The sensor layer collects the health information of the patients and transmits it to the healthcare service layer via the network layer."

"The network layer is responsible for collecting the information from the sensors and relaying it to the healthcare services layer and vice versa."

"In the semi honest model, the data collector follows the protocol for privacy preserving data collection however due to curiosity he may try to identify the individual."

"To ensure privacy-preserving data collection, the client-server-to-user model is employed."

"The untrusted server is accountable for transmitting the healthcare data from the patients' to the untrusted data collector."

"The client generates anonymous data which is transmitted to the collector to obtain healthcare services."

"The aggregation node is a locally trusted node that acts as an anonymization layer where each client data is collected, aggregated, and submitted to the server."

Related Works

"As healthcare records are information-rich sources, data analytics help us to extract the actionable insights about the data for healthcare monitoring, disease prediction, etc. However, there exists a privacy risk if the patient's identity is revealed during data collection, data publication, or data processing."

"A privacy-preserving data collection scheme for healthcare data is proposed by Li and others [30]."

"Kim and others [31] developed a privacy-preserving data collection mechanism for healthcare data streams."

"A k-anonymity based dynamic privacy-preserving data collection protocol is proposed [32] to reduce the leader collusion attack, which is a privacy attack."

"There have been several anonymization-based privacy-preserving approaches and cryptography-based approaches have also been proposed."

"Based on the inferences of the literature survey, we propose a clustering-based anonymization scheme that is resilient to various privacy attacks and to provide a better tradeoff between the data privacy and data utility."

Preliminary

"We define a taxonomy tree for categorical attributes and various information loss metrics that are used in the paper to evaluate our model."

"Personal data consists of various attributes, out of which this paper focuses only on three major attributes; Identifiers, Quasi Identifiers (QI) and Sensitive attributes (SA)."

"The average dissimilarity between the clusters is calculated by the formula The generalization of categorical attributes is achieved through taxonomy trees."

"Consider the representative vectors of two clusters as i and j. Then the distance between the vectors can be calculated using the following formula, Then the distance between the cluster tuples t_l and t_2 is as followsWhere $[T_{i,u}, T_{i,l}]$ are the quasi identifiers corresponding to the numerical attributes and H_i and CAT_i represents the categorical attributes of QI."

"The discernibility metric is defined as follows DM measures the information loss through the number of equivalent classes formed after generalization."

Privacy-Preserving Data Collection Scheme

"The proposed clustering-based anonymization scheme for privacy-preserving data collection consists of initialization, client-side, and server-side anonymization stages."

"In the client-side anonymization stage, clustering-based anonymization model is utilized to anonymize the QI attributes."

"In the server-side anonymization stage, the UPGMA based cluster combination method is utilized to compress the clusters to reduce the communication cost and to achieve better privacy."

"In the client-side anonymization stage, clustering operations are performed to generate clusters of anonymous tuples satisfying the clustering-based anonymity model."

"The clusters generated for every vector in the initialization stage are divided using a bottom-up clustering approach to form new clusters with anonymous tuples."

"In the server-side anonymization stage, to reduce the communication cost, the k_1-anonymous clusters formed at the client-side are compressed."

"UPGMA is a cluster combination method that is utilized to merge the clusters formed during the client-side anonymization stage."

Discussion

"In our proposed scheme, we employ the k-anonymity based model to resist the identity disclosure attack."

"Clustering based k anonymity property of the proposed scheme ensures that each cluster has at least k records similar to each other."

"Attribute disclosure attack is nullified in the proposed scheme."

"Shannon's conditional entropy [33] can be defined as follows Where, Then the conditional entropy of H(Y|X) is Let Z be the conditional probability of a patient's choice being $z = s$, with a particular set of random observations q such that, To calculate the uncertainty of an attacker knowing q is defined as entropy H(Z), and is calculated as follows Therefore the conditional probability of H(Y|X) is expressed as Finally, the minimum level of privacy can be calculated as In the proposed method the anonymization in both client-side and server-side satisfies the clustering-based anonymity model."

Experiments

"The experimental analysis of the results is analyzed based on (1) the size of records, (2) the size of clusters k, (3) sensitive values a, and (4) the number of QI attributes."

"The experimental analysis of the proposed privacy-preserving data collection scheme is carried under two aspects (i) Information Loss (ILoss) and (ii) Execution time."

"It is evident in both the graphs that the ILoss is decreasing when the value of k, α, and record size increases."

"The reason behind this ILoss decrease is when the value of k and α increases, the size of the cluster increases which in turn adds more records to the clusters."

"It is also showing that the information loss varies with the varying size of record and k value."

"The main reason is when the k value increases, the number of indistinguishable records within the cluster also increases, thereby leading to information loss."

Conclusion

"Data collection process are prone to privacy attacks aims to disclose the identity and sensitive information."

"A novel privacy-preserving data collection scheme for IoT based healthcare service systems has been proposed to anonymize the original records of the patients in the healthcare data collection process."

"A client-server-to-user model is analysed to identify privacy threats at different levels of data collection."

"The performance evaluation methods employed show that the proposed scheme has an edge over the state-of-the-art privacy-preserving data collection schemes."

5.2.4 Privacy by Designers: Software Developers' Privacy Mindset [34]

This is a machine-generated summary of:

Hadar, Irit; Hasson, Tomer; Ayalon, Oshrat; Toch, Eran; Birnhack, Michael; Sherman, Sofia; Balissa, Arod: Privacy by designers: software developers' privacy mindset [34]

Published in: Empirical Software Engineering (2017)

Link to original: https://doi.org/10.1007/s10664-017-9517-1

If you want to cite the papers, please refer to the original.

For technical reasons we could not place the page where the original quote is coming from.

Abstract-Summary

"Privacy by design (PbD) is a policy measure that guides software developers to apply inherent solutions to achieve better privacy protection."

"For PbD to be a viable option, it is important to understand developers' perceptions, interpretation and practices as to informational privacy (or data protection)."

"Our findings indicate that developers use the vocabulary of data security to approach privacy challenges, and that this vocabulary limits their perceptions of privacy mainly to third-party threats coming from outside of the organization; that organizational privacy climate is a powerful means for organizations to guide developers toward particular practices of privacy; and that software architectural patterns frame privacy solutions that are used throughout the development process, possibly explaining developers' preference of policy-based solutions to architectural solutions."

"Further, we show, through the use of the SCT schema for framing the findings of this study, how a theoretical model of the factors that influence developers' privacy practices can be conceptualized and used as a guide for future research toward effective implementation of PbD."

Introduction

"Our main research question is: What are the perceptions of privacy among developers involved in the design of software systems?"

"The findings shed a new light onto developers' privacy perceptions, interpretations and practices when designing new technologies and the interplay between the different forces affecting and affected by their privacy decision making."

"Based on the research findings and the classification of the data according to the schema of Social Cognitive Theory (SCT) (Bandura [35]), we propose a conceptualization of the factors that influence, and are influenced by developers' privacy practices, which can be used as a guide for future research toward effective implementation of PbD. The paper is organized as follows: The next section reviews related literature."

Related Research

"While these research works focus on addressing privacy concerns during the processes of development and maintenance of software, it is important to examine how the developed software will function and manage private data after its deployment, throughout its usage lifecycle."

"Analyzing the way in which privacy and data-utilization thinking is embedded in today's design processes is important for understanding how privacy can be effectively implemented by engineering practices."

"As the awareness about the need to consider privacy in software design increases, research has recently emerged on the perceptions and interpretations of software development professionals regarding privacy."

"Such theory would not only assist in explaining the gap between the privacy requirements and the privacy perceptions and interpretations of developers found in the above reviewed and the current research works, but could also point to means for bridging this gap."

Research Method

"We decided to use interviews rather than surveys, despite the inevitable result of a lower number of participants, as we were interested in in-depth exploration of developers' perceptions and interpretations of privacy and related concepts, and interviews are considered a key way of accessing the interpretations of informants in the field (Walsham [36])."

"The data-analysis procedure included open, axial, and finally selective coding (Strauss and Corbin [37, 38]) to determine the categories regarding the tension between developers' perceptions and interpretations of privacy and the law The open coding was conducted iteratively with the continuation of data collection, and included exploration for recurring themes."

"After identifying a central code (Strauss and Corbin [38]), namely, the interplay between developers' perceptions of privacy, characteristics of their work environment, and their actual practice, we selectively re-coded the data focusing on the schema of SCT and on the theory of organizational climate."

Findings

"We identified organizational climate as a central force representing the influence of the environment on developers' cognitive factors and behavior related to privacy."

"Ten of the participants provided examples and statements reflecting a positive organizational privacy climate, including organizational procedures governed by organizational policy, and in some cases additional strategies, such as a designated position within the organization responsible for privacy concerns, namely a Chief Privacy Officer (CPO), or means for educating developers to handle privacy concerns."

"While the role of organizational privacy policy is to guide behavior within the organization regarding privacy concerns, we found a pattern of statements reporting on organizational norms and practices that led to participants' impression that they were expected by their organization to overlook privacy concerns, including the organization's own privacy policy and regulations."

"The participants' answers distributed as follows: three examples of privacy-by-architecture solutions, seven examples of privacy-by-policy, and six examples of security solutions."

Discussion

"We detail the environmental mechanisms that influence, and are influenced by, developers when handling privacy concerns, identifying organizational privacy climate as a powerful means for organizations to guide developers toward particular interpretations of privacy, and specifically their perceptions as to how they are expected to act upon privacy concerns, which can either be aligned or conflict with the stated policy."

"Such measures would be instrumental for managing and controlling different mechanism for increasing positive organizational privacy climate toward improving awareness, interpretation, and behavior of developers in the context of addressing privacy concerns in the developed software systems."

"Based on the findings and related research on organizational climate, we propose forming a strong positive organizational privacy climate and appropriate education to familiarize developers with privacy solutions beyond those addressing security concerns, as a future direction for improvement mechanisms."

Limitations

"Some limitations should be considered in interpreting the findings of this study."

"This setting is different from real-life software development practice, presenting less time and performance pressure."

"While this can be considered a limitation as well, in the context of the findings of this study—when developers were specifically asked to discuss privacy and so were highly attuned to this topic—we expect their answers to reflect more attention and consideration of privacy than in the complexity and tight schedule of real-life situations."

"Future research in additional domains, and using quantitative research methods, could contribute to the generalization of these finding and the validation of the theoretical model, its components and the relationship between them, as identified in this study."

Conclusion

"The findings of this study suggest that organizational privacy climate highly influences developers' privacy interpretation and behavior; thus, it may potentially serve as an effective mechanism to bring about the required change in the privacy mindset and practices as to informational privacy, starting with the adaptation of organizational policy to the principles of FIPPs and followed by the diffusion of this policy into the organizational climate."

"Other findings, notably developers' high familiarity with security solutions, as opposed to solutions of other privacy-related concerns, as well as developers' preference to use privacy-by-policy rather than privacy-by-architecture solutions, indicate that developers lack the required knowledge for effectively design privacy preserving technologies."

"Providing developers with knowledge, by means of education, as well as motivation, by means of positive organizational privacy climate, could potentially create the mindset required for designing privacy-preserving solutions."

5.2.5 Understanding Developers' Privacy and Security Mindsets via Climate Theory [39]

This is a machine-generated summary of:

Arizon-Peretz, Renana; Hadar, Irit; Luria, Gil; Sherman, Sofia: Understanding developers' privacy and security mindsets via climate theory [39]

Published in: Empirical Software Engineering (2021)

Link to original: https://doi.org/10.1007/s10664-021-09995-z

If you want to cite the papers, please refer to the original.

For technical reasons we could not place the page where the original quote is coming from.

Abstract-Summary

"Privacy and security by design are policy measures that guide software developers to engineer Privacy and security solutions inherently into the software systems they develop."

"We aimed to explore the potential of using organizational climate theory for attaining a better understanding of developers' perceptions and behaviors and the underlying forces affecting them, and to unveil the constructs that compose organizational privacy and security climates."

"Our findings indicate that software developers are faced with inconsistent and confusing cues conveyed by management and other parties in their work environment, many of which indicate that these facets are of relatively low priority, leading to perceptions and behaviors that are not in line with those expected and recommended by policy makers."

"Further, we show how these perceptions and behaviors can be explained by constructs of the organizational climate theory and how, based on our findings, organizational climate mechanisms can be used to go beyond understanding developers' current privacy and security mindsets toward improving them, thereby leading to an effective implementation of privacy and security by design."

Introduction

"Many engineers are under the impression that their organizations do not give them enough time, and this limitation leads them to think that the decision whether to incorporate privacy and security mechanisms into a system design rests with them (Spiekermann et al. [40])."

"Given the findings about the privacy and security-related behaviors of developers and the role played by the organizational environment, our working assumption is that the study of organizational privacy and security climate would illuminate not only developers' privacy and security mindsets but also the organizational factors from which they stem."

"Motivated by previous findings about the potential effects of the organizational environment in this context (Hadar et al. [41]; Spiekermann et al. [40]) and informed by the interplay and dependencies between privacy and security, the current research study was aimed to shed new light on developers' privacy and security perceptions and behavior via a focused investigation of organizational privacy climate and organizational security climate."

Theoretical Background

"If we consider organizational climate as a facet-specific construct, its perceived priority by comparison with other relevant demands constitutes a proximal indicator informing employees of the kinds of role behavior that are likely to be rewarded (Zohar and Polachek [42])."

"We review key climate themes from the organizational climate theory, which served us in designing and focusing our research study, including importance, communication, rewards, consequences of omission, training, management commitment, manager support, and practices, policies, and procedures."

"While part of management commitment is its investment in HR practices, such as training as rewards, the literature of organizational climate makes a distinction between this overall perception of employees about management commitment and their perception of specific practices of HR (see, e.g., Christian et al. [43] and Neal and Griffin [44])."

Method

"And in line with the principles of organizational climate theory, the study was guided by the following research question: What climate-related themes for privacy and security are manifested in developers' working environment and how?"

"As this is a first exploration of organizational privacy and security climates, a qualitative approach is appropriate for understanding developers' perceptions of these facets in their full complexity and for conceptualizing emergent themes in correspondence with the organizational climate theory."

"Data analysis was performed according to the principles of provisional coding (Miles et al. [45]), initially informed by key themes of the theory of organizational climate and used for coding the collected data."

"During data analysis we identified the recurring pattern of discourse taking place between the participants and different role holders in the organization, leading us to define codes such as discourse with direct manager about security, discourse with team members about security, discourse with peers from other organizational units about security, and discourse with security expert."

Findings

"An important part of the analysis of the statements was related to the level of climate indicated in each statement, reflecting either high or low climate level, namely, whether the emphasis on the facet (i.e., security or privacy) is perceived as high or low within the organization."

"Statements that were classified as indicating a high level of this climate theme described behavior that encourages employees to behave in a way that promotes secured or privacy protecting code."

"It is important to note that, while the statements classified into this theme refer directly to the efforts and investments on the part of higher management, this theme is actually intertwined with other themes presented here, such as training, rewards, personal- and unit-level consequences of omission, and the assignment of experts supporting security and privacy in the organization, all being manifestations of management commitment."

Discussion

"The research findings suggest the following insights. (a) Both facets are reflected in diverse climate themes, demonstrating the adequacy of organizational climate theory for investigating privacy and security perceptions and behaviors in their full complexity. (b) The security facet shows a more widely spread representation across the different themes than does the privacy facet, implying a higher presence of security than of privacy in the discourse and awareness of developers. (c) Both facets exhibit indications that the managements of the organizations employing our participants may not create organizational environments that consistently convey the importance of security, and even less so of privacy, when developing software."

"Additional indications of cues leading to a low level of privacy and security climates include the findings related to the themes discussed in subsection (a), namely, the indications of the low relative importance of security and privacy when they compete with other facets, indications of perceived low managers' support with regard to these facets, and indications of sporadic communication with supervisors and peers."

Limitations

"Credibility is analogous to internal validity in positivist terms and addresses the "truth" of the research findings."

"The credibility of the findings, in our context, is inherently threatened by the use of interviews as a data source for practitioners' perceptions and behaviors, which could be affected by participants' reluctance to cooperate with researchers with whom they have not established familiarity and trust, inaccurate recollection, or social desirability bias."

"To ensure the credibility of the data analysis and interpretations of the findings, we established structural coherence (Guba [46]), that is, testing of every datum and interpretation against all others to be certain that no internal conflicts or contradictions exist."

"The research team consulted privacy and security experts throughout the different phases of the research and, in particular, during the design of the interview guide and the interpretation of the data."

Conclusion

"We aimed to learn about developers' privacy and security mindsets via organizational climate theory."

"We applied and adapted the theory of organizational climate to the facets of privacy and security in the context of software development."

"We identified the following climate themes affecting the emergence of organizational privacy and security climates: communication, with four sub-themes thereof according to the parties with whom the developer communicates, practices, policies, and procedures, importance, with a distinction between absolute importance and importance relative to other competing facets (e.g., functionality), rewards and consequences of omission, extending beyond their typical effect on the individual circle to team, company, and customers, manager support, training, and management commitment."

"The analysis of these mindsets through the lens of organizational climate theory reveals that developers' work environment, which has a critical role in shaping their mindsets, sends them inconsistent and confusing cues as to the importance of privacy and security in their daily work, with many of these cues in fact suggesting that the priority of these facets is relatively low."

5.2.6 Interdisciplinary Perspectives on Privacy Awareness in Lifelogging Technology Development [47]

This is a machine-generated summary of:

Wilkowska, Wiktoria; Offermann, Julia; Colonna, Liane; Florez-Revuelta, Francisco; Climent-Pérez, Pau; Mihailidis, Alex; Poli, Angelica; Spinsante, Susanna; Ziefle, Martina: Interdisciplinary perspectives on privacy awareness in lifelogging technology development [47]

Published in: Journal of Ambient Intelligence and Humanized Computing (2022).

Link to original: https://doi.org/10.1007/s12652-022-04486-5

If you want to cite the papers, please refer to the original.

For technical reasons we could not place the page where the original quote is coming from.

Abstract-Summary

"The consequences of population aging can be mitigated to some extent by assisting technologies that can support the autonomous living of older individuals and persons in need of care in their private environments as long as possible."

"A variety of technical solutions are already available on the market, but privacy protection is a serious, often neglected, issue when using such (assisting) technology."

"In a three-year project PAAL ('Privacy-Aware and Acceptable Lifelogging Services for Older and Frail People'), researchers from different disciplines, such as law, rehabilitation, human-computer interaction, and computer science, investigated the phenomenon of privacy when using assistive lifelogging technologies."

"A user-centered empirical approach was applied to the lifelogging technologies, investigating the perceptions and attitudes of (older) users with different health-related and biographical profiles."

"Partners of the PAAL project present insights gained from their cross-national, interdisciplinary work regarding privacy-aware and acceptable lifelogging technologies."

Introduction

"The use of assisting technologies in care environments can serve to address current challenges, such as the lack of care personnel and the increasing numbers of people in need of care, by providing relief and support in tasks of the daily care routine (Rashidi and Mihailidis [48])."

"In order to implement a broad spectrum of assisting lifelogging systems—fulfilling diverse supporting and relieving functions, as well as being both privacy-aware from a legal perspective and sustainably accepted by their future users—an interdisciplinary collaboration of diverse technical, legal, and social disciplines is needed."

"The current work describes a multidisciplinary view on using lifelogging technologies for assistance in the everyday life which was the primary goal of the European project PAAL ('Privacy-Aware and Acceptable Lifelogging services for older and frail people')."

Privacy by Design as a Legal Requirement

"The Council of Europe adopted the Convention 108 which secures the right to privacy as enshrined in Article 8 of the ECHR in regard to automatic processing of data by safeguarding the individual against the unjustified collection, processing, use, storage, and dissemination of their personal data."

"Organizational measures include matters, like conducting privacy impact assessments, documenting data processes that contain personal data, and appointing a Data Protection Officer (DPO)."

"Implementing technological and organizational safeguards to guarantee the protection of personal data is a critical legal requirement, especially because these devices process highly sensitive personal data like key human biological signals that, if lost or stolen, will not only put an individual's reputation at risk, but can also threaten his or her health or wellbeing."

Realization of PbD Approach Using Exemplary Lifelogging Applications

"Considering the risk of the potential exposure of personal privacy that can lead to breach or misuse of sensitive data, an alternative is to use less privacy intrusive sensors that provide reminders and monitor changes in a person's health status."

"An alternative way to address the problem of protecting the user's privacy, which is being investigated in the framework of the PAAL project, focuses on identifying those features inherently present within the collected signals, like in the case of the acceleration measured from a wrist-worn device, with the aim of removing those features, which may expose personal details about the person, without hindering the performance of automatic ADL classification algorithms (Poli et al. [49])."

"At the user level, personal privacy and data protection are perceived as top priorities of the use."

Discussion

"The collaborative project PAAL demonstrated such an approach, (1) providing legal regulations and socio-ethical recommendations in order to accordingly protect the user's privacy from abuse, (2) appropriately considering the real user's needs, perceived benefits, barriers, and conditions of an accepted use as revealed in empirical studies, and (3) consequently implementing the gained knowledge into the technical realization."

"In an effort to approach this research gap, the interdisciplinary collaboration has provided a number of examples as to how privacy can be methodologically integrated into various levels of lifelogging systems, particularly taking technological, legal, and social perspectives into account."

"While the multi-leveled technical approach for identifying the PbD issues in lifelogging may be generalized to different technologies, ranging from VAs to video-based systems and from wearable to ambient sensor-based technologies, to implement the actual privacy countermeasure is strongly solution-specific."

Conclusion and Outlook

"In the light of some serious privacy concerns that threaten the potential of lifelogging applications to improve efficiency and care in the healthcare settings, there is a high demand for a robust privacy protection framework which requires the active involvement of governments, organizations, developers, and users."

"The collaboration in the presented project PAAL provides a framework combining legal, social, user-centered, and technical requirements for the design of lifelogging technologies, being used for the daily assistance of older and frail people as well as their private and institutional caregivers."

"Besides, the presented empirical findings and the interdisciplinary exchange reassert that privacy concerns can pose a serious deterrent to the technology adoption."

"An interdisciplinary and multi-technological approach in the development enables thus more promising, competitive, and sustainable technology solutions for the users in the future."

5.2.7 Privacy Preserving Electronic Petitions [50]

This is a machine-generated summary of:

Diaz, Claudia; Kosta, Eleni; Dekeyser, Hannelore; Kohlweiss, Markulf; Nigusse, Girma: Privacy preserving electronic petitions [50]

Published in: Identity in the Information Society (2008).

Link to original: https://doi.org/10.1007/s12394-009-0012-8

If you want to cite the papers, please refer to the original.

For technical reasons we could not place the page where the original quote is coming from.

Abstract-Summary

"The design of a secure and privacy preserving e-petition system that we have implemented as a proof-of-concept demonstrator."

"Our system ensures that duplicate signatures are detectable, while preserving the anonymity of petition signers."

Introduction

"To prevent this, some e-petition servers check the IP address of the signer and allow only one signature per IP address."

"We propose using the existing PKI-based electronic IDs cards (De Cock [51]) in combination with anonymous credential protocols to ensure that (i) signatures are legitimate, (ii) each citizen can sign a petition at most once, and (iii) petition signers are anonymous."

"Our e-petition design and implementation uses the Belgian e-ID for initial authentication, and then allows the user to obtain an anonymous credential (Chaum [52]; Camenisch and Lysyanskaya [53]; Camenisch et al. [54]) that is used to electronically sign petitions on a server."

"The citizens use the anonymous credential obtained from the credential issuer to interact with the e-petition server."

"Our design for the e-petition server aims at a notable advance in protecting user privacy, by shielding off any and all identifiable information about the users through the use of anonymous credentials and anonymous communications channels."

Requirements

"This requirement includes private key material held by the user that is not protected by the tamper resistance of the e-ID."

"Data integrity: No entity should be able to modify the data exchanged between the citizen and the e-government servers."

"Confidentiality: All the data exchanges between a citizen and the e-government servers must be kept confidential from other entities."

"Traffic analysis protection is required so that external observers are not able to determine that a citizen is accessing the e-petition server (or a particular petition in the server)."

"Signer anonymity: The e-petition server (even in collusion with the credential issuer) must not be able to identify the citizens who have signed the petitions."

"Multiple signing prevention: An e-petition application has to be designed in a way it can properly detect and rectify citizen attempts to sign a single e-petition multiple times."

Anonymous Credentials

"Today's anonymous credential protocols (Camenisch and Lysyanskaya [53]) rely on zero-knowledge proofs (Goldwasser et al. [55]) to reduce to the minimum the amount of information disclosed about their owners."

"Such credentials can be realized based on the so-called epoch number of direct anonymous attestation (Brickell et al. [56])."

"By binding a different tag to every context in which a credential is shown, k-times anonymous authentication (Teranishi et al. [57]) can be used to create unique serial numbers."

"The schemes in Damgård et al. [58] and Camenisch et al. [54] even support the identification of the owner of a credential that was shown twice with the same serial number."

"For the purpose of this paper it is sufficient to note that assertion-based signatures for various assertion types can be obtained by applying the Fiat-Shamir heuristic (Fiat and Shamir [59]) to various privacy enhanced authentication systems, such as existing anonymous credentials systems, group signatures, and the e-token scheme used by us."

Privacy-Preserving e-petition Design

"The demonstrator extends 'identity mixer' and implements a privacy preserving protocol between a client and two web server applications: the issuer of the anonymous credential and the e-petition server."

"In our demonstrator, the petition signing and petition signature publication processes are managed by the e-petition server, and the user is able to (unlinkably) sign as many distinct petitions as desired with the same credential."

"The e-petition server publishes the protocol transcripts of every successful petition signature (step 8), so that anyone can verify that their signature has been counted, and additionally check that all transcripts correspond to signatures made by users who possess valid credentials."

"We use a special assertion-based signature scheme for signing petitions: Camenisch and others have proposed a protocol for periodically spendable e-tokens (Camenisch et al. [54])."

Security Analysis

"In our design, the initial authentication of the citizen is achieved through the Belgian e-ID card, which provides 2-factor strong authentication (physical possession of the card and knowledge of its PIN code)."

"The e-government servers also authenticate themselves towards the user to prevent malicious servers from impersonating legitimate ones."

"The authentication of the e-government servers is provided through digital signatures and relies on the security of private keys."

"To prevent citizen identification using the IP address, the citizen communicates to the servers through an anonymous communication infrastructure (e.g., Tor Dingledine et al. [60])."

"Our protocols provide mechanisms to detect and remove signature duplicates, without compromising signer anonymity."

"In our design, the e-petition server publishes the protocol transcripts of every signature."

Implementation

"The middleware facilitates the 2-factor authentication of users towards the e-petition Web server and retrieval of user specific credential attributes."

"The demonstrator consists of two simple Web interfaces for issuing anonymous credential and signing e-petitions."

"Using the issuing server Web site, users can request and save anonymous e-petition credentials into secondary memories."

"Once they get their credentials, they can consult the e-petition signing server Web site."

Data Protection Implications of e-petition

"In the current design of the system, the user is fully identifiable by the credential issuer, who in this case is processing personal data of the user in order to ensure authentication and authorization and to issue the anonymous credential that will be used for the e-petition signing."

"The credential issuer is thus rendered controller of the data; i.e., the one that determines the purposes and the means of the processing of personal data, and has to fulfil the obligations that the data protection legislation foresees for the data controller."

"In our e-petition system, it shall be examined whether the data that relate to the anonymous credential are anonymous or whether the user is just pseudonymous towards the e-petition web server and thus the latter processes personal data."

"The data that are processed by the e-petition web server are anonymous and the data protection legislation will not apply."

Conclusions

"Our design shows that anonymous credential protocols can be used to extend existing PKI-based national e-ID infrastructures, to achieve a degree of security and privacy protection that PKI alone cannot provide."

"Of directly signing petitions with the existing Belgian e ID card, we use it to provide initial strong authentication, and ensure that each user obtains at most one anonymous credential."

"The use of the anonymous credential in the e-petition signing process ensures maximum privacy protection through data minimization."

"We have introduced anonymous credentials, studied the privacy and security requirements of the e-petition application, presented an overview of our design's architecture and protocols, and discussed the applicability of data protection legislation."

5.2.8 Multi-level Privacy Analysis of Business Processes: The Pleak Toolset [61]

This is a machine-generated summary of:

Dumas, Marlon; García-Bañuelos, Luciano; Jääger, Joosep; Laud, Peeter; Matulevičius, Raimundas; Pankova, Alisa; Pettai, Martin; Pullonen-Raudvere, Pille; Toots, Aivo; Tuuling, Reedik; Yerokhin, Maksym: Multi-level privacy analysis of business processes: the Pleak toolset [61]

Published in: International Journal on Software Tools for Technology Transfer (2021)

Link to original: https://doi.org/10.1007/s10009-021-00636-w

If you want to cite the papers, please refer to the original.

For technical reasons we could not place the page where the original quote is coming from.

Abstract-Summary

"Privacy-enhancing technologies (PETs), such as secure multi-party computation and differential privacy, provide mechanisms to perform computations over private data and to protect the disclosure of private data and derivatives thereof."

"When PETs are used to protect individual computations or disclosures, their privacy properties and their effect on the utility of the disclosed data can be straightforwardly asserted."

"This article presents a multi-level approach to analyze privacy properties of business processes that rely on PETs to protect private data."

"The approach is embodied in an open-source toolset, Pleak, that allows analysts to capture privacy-enhanced business process models and to characterize and quantify to what extent the outputs of a process leak information about its inputs."

Introduction

"In the context of business processes, this principle implies that workers involved in the execution of a business process should only have access to private data to the extent it is required to perform their tasks."

"Some of the private data processing tasks and data exchanges may be protected by Privacy-Enhancing Technologies (PETs) such as encryption, multi-party computation, and differential privacy."

"The [62, 63], we presented an extension of the BPMN notation, Privacy-Enhanced BPMN (PE-BPMN), for capturing business processes in which PETs are used to protect and control private data processing and disclosure."

"We also introduced basic techniques for analyzing PE-BPMN models in order to determine which parties in the process may get access to private data."

"This toolset allows users to analyze both plain and privacy-enhanced BPMN models in order to characterize and quantify to what extent the outputs of a process leak information about its inputs."

Background

"Out of all PETs we are mainly interested in the ones providing data confidentiality, privacy preserving data processing and secure networking as these are crucial for capturing privacy leakages."

"We review previous studies that have addressed problems related to modeling and analyzing security and privacy aspects of business processes."

"Various studies have addressed the question of how to capture security and privacy in business process models."

"These previous studies do not consider the problem of analyzing a (privacy-enhanced) business process model in order to identify and to quantify potential leakages."

"We expand this principle and illustrate how privacy leakage in business processes could be determined using other PETs (e.g., for confidentiality protection, entity authentication and confidential input) systematically adapted to the BPMN modeling language."

Architecture

"The first editor, the PE-BPMN & Leaks-When Editor, allows users to design and to analyze two types of process models: PE-BPMN process models (as defined above) and data processing workflows—also known as SQL workflows."

"A SQL workflow is a BPMN process model in which each task is specified by means of a SQL query that maps one or more input tables into an output table."

"The PE-BPMN & Leaks-When editor implements the top two privacy analysis levels mentioned in Sect. 5.1: the disclosure level (which data objects are disclosed or leaked to which party?) and the qualitative level (which attributes or functions are disclosed in a way that potentially violates access policies?)."

"For SQL-based analysis clicking on the task displays the SQL queries and clicking on the input data objects (e.g., Port or Ship) shows the database schema for these inputs."

Disclosure Analysis of PE-BPMN Models

"The PE-BPMN disclosure analyzer takes as input a PE-BPMN model and produces a tabular report indicating which data objects are disclosed to each party in the process, and what privacy protections, if any, are applied to the disclosed data."

"The Visible marker indicates that a data object is disclosed to a party without any protection."

"When the visibility marker for a data object is hidden (H marker) or not visible (dash marker), the dependency marker is included in the disclosure report to indicate that a derivative of the data object may be disclosed to that party during the execution of the process."

"Given a data object A and a party P, the dependency marker D indicates that a direct derivative of data object A may be disclosed to party P during the execution of the process."

Leaks-When Analysis

"A leaks-when report is a directed acyclic graph that captures, for each attribute of an output object disclosed by the SQL workflow: (i) from which attributes of the input objects it is derived and (ii) how and under what condition this disclosure occurs."

"Each statement in a disclosure policy grants a right to run 'select' queries over one or more attributes of an input data objects to one party (pool) in the SQL workflow."

"In the running example, a global policy statement over the parameters object may grant access to this input data object to a given party (pool) for the full execution of the SQL workflow."

"Given a disclosure policy, the leaks-when analyzer tags each node and flow in the leaks-when report with one of three tags: Directly policy-violating (represented in red in the graphical interface) indicates that the SQL workflow directly accesses an attribute of an input data object that should not be disclosed according to the policy."

Sensitivity Analysis

"If x is some input (e.g., an attribute of some table), and y is the output of the SQL workflow (e.g., an aggregated statistic), we may want to know how much a potential attacker who observes y may learn about x. We assume that the attacker only observes the output and does not interfere with the workflow."

"Taking into account attacker's background allows to achieve reasonable privacy even without inserting any noise into the workflow (studying different types of attacker for different aggregated statistics has been done e.g., in [64])."

"In Pleak, we can analyze certain noise addition mechanisms, learning how much noise needs to be added to ensure desired level of privacy."

"Such a norm ensures differential privacy w.r.t. ship location even if different ships can be correlated."

Guessing Advantage Analysis

"We want to know how much the attacker's success probability increases after seeing a SQL workflow output, compared to success probability before seeing the output. (Guessing advantage) Let X be an input space."

"To estimate the noise magnitude needed to achieve desired upper bound on guessing advantage, we need a SQL query, a database schema, and a description of the attacker's goal."

"Let us say that the attacker wants to guess the ship location within 5 units w.r.t."

""worst instance" means that, if there are several ships whose location the attacker is trying to guess, then the reported numbers correspond to the one that requires the most noise for its protection."

"If the noise magnitude is fixed, the more precise we require a guess from the attacker to be, the less advantage he has in succeeding."

Conclusion

"The Pleak toolset implements a collection of techniques to detect and analyze privacy leakages arising from the execution of business processes based on both classical process models (in the BPMN notation) and privacy-enhanced process models (in the PE-BPMN notation)."

"The first level of abstraction (the so-called Boolean level) allows us to see which input data sources are directly or indirectly disclosed to each party in the process."

"The toolset has also been validated using three other privacy-enhanced business processes defined in the context of the DARPA Brandeis programme on privacy technologies."

"In its current implementation, the Pleak toolset supports the quantitative analysis of privacy-enhanced business processes in which the computations are specified in the SQL query language."

"Another avenue for future work is to extend the set of techniques in the Pleak toolset with program analysis techniques that would allow it to handle other languages."

5.2.9 Challenges in Enabling Privacy Self Management [65]

This is a machine-generated summary of:

Vidhani, Kumar; Banahatti, Vijayanand; Lodha, Sachin: Challenges in enabling privacy self management [65]

Published in: CSI Transactions on ICT (2021)

Link to original: https://doi.org/10.1007/s40012-021-00335-6

If you want to cite the papers, please refer to the original.

For technical reasons we could not place the page where the original quote is coming from.

Abstract-Summary

"Newer privacy regulations demand enterprises to inculcate privacy-by-design principles."

"They are mandating enterprises to implement privacy self management paradigm that empowers data subjects with control and rights over their data."

"This paradigm shift has brought in a major disruption and created multiple challenges for enterprises and data subjects."

"We first highlight the challenges faced by enterprises with respect to data, applications and technology."

"We discuss issues in bringing enterprise legacy systems up to speed as well as some potential challenges for greenfield applications while following the privacy-by-design principles."

"We briefly outline four core principles that could serve enterprise to create an effective, rewarding and futuristic privacy program."

Introduction

"The European Union took a different path for data privacy and introduced cross sectoral Data Protection Directive (DPD) which served as guidelines for the enterprises towards the processing of European citizens' personal and sensitive data [66]."

"Data collection is a critical activity from the enterprise perspective as there are services offered by the enterprise that cannot be executed without customers sharing their sensitive data with them (e.g., sharing medical history with hospital)."

"Shift in the privacy management paradigm from merely an accountability to some regulations and principles to data subjects becoming an active stakeholder of enterprise information system have a huge impact on enterprises."

"While these new privacy regulations have put forth multiple challenges for enterprise; they have also made a data subject an active stakeholder in the model and tasked him with additional responsibilities."

Enterprise Challenges

"To address this problem, enterprises have invested in the Master Data Management (MDM) technology that provides a single view of enterprises' customers, suppliers, products, etc. With the introduction of the privacy self management paradigm, MDM philosophy is likely to face challenges for the same reasons stated above."

"While the technology for relational databases is mature and makes it easy to extract information pertaining to an individual through Structured Query Language (SQL); for the data stored in unstructured form (e.g., files) the extraction task poses a big challenge to enterprise."

"For the greenfield enterprise applications, it is relatively easier to enable privacy self management by eliciting expectations from data subject in requirements gathering phase and implementing the consent-notice model of privacy regulation as per the elicited expectations."

"Automatic decision making supported by AI seriously challenges the privacy self management paradigm as a new purpose may get generated during data processing without involvement of the data subject."

Data Subject Challenges

"The challenge inappropriate consent demands alertness from data subject and regulators."

"Given the fact that most data subjects do not have adequate technical or legal background, this makes it near impossible for data subjects to relate, compare or differentiate between two privacy policies."

"Once privacy self management kicks in, a data subject has to provide or revoke consent for multiple purposes even for a single online service."

"New privacy laws (e.g., GDPR) demand clear articulation of purpose before obtaining consents from data subjects."

"It has led to further ballooning up of privacy policy and purpose text that acts more as a deterrent rather than an enabler for data subjects."

"Unethical enterprises may use them to make data subjects in giving consents [67]."

"Data subjects needs to be very alert and attentive while making consent decisions."

Direction for Enterprise

"This privacy program could be built around a set of foundational principles that respect privacy of data subjects and make privacy an integral part of the enterprise culture."

"Such a principles driven response would be more sustainable as well as rewarding for both enterprises and data subjects."

"Transparency: Transparency in the enterprise operations that process sensitive data of data subjects should be a founding principle of the enterprise privacy program."

"Transparency in the data collection processes and unambiguous explanation of purposes for data processing will build trust between customers and enterprise."

"Control: Enterprise should give adequate control to data subjects over their data."

"Enterprise should share the value derived from processing customers' data back with them."

"Accountability: Newer privacy regulations hold enterprise accountable for preserving privacy of data subjects, then be it as a data controller or its processor."

Conclusion

"It also has thrown up many challenges to enterprise."

"We have discussed major enterprise challenges from the perspectives of data, application and technology."

"We have also discussed certain challenges that a data subject is likely to face in the new privacy self management paradigm."

5.2.10 Releasing Survey Microdata with Exact Cluster Locations and Additional Privacy Safeguards [68]

This is a machine-generated summary of:

Koebe, Till; Arias-Salazar, Alejandra; Schmid, Timo: Releasing survey microdata with exact cluster locations and additional privacy safeguards [68]

Published in: Humanities and Social Sciences Communications (2023)

Link to original: https://doi.org/10.1057/s41599-023-01694-y

If you want to cite the papers, please refer to the original.

For technical reasons we could not place the page where the original quote is coming from.

Abstract-Summary

"To safeguard the respondents' privacy, micro-level survey data is usually (pseudo)-anonymized through deletion or perturbation procedures such as obfuscating the true location of data collection."

"This, however, poses a challenge to emerging approaches that augment survey data with auxiliary information on a local level."

"We propose an alternative microdata dissemination strategy that leverages the utility of the original microdata with additional privacy safeguards through synthetically generated data using generative models."

Introduction

"This leaves data producers in a difficult situation: finding the balance between disclosing unit-level data (i.e., containing individual- or household-specific information, commonly described as 'microdata') to foster use and to support knowledge generation and complying with relevant regulations."

"In that regard, advances in synthetic data generation have introduced new ways to narrow the void between information loss and privacy protection."

"Going further, the challenge for data producers is to define adequate microdata dissemination strategies that allow users to satisfy their needs, i.e., release survey microdata that can be used for statistical analysis and that are compatible with other sources of information allowing to answer new and more detailed research questions and—at the same time—it must be ensured that the identities of the respondents are protected."

"The alternative microdata dissemination strategy obviously conserves data utility for analysis on the representative level via the first dataset, while the second dataset allows for the accurate capture of proximity-related information."

Unsatisfied Basic Needs in Costa Rica

"As our reference dataset in this project, we use data from Costa Rica—notably the Xth Population and VIth Housing Census of Costa Rica, 2011 (Censo Nacional de población y Viviendas de Costa Rica 2011)—to produce three different data file types: First, we draw survey samples from a census population using a stratified two-stage cluster sample design without applying any statistical disclosure control mechanisms."

"To exemplify the importance of an alternative microdata dissemination strategy, we select the Unsatisfied Basic Needs index (Necesidades Básicas Insatisfechas (NBI))—a composite indicator similar to the multidimensional poverty index (MPI) (Alkire et al. [69], Méndez and Bravo [70]) used as a key statistical indicator in Costa Rica—as our target variable for survey augmentation in section "Utility for survey augmentation"."

"District-level NBI estimates have been derived from household sample surveys only, thus leaving out-of-sample districts without recent data and therefore less suitable to inform targeted policy interventions."

Results

"The obfuscated zip code is denoted by $Y_{zip} \in y$. Following our proposed data dissemination strategy, we further define the true survey without small-area geographic identifier as 'No Zip Code' survey $D_{no} := (X_2, \ldots, X_m)$ given that $X_1 \leftarrow X_{zip}$."

"The strategy proposed in this paper considers two datasets for dissemination: (1) the 'Synthetic' survey D_{syn} with original zip codes and remaining attributes being synthetically generated using a copula-based approach, and (2) the true original survey microdata stripped of geographic identifiers below the strata-level—the 'No Zip Code' survey D_{no}."

"Recalling that roughly 70% of the displaced clusters stay within the same zip code in the geomasked survey, even predicting the sensitive attribute for strata hosting as little as two small areas, average population uniqueness in the synthetic data would not exceed much the 50/50-chance of the random guess, thus providing better privacy protection in the re-identified original survey than the geomasked alternative."

Discussion

"We proposed and evaluated an alternative data dissemination strategy for micro-level survey data that improves the trade-off between privacy risk and data utility."

"We showed that by publishing two datasets, namely the original survey data with limited geographic identifiers and a synthetically generated survey dataset with the true cluster locations, re-identification risks can be reduced significantly vis-à-vis popular geomasking approaches without incurring additional losses in terms of data utility for survey augmentation."

"Even though intuitive as the sampling variance naturally decreases in n, at some point it may become misleading with potentially negative effects on the model performance as the synthetic data generating process still relies on the same information conveyed in the true survey with sample size n. However, in our experiment the adjusted R^2 does not exhibit a bump, but increases monotonically, thus hinting at little additional explanatory power of our satellite-derived covariates vis-à-vis the area-level direct survey estimate for the in-sample areas."

5.2.11 Visual Privacy Attacks and Defenses in Deep Learning: A Survey [71]

This is a machine-generated summary of:

Zhang, Guangsheng; Liu, Bo; Zhu, Tianqing; Zhou, Andi; Zhou, Wanlei: Visual privacy attacks and defenses in deep learning: a survey [71]

Published in: Artificial Intelligence Review (2022)

Link to original: https://doi.org/10.1007/s10462-021-10123-y

If you want to cite the papers, please refer to the original.

For technical reasons we could not place the page where the original quote is coming from.

Abstract-Summary

"Visual privacy in the context of deep learning is now an important and challenging topic."

"In this survey, we discuss algorithms of visual privacy attacks and the corresponding defense mechanisms in deep learning."

"We analyze the privacy issues in both visual data and visual deep learning systems."

"We show that deep learning can be used as a powerful privacy attack tool as well as preservation techniques with great potential."

"By thoroughly investigating the relationship of visual privacy and deep learning, this article sheds insights on incorporating privacy requirements in the deep learning era."

Introduction

"Many images and videos are used in the training and inference phase of deep learning algorithms, which may meet privacy violations (Oh et al. [72])."

"Some surveys covered only on privacy-preserving algorithms in deep learning systems (Tanuwidjaja et al. [73]; Boulemtafes et al. [74]) or on machine learning as a service (MLaaS) (Tanuwidjaja et al. [75])."

"Ha et al. [76, 77] and Liu et al. [78] have discussed not only on privacy, but also some methods on security in deep learning systems."

"After reviewing the literature from high-quality journals and conferences, we categorize some deep learning algorithms and technologies into attack methods and corresponding defense mechanisms concerning visual privacy."

"We summarize the attack methods for visual privacy in deep learning."

"We review the defense mechanisms corresponding to the attack methods for visual privacy in deep learning."

"Our survey gives a comprehensive review of visual privacy and the corresponding attack and defense methods in deep learning."

Preliminaries

"As deep learning models are trained by image/video datasets, these visual datasets and models can also be considered as privacy."

"Differential privacy The algorithms of differential privacy involve adding noises to the data or the deep learning model until the attacker cannot distinguish an individual data sample from the dataset or cannot recover certain private information from the model."

"Privacy preservation targets The targets for privacy preservation can be divided into two different groups: visual data (image and video data samples) and visual deep learning systems (deep learning systems focused on visual data)."

"In these circumstances, deep learning systems should be considered as private data to be protected, including the deep learning model and the whole training/inference phase."

"People who develop deep learning models can see the data directly in the dataset."

"The algorithms are then divided into visual data privacy and visual deep learning system privacy."

Attack Methods

"When the attackers have deep learning model-based attack tools, they can detect, extract, or retrieve sensitive information in visual data."

"With sensitive information in the training dataset, the deep learning models are vulnerable to various attacks."

"There are more papers extending the scenarios to attacks against other deep learning models, including semantic segmentation He et al. [79] and object detection Park and Kang [80]."

"Discussion There are three main reasons for the existence of membership inference attacks: the overfitting feature of the deep learning models, the data distribution between member and non-member samples, and the influence of data augmentations."

"Salem et al. [81] discovered that a more overfitted deep learning model would be more vulnerable to membership inference attacks."

"Discussion Deep learning models are more vulnerable when they are overfitted, causing the success of model inversion attacks, which is similar to membership inference attacks (Yeom et al. [82])."

Defense Mechanisms

"Chen et al. [83] presented a face data privacy-preserving approach using GAN by generating fake images of faces."

"Applying differential privacy in visual data needs to prevent identity information leakage in any image while preserving image quality, as these images are for browsing in social networks (Liu et al. [84])."

"Phan et al. [85] provided differential privacy preservation of deep learning models with an adaptive Laplace mechanism."

"Many research paper showed that differentially private deep learning models provided privacy protection against membership inference attacks, while offering reduced model utility at the same time (Rahman et al. [86]; Truex et al. [87]; Chen et al. [88]; He et al. [79]; Leino and Fredrikson [89]; Park and Kang [80]; Choquette-Choo et al. [90])."

"Nasr et al. [91] proposed an adversarial regularization method, which was an optimization privacy-preserving mechanism, by making the victim model's prediction in the training data indistinguishable from other data samples of the same distribution."

Future Directions

"We have reviewed recent research achievements on attack methods and corresponding defense mechanisms for visual data privacy and visual deep learning system privacy."

"Attacks on visual deep learning system privacy in both hite-box and black-box settings The white-box and black-box settings for privacy attacks mean how much access the attacker can get to the deep learning model."

"Applying adversarial perturbations to an image privacy-preserving framework, privacy in visual data can be well protected against the attacker while keeping the application's utility at the same time."

"Trade-off between model privacy and utility when using differential privacy It is an ongoing challenge to develop new deep learning frameworks and models that accommodate the needs of visual privacy preservation."

"Adversarial examples are developed for attacking deep learning models, which on the other hand, are adopted in preserving privacy in visual data."

"More deep learning technologies will be utilized by visual privacy attacks and defenses."

Conclusion

"In this survey, we have discussed visual privacy attacks and defenses in the context of deep learning."

"We have divided papers into attack methods and corresponding defense mechanisms, and then further into those focus on visual data or visual deep learning systems as the privacy preservation target."

"Major research trends on defense include: personalized privacy detection, leveraging newly developed deep learning methods in privacy preservation, balancing model privacy and utility, evaluating larger datasets for visual privacy, utilizing multiple technologies in one task and utilizing one kind of technology in multiple tasks, breaking the never-ending attack-defense cycle."

"Visual privacy attacks and defenses require different technologies that can be applied to images and videos."

"We believe our timely study will shed valuable light on the research problems associated with visual privacy attacks and defenses."

5.2.12 Information Privacy Assimilation in IT Organizations [92]

This is a machine-generated summary of:

Attili, V. S. Prakash; Mathew, Saji K.; Sugumaran, Vijayan: Information Privacy Assimilation in IT Organizations [92]

Published in: Information Systems Frontiers (2021)

Link to original: https://doi.org/10.1007/s10796-021-10158-0

If you want to cite the papers, please refer to the original.

For technical reasons we could not place the page where the original quote is coming from.

Abstract-Summary

"Following the neo-institutional theory, this study develops a model for understanding the mechanism of information privacy assimilation in Information Technology (IT) organizations."

"This study treats information privacy as a distinct dimension separate from security."

"After analyzing a sample survey data of 214 respondents from the IT industry, privacy capability and organizational culture emerged as influencing factors with a statistically significant influence on information privacy assimilation."

"The findings from this study support the mediating role of senior management participation between the external coercive forces and privacy-related business strategy."

"Business strategy also plays a mediating role between coercive/normative forces and privacy-related activities within an organization."

"The mimetic forces show a direct influence on privacy-related activities."

"A positive moderating effect of organizational culture on normative forces and privacy-related activities relationship; and a negative moderating effect of privacy capability on mimetic forces and privacy-related activities relationship are observed."

"These findings could enable senior managers to respond to institutional pressures by focusing on appropriate factors within an organization for developing effective privacy strategies and actions."

"Based on the lessons learned from the pilot study, we have updated the research model and hypotheses with a focus on privacy-related business strategy and activities."

Introduction

"By 2010, with the rise of SMAC (Social-Mobile-Analytics-Cloud) as a new information technology stack, academic researchers have focused on information privacy as a distinct construct different from information security (Attili et al. [93]; Conger et al. [94]; Kim et al. [95])."

"In the United States, apart from strengthening the regulations like HIPAA (Health Insurance Portability and Accountability Act), COPPA (Children's Online Privacy Protection Act), a focus on state-specific data privacy laws including the California Consumer Privacy Act (CCPA) have been enacted particularly post GDPR."

"In the backdrop of a growing data-centric world, organizations need to comply with information privacy regulations globally and also listen to customer concerns."

"How do organizations assimilate privacy in the strategies and actions that they undertake?"

"We address this gap by analyzing information privacy assimilation in IT organizations through a sample survey of the industry."

"Based on the lessons learned from the pilot study, we have updated the research model and hypotheses with a focus on privacy-related business strategy and activities."

Literature Review

"Information privacy focus for organizations is viewed by the American Institute of Chartered Public Accountants (AICPA) as the responsibility of organizations as well as individuals in collecting, using, disclosing and retaining Personally Identifiable Information (PII) (Johnson [96])."

"Much of the research focused on individuals; however, Information privacy is not just an individual responsibility, organizations also have a key role to play."

"A recent study by Belanger and James [97] has developed a Theory for Multi-level Information Privacy (TMIP), recognizing information privacy at an individual, group, and organizational levels."

"Attili et al. [93] studied the concept of assimilation pertaining to information privacy in IT organizations."

"In the current era when privacy regulations are evolving globally, this research attempts to carve out the privacy from security in the context of IT organizations and empirically investigates how external forces influence organizational information privacy assimilation in the strategies and actions of IT organizations."

Research Model and Hypotheses Development

"H1a: Coercive forces pertaining to information privacy will positively influence senior management participation."

"H1b: Normative forces pertaining to information privacy will positively influence senior management participation."

"H1c: Mimetic forces pertaining to information privacy will positively influence senior management participation."

"H2a: Senior management participation has a mediating effect on the relationship between coercive forces and privacy-related business strategy."

"H2b: Senior management participation has a mediating effect on the relationship between normative forces and privacy-related business strategy."

"H2c: Senior management participation has a mediating effect on the relationship between mimetic forces and privacy-related business strategy."

"With a social learning environment, it is reported that the external institutional forces are influencing middle-level management, who are responsible for privacy-related activities within the organization."

"H3b: Privacy-related business strategy has a mediating effect on the relationship between normative forces and the privacy-related activities of an organization."

Research Method

"In order to test the research model developed in the previous section, we undertook a sample survey."

"The research design included a survey method for data collection and statistical techniques such as PLS path modeling for hypotheses testing."

"This sample of 194 valid responses were analyzed to ensure the validity of the instrument and constructs."

"Although probability sampling is ideal for statistical inference, we faced the following constraints in following this approach: (a) access to members of a sample frame of the Indian IT industry is very restrictive due to confidentiality, (b) data security and privacy policies being nascent in India, getting key informants with valid experience was a huge challenge."

"The sample is a set of software professionals and leaders across IT organizations, who have 10+ years of IT experience and were aware of information privacy aspects in their current roles within their organization."

Data Analysis and Results

"In order to further determine the significance of the PLS path modeling estimates, bootstrapping method was used (Henseler et al. [98])."

"After assessing the direct effect, we included the mediator variable in the PLS path model and assessed the significance of the indirect path."

"The indirect effect was assessed after running the bootstrapping procedure and if it was found significant then mediator absorbs some of the direct path."

"To assess how much of the direct path was absorbed, we calculated the variance accounted for (VAF), for example, indirect effect/total effect."

"It shows that for coercive force, the direct effect ($\beta = 0.086$, $p = 0.232$) is not significant but indirect effect ($\beta = 0.074$, $p = 0.030$) is statistically significant."

"In order to evaluate the significance of the hypothesized relationships, bootstrapping was conducted using the product indicator scheme (Hair et al. [99]; Ringle et al. [100])."

Discussion of Results

"Our study focused on the mechanism by which external institutional forces influence privacy assimilation in IT organizations with a focus on business strategy and privacy related organizational activities."

"Our findings show that senior management support is a key factor that mediates the impact of external forces on privacy related business strategy within organizations."

"The influence of coercive forces on privacy business strategy is fully mediated by senior management participation."

"Statistically significant, the relatively less influence of coercive forces ($\beta = 0.166$, $p = 0.017$) highlights the scope for more collaboration between the legal framework and IT organizations in the area of privacy."

"Organizational culture also emerged as a significant influencing factor ($\beta = 0.322$, $p = 0.000$) on privacy related activities at the organization."

Implications

"The purpose of this study was two-fold: (a) to develop a conceptual model to understand information privacy in organizations and validate the model analytically, and (b) to test the model with quantitative data for generalization."

"The main contribution of this study is in developing a theory for information privacy assimilation, building on the concepts derived from neo-institutional theory using primary data collected through survey questionnaire."

"This study distinctly contributes to the body of knowledge in information privacy research, particularly in theory development and testing at the organizational level."

"This research addresses information privacy as a separate phenomenon from data security and develops conceptual elements and relationships specific to information privacy."

"This study develops a theory for information privacy at the organizational level, based on institutional theory which is further tested in the context of information privacy and IT domain using sample survey data."

Conclusion

"The findings from this study support the mediating role of senior management participation between the external coercive forces and privacy related business strategy."

"Business strategy plays a mediating role between coercive/normative forces and privacy related activities within organization."

"A negative moderating effect of privacy capability on mimetic forces impacting privacy assimilation is also observed."

"The study helps to identify appropriate privacy measures from an IT organization's view point, the influence of the business domain (health care, banking) was not considered."

"The quantitative sample used in the study is limited to US organizations with operations expanding to India and Indian organizations predominantly working within US regions."

"Lack of organizations from the European region in the sample geographically limits the study."

5.2.13 Data Sharing Platforms: Instruments to Inform and Shape Science Policy on Data Sharing? [101]

This is a machine-generated summary of:

Devriendt, Thijs; Shabani, Mahsa; Lekadir, Karim; Borry, Pascal: Data sharing platforms: instruments to inform and shape science policy on data sharing? [101]

Published in: Scientometrics (2022)

Link to original: https://doi.org/10.1007/s11192-022-04361-2

If you want to cite the papers, please refer to the original.

For technical reasons we could not place the page where the original quote is coming from.

Abstract-Summary

"The lack of incentives for data sharing has been conceptualized in both scientific literature and policy documents as a problem of science policy."

"It is shown how the design of platforms may help in addressing policy barriers to data sharing in the long-term."

"Platforms can be made into policy instruments that generate information on the data sharing process and the functionality of data access committees."

"This allows platforms to be used to inform science policy development, to monitor data sharing practices and to steer funding prioritization for cohorts and data infrastructures themselves."

Introduction

"Platforms that aim to facilitate the sharing of research cohort data are currently under development."

"Some platforms to connect research cohort data already exist within specific disease areas."

"The European Medical Information Framework (EMIF) platform, which ran until 2018, brought together electronic health record data and research cohort data sources, focusing primarily on Alzheimer's disease (Oliveira et al. [102])."

"Such concerns that the academic reward system does not sufficiently incentivize open science practices, including data sharing, are also captured within many policy documents (Altmetrics Working Group of the Open Science Policy Platform [103]; Leonelli [104]; Working Group on Rewards under Open Science [105]; Wouters et al. [106])."

"The CRediT taxonomy and data authorship both add detail to the type of contributions of researchers, and enable better recognition for data production, management and sharing (Allen et al. [107]; Bierer et al. [108])."

"As platforms only facilitate data sharing through technical means, they might fall short of addressing policy-related barriers related to the lack of recognition."

Tracing Information on Cohort Usage and Data Access Committee Functionality Through Platforms

"If platforms employ central portals for sending out data access requests (e.g., via data catalogues) using common templates for a range of cohorts, platforms may provide useful information on cohort data usage and DACs."

"The use of common templates for data usage requests featuring a minimum of standardized categories over platforms would enable collecting this information over all platforms."

"The data access management systems used within platforms should integrate mechanisms to transfer this information to one central place, where it can be made available to the research community, science policy makers and funding agencies."

"In analogy with web of science's data citation index (SCI), this central place can be considered an index of cohort data usage, where records on data access requests, information on DAC functionality and outcome publications for each cohort and platform are brought together."

Meta-Research on Data Sharing Practices

"By collecting information on data access requests and the functioning of DACs, the relationship between collected indicators and research outputs can be studied."

"Fully open data sharing would be characterized by a high degree of acceptance of access requests and lack of cohort affiliates on outcome publications."

"The scope of the collected information exceeds that of co-author network analyses as instances where data is shared for non-academic purposes (e.g., with industry) or where collaborations did not result in publications become visible."

"By tracking the process from incoming data access request to publication, it may become possible to study through quantitative methods where impediments occur."

"In that case, the impact of centralization of DACs on the processing time of data access requests, requests for scientific revisions, approval rate and non-responsiveness could be studied."

Developing Science Policies Suitable for Data-Intensive Research

"The advantages of collecting information on the data sharing process systematically over all cohorts and platforms will be outlined in terms of (a) gathering insights to develop new policies; (b) monitoring compliance with existing policies; (c) observing the effects of novel policies; (d) impacting the allocation of funding to data producers; and (e) reviewing the evolution of data sharing platforms (for funding prioritization)."

"The introduction of DMPs by many funders of science in recent years has prompted researchers to contemplate their data sharing practices."

"We argue that the system described here could have the same function for privacy-sensitive data in platforms, if information of the cohort data usage index is fed back into software programs used by funders of science to monitor the impact of research outputs (e.g., by linking funder PIDs to cohorts)."

Potential Obstacles to Establish Monitoring Mechanisms

"If common or fewer data access management systems would be used, the adoption of these can be foreseen at the conception of the infrastructure."

"Cultural barriers may impede adoption of data access management systems and/or common templates for data reuse."

"Researchers may also not be committed to creating common templates for data reuse."

"To overcome this barrier, the advantages of common templates for requesting data access, such as the diminished burden for the data requesting party and potential information extracted (as outlined here) need to be emphasized."

Scope of Proposal

"The "data sharing platforms" that are described in this article consist out of multiple subcomponents that facilitate data sharing, including decentralized data repositories, data catalogues, data management systems, and virtual research environments."

"The only two relevant components for this manuscript are the data access management system and the data catalogue."

"In all scenarios, it is possible to track decision-making of DACs through the data access management system."

"The only requirements are that (1) the DAC is formally inputted in the data access management system; (2) the DAC is linked to the place where metadata is stored (i.e., the data catalogue); (3) DACs input their decisions on data access in the data access management system."

Conclusion

"Data sharing platforms can collect information on the usage of available cohorts and the functionality of DACs."

"This increases the transparency over cohort and platform usage and the "invisible" processes of DACs."

"Systematically collecting indicators in standardized fashion allows setting up meta-research programs that increases knowledge on data sharing practices to the benefit of the research community, policy makers and funders of science."

"The infrastructures themselves may be able to partly remediate policy-related barriers for data sharing."

5.2.14 Comparative Sensitivity of Social Media Data and Their Acceptable Use in Research [109]

This is a machine-generated summary of:

Hemphill, Libby; Schöpke-Gonzalez, Angela; Panda, Anmol: Comparative sensitivity of social media data and their acceptable use in research [109]

Published in: Scientific Data (2022)

Link to original: https://doi.org/10.1038/s41597-022-01773-w

If you want to cite the papers, please refer to the original.

For technical reasons we could not place the page where the original quote is coming from.

Abstract-Summary

"Social media data offer a rich resource for researchers interested in public health, labor economics, politics, social behaviors, and other topics."

"Scale and anonymity mean that researchers often cannot directly get permission from users to collect and analyze their social media data."

"We compare individuals' perceptions of acceptable uses of other types of sensitive data, such as health records and individual identifiers, with their perceptions of acceptable uses of social media data."

"Our survey of 1018 people shows that individuals think of their social media data as moderately sensitive and agree that it should be protected."

Introduction

"A basic principle in research called respect for persons can bring clarity to how researchers can think about ethical social media data use practices."

"The scale and anonymity of social media mean that researchers often cannot directly elicit individuals' perceptions about the acceptable uses of data they generate on social media."

"To understand how we can realize respect for persons in the particular context of social media data research, we look to scholarship on users' perceptions about acceptable uses of other types of sensitive yet widely available data."

"Comparing individuals' acceptable use perceptions for their social media data with these other types of sensitive data opens opportunities for us to learn from existing respect-for-persons best practices aside from explicit consent developed to support sensitive data."

"Our survey study thus addresses the following research questions: RQ1: How do participants perceptions of acceptable social media data use compare to other types of sensitive data about them?"

Factors Affecting Perceptions of Acceptable Use

"Literature suggests that individuals' perceptions of acceptable use also vary across data use purposes like health research or marketing."

"Beyond general use purpose (e.g., for public health research, for marketing, for public awareness, etc.), people's understanding of exactly how their social media data will be used also affects their perceptions of its acceptable use."

"Research reports that perceptions of acceptable data use also vary based on a data sharer's personal characteristics."

"Social media users' perceptions of whether using their data is acceptable may be mediated by how sensitive they perceive their data to be, the data analyst, data use purpose, data type, and their personal characteristics."

"Our study draws from existing scholarship's insights on sensitive data and acceptable use to learn specifically about how people feel about their social media data's use relative to other data types that people have indicated is sensitive to them like health records [110] and location [111]."

Results

"Respondents with higher scores on the digital privacy concern scale were less likely to agree that it their data's use is acceptable (OR = 0.892)."

"Those who had higher institutional trust scores (OR = 1.041) and privacy behavior scores (OR = 1.112) were more likely to indicate that using their data was acceptable."

"According to ANOVA analyses, we found that the best model included data type, use purpose, data analyst, behavioral scales, and demographics but not sensitivity as predictors of respondents finding their social media data's use by researchers acceptable for social/behavioral research."

"Individuals with concerns about digital privacy (e.g., they hesitate to provide information when its requested) were less likely to be accepting of their data being used."

Discussion

"Our results suggest that individuals are generally accepting of academic researchers using social media data and that online screen names are less sensitive than many other types of data, including individual identifiers such as driver's license numbers and medical histories, but more sensitive than height, weight, race, and religion."

"We found that sensitivity is not a good predictor of whether individuals thought it was acceptable for researchers to use social media data."

"Our work shows that beyond considering individuals' preferences for their data's use under a respect for persons framework, social media researchers must also attend to the contexts in which user data are generated and used."

"Through surveying individuals, our study offers a benchmark of individuals' perceptions of their social media data's sensitivity and acceptable use from which social media data researchers can cultivate general respect-for-persons practices."

"People generally find it acceptable for researchers to use their social media data but prefer that researchers clearly articulate the benefits of their work."

Methods

"To understand participants' perceptions about acceptable social media data use, we surveyed 1018 people through Qualtrics panels and Mechanical Turk [112]."

"We used statistical analyses to identify patterns in survey responses, specifically regarding (a) the sensitivity of individuals' online identifiers relative to other personal identifiers, and (b) whether or not participants perceive a particular data analyst using a specific type of data for a purpose as acceptable or not."

"Our survey instrument measured relationships between acceptable use and relevant constructs like data sensitivity, and personal characteristics like trust in institutions."

"Their study evaluated respondents' relationships to a specific kind of social media data: the online screen name."

"Evaluating respondents' relationships to their online screen name does not encompass all types of social media data."

"Had we included the specific data types we asked about in scenarios (e.g., social media posts) in the sensitivity questions, we may have gotten different results."

5.2.15 Comparative Legal Study on Privacy and Personal Data Protection for Robots Equipped with Artificial Intelligence: Looking at Functional and Technological Aspects [113]

This is a machine-generated summary of:

Ishii, Kaori: Comparative legal study on privacy and personal data protection for robots equipped with artificial intelligence: looking at functional and technological aspects [113]

Published in: AI & SOCIETY (2017)

Link to original: https://doi.org/10.1007/s00146-017-0758-8

If you want to cite the papers, please refer to the original.

For technical reasons we could not place the page where the original quote is coming from.

Abstract-Summary

"This paper undertakes a comparative legal study to analyze the challenges of privacy and personal data protection posed by Artificial Intelligence ("AI") embedded in Robots, and to offer policy suggestions."

"After identifying the benefits from various AI usages and the risks posed by AI-related technologies, I then analyze legal frameworks and relevant discussions in the EU, USA, Canada, and Japan, and further consider the efforts of Privacy by Design ("PbD") originating in Ontario, Canada."

"The unpredictability arising from the AI machine learning function poses further difficulties, which have only been partially addressed by legal frameworks in the aforementioned jurisdictions."

"The first priority is adopting PbD as the most flexible, soft-legal, and preferable approach toward AI-oriented issues."

"Implementing PbD will protect individual privacy and personal data without specific efforts, and achieve both the development of AI and the advancement of privacy and personal data protection."

"If AI surpasses human intelligence, a terminating function, such as a "kill switch" will be the last resort to preserve individual choice."

Introduction

"This paper's purpose is to undertake a comparative legal study to analyze the challenges of privacy and personal data protection posed by robots equipped with AI, and to offer policy suggestions."

"Robots do not necessarily require AI, the incorporation of sophisticated functions of AI into robots greatly affects privacy and personal data protection."

"To consider the risks and challenges on privacy and personal data protection, we need to recognize the current state of AI."

"AI technology is rapidly advancing, which might entail unpredictable effects on privacy and personal data in the long run, in particular, from its machine learning function."

"From a legal perspective, the Ministry of Internal Affairs and Communications ("MIC") in Japan, co-operating with academics, has analyzed the impacts, risks, and challenges posed by AI, including privacy and personal data protection (Institute for Information and Communications Policy [114])."

Usages, Benefits, and Risks of AI Robots

"AI robots for military use need to be carefully considered, they are able to protect citizens by detecting enemies ahead of attacks."

"AI robots for disaster response will help evacuations of victims in dangerous areas that human rescuers cannot reach."

"Individuals should have the right to know about what happens when their data are processed within an AI robot."

"The sensor function enables robots to collect and record a wide range of rich personal data through continual monitoring of an individual's lifestyle and activities."

"Human-resembling robots psychologically change people's perceptions of robots based on their appearances, which lead to the disclosure of sensitive personal data."

"The more robots become human-like, the more people would accept them as quasi-humans or their clones, rather than tools; this could lead to people sharing rich details of their intimate private lives and sensitive personal data with robots."

Governing Privacy and Personal Data Protection for AI and Robots: Laws, Policies, and Other Approaches

"The Privacy Act of 1983 (R.S.C., 1985, c. P-21) covers the personal data handling practices of federal government institutions, and the Personal Information Protection and Electronic Documents Act ("PIPEDA") of 2000 (S.C. 2000, c. 5) covers the equivalent practices of private sector organizations in the course of commercial activities across Canada."

"SmartData is expected to proactively build privacy and security right from the beginning and incorporate the Fair Information Principles into the technology itself, thereby establishing an individual's control over their personal data."

"It is thought to be necessary to consider the legislation on privacy and personal data corresponding to AI Networks, considering the balance between protection and use" (Institute for Information and Communications Policy [114], p. 55)."

Discussion

"While I consider PbD to be the best approach to protect privacy as the development of AI robots progresses into the future, the technical methods to implement PbD are a future research topic that needs to be examined."

"After examining the role of consent in Canada and other jurisdictions, challenges of meaningful consent, the OPC identified several possible solutions: Enhancing informed consent which includes mechanism of PbD, alternative solutions that might introduce certain limited permissible uses without consent or introduce certain "no-go zones" of prohibited uses; stronger accountability mechanisms for organizations to demonstrate compliance with their current legal obligations; new accountability mechanisms that introduce broader notions of fairness and ethics; and strengthening regulatory oversight to ensure that proposed solutions are effective in protecting privacy (Office of the Privacy Commissioner of Canada [115])."

"Privacy laws in Canada and the Personal Information Protection Act in Japan also do not provide effective protections against AI risks."

Conclusion

"I have tried to analyze and offer policy suggestions on the challenges on privacy and personal data protection posed by AI and robots from a comparative legal perspective, focusing on the EU, USA, Canada, and Japan."

"While AI robots offer many benefits in the field of transportation and logistics systems, traffic, public safety, disaster response, and medical care, etc., several issues need to be considered from the perspective of privacy and personal data protection, mainly raised by the unpredictability of AI."

"The current legal protections in the EU, USA, Canada, and Japan are partially applicable to AI-oriented issues, including discriminatory decisions caused by profiling, impeding transparency, and deterring individual consent, jurisprudence has not yet clearly formulated the best approach."

"AI challenges not only privacy and personal data but also the overall current legal framework itself."

5.2.16 *Toward an Understanding of the Antecedents to Health Information Privacy Concern: A Mixed Methods Study [116]*

This is a machine-generated summary of:

Fox, Grace; James, Tabitha L.: Toward an Understanding of the Antecedents to Health Information Privacy Concern: A Mixed Methods Study [116]

Published in: Information Systems Frontiers (2020)

Link to original: https://doi.org/10.1007/s10796-020-10053-0

If you want to cite the papers, please refer to the original.

For technical reasons we could not place the page where the original quote is coming from.

Abstract-Summary

"As personal health information is digitized and entrusted to healthcare professionals and the technology vendors that manage health information systems (e.g., electronic health records), questions continue to arise regarding how this information is used and protected."

"By understanding what factors shape people's health information privacy concerns (HIPCs), organizations can better manage reactions and concerns regarding the use of new technologies and guidance can be produced to help people better protect their health information."

"We also show that users who report high HIPC are less likely to allow their health information to be included in an electronic health record system."

Introduction

"Privacy concern in the health context remains relatively under examined in the IS literature with the exception of a small number of studies [e.g., (Anderson and Agarwal [117]; Bansal et al. [118]; Kordzadeh and Warren [119]; Kordzadeh et al. [120]; Xu et al. [121]; Zhang et al. [122])]."

"To contextualizing the information privacy concern (IPC) construct for health information, we also identify and test factors that give rise to HIPCs through a mixed methods study and confirm the influence of HIPC on a health privacy-related behavior."

"A few studies have modified these privacy concern constructs to specify the industry or context of the organizations handling individuals' information, for example, health care (Angst and Agarwal [123]; Li and Slee [124])."

"We examine factors that shape individuals' HIPCs in addition to confirming the influence of HIPCs on a health privacy-related outcome."

Theoretical Development

"Borrowing from the definition of Hong and Thong (Hong and Thong [125]), we define health information privacy concern (HIPC) as an individual's perception of his or her concern for how personal information is handled by health entities."

"Their study examined several distinct pieces of medical information (e.g., physicals, cancer, depression) and interpreted their finding of higher privacy concern among older adults (>45 years of age) as being a result of more health issues as age increases."

"Bansal and others (Bansal et al. [126]) demonstrated that prior privacy invasion experiences increased privacy concern for health information specifically."

"H3a: As experience with health information privacy invasion increases, an individual's HIPC will increase."

"H3b: As exposure of health information privacy media coverage increases, an individual's HIPC will increase."

"Privacy concern has also been found to indirectly influence, through perceived privacy risk, individuals' intentions to use a personal health record system (Li et al. [127])."

Methodology

"The reasoning expressed by these interviewees can be largely tied back to their perceptions of how each party would use the data, with several interviewees (n = 5) stating that health professionals would use the data solely for the patient's benefit, whereas they believed that technology companies would use data for their own commercial goals."

"Many interviewees did not believe technology vendors would uphold high ethical standards due to the commercial aims of these companies, and the view that they only seek health data for monetary reasons, whereas health professionals require this information to administer treatment."

"A small number of interviewees (n = 5) expressed the view that their health data belonged to the health professional or professionals who created the data but felt they should own the information."

Discussion

"The results suggest that companies working with sensitive health data need to make an effort to assuage individuals' concerns regarding its use in order to implement information systems that will store and use the data."

"It also suggests that it is important to reduce highly publicized breaches of health data to encourage privacy-sensitive individuals to allow their data to be included in EHR systems."

"Our results point to some areas of focus to ease this transition; including, building trust in the technology vendors who will implement them, implementing and increasing awareness of legislation that protects health data, gaining an understanding of ownership, and clearly conveying what shared ownership means in the patient-healthcare provider relationship."

"Our quantitative results indicate that as perceptions of information ownership increase, so does HIPC, and our qualitative findings suggest that people may have distinctly different views on who owns their PHI."

Conclusion

"We conducted a mixed methods study of HIPC using a sample of individuals from the Republic of Ireland."

"Individual perceptions of risk, trust, information sensitivity, and information ownership were all important in shaping HIPC."

"Awareness of health privacy events from the media was also a driver of HIPC."

"We find that increased HIPCs decrease individual's intention to allow their information to be included in an EHR system."

5.2.17 Security and Privacy Requirements for a Multi-institutional Cancer Research Data Grid: An Interview-Based Study [128]

This is a machine-generated summary of:

Manion, Frank J; Robbins, Robert J; Weems, William A; Crowley, Rebecca S: Security and privacy requirements for a multi-institutional cancer research data grid: an interview-based study [128]

Published in: BMC Medical Informatics and Decision Making (2009)

Link to original: https://doi.org/10.1186/1472-6947-9-31

If you want to cite the papers, please refer to the original.

For technical reasons we could not place the page where the original quote is coming from.

Abstract-Summary

"Grid-based systems—such as the cancer Biomedical Informatics Grid (caBIG)—seek to develop new mechanisms to facilitate real-time federation of cancer-relevant data sources, including sources protected under a variety of regulatory laws, such as HIPAA and 21CFR11."

"These systems embody new models for data sharing, and hence pose new challenges to the regulatory community, and to those who would develop or adopt them."

"For quantitative analysis, data was aggregated at the individual or institutional unit of analysis, depending on the specific interview question."

"Results were grouped by broad themes, including among others: governance, legal and financial issues, partnership agreements, de-identification, institutional technical infrastructure for security and privacy protection, training, risk management, auditing, IRB issues, and patient/subject consent."

"The findings suggest that with additional work, large scale federated sharing of data within a regulated environment is possible."

"These include: (1) the necessity to construct separate legal or corporate entities for governance of federated sharing initiatives on this scale; (2) consensus on the treatment of foreign and commercial partnerships; (3) the development of risk models and risk management processes; (4) development of technical infrastructure to support the credentialing process associated with research including human subjects; (5) exploring the feasibility of developing large-scale, federated honest broker approaches; (6) the development of suitable, federated identity provisioning processes to support federated authentication and authorization; (7) community development of requisite HIPAA and research ethics training modules by federation members; (8) the recognition of the need for central auditing requirements and authority, and; (9) use of two-protocol data exchange models where possible in the federation."

Background

"Although neither comprehensive nor exhaustive, this study developed valuable primary data on the attitudes of those involved in regulatory decision-making relevant to the development and functioning of a large-scale and federated, biomedical data grid."

"CaBIG includes identity management processes in its federation model to provide the needed authentication on which authorization decisions ultimately rely."

"If caBIG or any federated biomedical data grid is to meet the needs of all relevant parties, those needs must be known—especially those of the often non-technical staff charged with overseeing data integrity and privacy."

"There are various federal regulations and standards governing federal employees' and contractors' use of electronic equipment, such as the Federal Information Processing Standards 201-1 (Personal Identity Verification requirements), that will have some impact on caBIG."

Methods

"Our approach was to develop structured elicitation interviews of key regulatory personnel at a subset of cancer centers involved in exchange of data using the caBIG system."

"Individuals who participated in the development of the instruments included representatives of the security project (7), members of the Data Sharing and Intellectual Capital Working Group (3) and Architecture Working Group (1), Institutional Review Board directors (3), external advisors (3), grid technologists (3), NCICB representatives (5), patient advocates (3), caTIES adopters (4), caTIES development team (2), and other stakeholders (4)."

"Participating cancer centers included all four current adopters of the caTIES System, the test-bed system described in the Interview Instrument."

"The interview scenarios were structured such that individual participants were asked a subset of the 87 questions across four scenarios, based on organizational role and expertise."

"For questions where each participant provided a single response, we show counts with interview as the unit of analysis."

"For questions where participants enumerated multiple items in response to a question, we use interview statements as the unit of analysis."

Results

"For right now, we would consider that our responsibility to ensure when that data goes out there, it is under an appropriate agreement limiting the use by that person to something that is either approved human subject research or not considered human subject research under the federal [guidelines]"—University and IRB Legal Counsel As articulated by one privacy officer, a potential problem with Data Use and Confidentiality Agreements in relation to use within a federated data-sharing environment is that they are typically specific to a project and executed between the researchers and the other institution: "I'd have to go through the [regulation] but there has to be some relationship then back to the researchers just not the institutions."

Discussion

"Most of the participants in the interviews had trouble understanding the concepts and implications involved with a federated environment, even though federation of identity and data sharing practices represent a model that retains and enhances local control."

"In a federation where the basic membership level is expressed at the institutional level, it may be beneficial to develop mechanisms that allow individual investigators who are not affiliated with member institutions to access or even contribute data to a federation in a secure fashion."

"Specific auditing functions this central group would be charged with overseeing include both technical and non-technical components, and consist of policy review, adherence to agreements, adherence to technical procedure and technical security architecture, adherence to data release only through protocol, incident aggregation, incident analysis, and communication of audit data back to the member institution."

"Specific tooling is needed to support both the centralized auditing functions proposed for the governing body, and the specific data required for trust development at the individual institutions."

Conclusion

"This study identified and explored security and privacy requirements for large-scale federated biomedical data sharing initiatives."

"These requirements are currently being used as a model for the development of the caBIG data sharing and security framework (DSSF)."

"These requirements also represent a general framework that can be used to inform the development of other large multi-institutional data sharing consortium."

"The findings, as well as the experiences of other large scale data sharing initiatives suggest that data sharing mechanisms will increasingly require strong central governance, and institutional commitment to the security procedures and policies of these organizations."

5.2.18 Alternatives for Protecting Privacy While Respecting Patient Care and Public Health Needs [129]

This is a machine-generated summary of:

DeCew, Judith Wagner: Alternatives for protecting privacy while respecting patient care and public health needs [129]

Published in: Ethics and Information Technology (1999)

Link to original: https://doi.org/10.1023/a:1010012018213

If you want to cite the papers, please refer to the original.

For technical reasons we could not place the page where the original quote is coming from.

Abstract-Summary

"I then examine three alternative approaches to protection of medical records: reliance on governmental guidelines, the use of corporate self-regulation, and my own third hybrid view on how to maintain a presumption in favor of privacy with respect to medical information, safeguarding privacy as vigorously and comprehensively as possible, without sacrificing the benefits of new information technology in medicine."

5.2.19 Monitoring Ethical, Legal, and Social Issues in Developing Population Genetic Databases [130]

This is a machine-generated summary of:

Austin, Melissa A; Harding, Sarah E; McElroy, Courtney E: Monitoring ethical, legal, and social issues in developing population genetic databases [130]

Published in: Genetics in Medicine (2003)

Link to original: https://doi.org/10.1097/01.gim.0000093976.08649.1b

If you want to cite the papers, please refer to the original.

For technical reasons we could not place the page where the original quote is coming from.

Abstract-Summary

"Purpose To characterize ethical, legal, and social issues unique to population genetic database research and to determine the relevance of international recommendations and guidelines for addressing these issues in the development of "genebank" projects globally."

"Results We identified and characterized five categories of ethical, legal, and social issues unique to genebank development: sponsorship and benefit-sharing, neutrality and regulatory power of ethics committees, public engagement, consent, and data protection."

"There is a growing need for more explicit, enforceable, and coordinated international guidelines relevant to the development and implementation of genebanks."

"By comparing ethical, legal and social issues as they arise in genebanks, researchers can better evaluate how to best use these projects to improve public health while protecting participating populations."

Main

"With the exception of the project in the Kingdom of Tonga, which was terminated due to ethical conflicts, development on all of these genetic databases has progressed significantly in the last year."

"[131] Although consent for most genetic research is obtained for a specific disease and research agenda, genebank projects obtain "generalized consent" for an unspecified set of common diseases and for an undefined period of study."

"While the characteristics of genebanks vary considerably, [131] all raise similar ethical, legal, and social issues that need to be examined as the projects develop."

"We identify and examine five general categories of ethical, legal, and social issues (ELSI) in genebank development: genebank sponsorship and benefit-sharing, ethics committees, public engagement, consent, and data protection."

"We illustrate specific issues using examples from established and proposed genebank projects, and evaluate the relevance of guidelines and recommendations provided by various international and regional ethics bodies in addressing these issues."

Methods and Results

"To identify ethical, legal, and social issues relevant and unique to genebank research, we undertook a comprehensive electronic search and literature review, using such search terms as biobank, genetic database, population genetics, and genetics and ethics, and terms specific to previously identified genebank projects."

"Because gene discovery within a population has implications beyond its study participants, current genebank ethics committees may not protect all of the individuals at risk from the research."

"Collecting the large number of samples necessary for genetic database research requires a significant recruitment effort and necessitates public awareness of the possible risks and benefits associated with participation in the genebank."

"Legislation on the IHD (Icelandic Health care Database) grants power to a three-person committee, with members appointed by the government, the national university, and deCode, to determine access to genebank data, "provided that the research may not be expected, in the view of the committee, to have an adverse effect upon the licensee's commercial interests.""

Summary and Conclusions

"We have identified and characterized five categories of such issues: genebank sponsorship and benefit-sharing, neutrality and regulatory power of ethics committees, public engagement, consent, and data protection, not all of which are included in current international policy regulations and guidelines."

"More empirical research examining specific issues, such as what constitutes a reasonable effort to protect privacy and the risks inherent in generalized consent, and a coordinated effort to develop comprehensive and widely applicable guidelines would be valuable."

"As genebank projects progress, there is a growing need for more explicit, enforceable, and coordinated international policy guidelines."

"A global and independent advisory group would monitor the development of genebanks, assist in identifying potential issues as they arise, and generally ensure that genebank projects meet their goals of improving public health while protecting the participating populations."

5.2.20 Exploring the Need for a Suitable Privacy Framework for mHealth When Managing Chronic Diseases [132]

This is a machine-generated summary of:

Jusob, Farad Rafique; George, Carlisle; Mapp, Glenford: Exploring the need for a suitable privacy framework for mHealth when managing chronic diseases [132]

Published in: Journal of Reliable Intelligent Environments (2017)

Link to original: https://doi.org/10.1007/s40860-017-0049-7

If you want to cite the papers, please refer to the original.

For technical reasons we could not place the page where the original quote is coming from.

Abstract-Summary

"This paper reviews and examines the challenges of preserving user privacy in the context of using mHealth to manage chronic diseases."

"The paper first discusses mHealth, its importance in managing chronic diseases, and the associated privacy concerns."

"The paper compares the existing privacy frameworks applicable to mHealth."

"The key principles gathered from the frameworks are analysed in the context of their suitability for enabling adequate privacy when using mHealth for managing chronic diseases."

"The paper argues that a new privacy framework is needed for mHealth in the context of managing chronic diseases."

Introduction

"In Europe, new challenges are being faced by healthcare systems, such as an increase in the elderly population (highly susceptible to chronic diseases) as well as budget cuts."

"mHealth services can be beneficial to managing chronic diseases by, for example, improving patient monitoring without the need to visit a health centre [133]."

"This is especially important in light of an increasing aging population, increases in chronic diseases, high cost of healthcare services, and the need to use limited resources effectively."

"The paper will then discuss the suitability of these frameworks and principles for safeguarding privacy when using mHealth to monitor chronic diseases."

"The paper will focus on the need for a new framework for mHealth in the context of managing chronic diseases."

mHealth

"Body sensors and mobile apps enable the collection of considerable medical, physiological, daily activity, and lifestyle data which is used to improve personalised treatment and medication for patients as well as enable users to manage their own health by self-assessment [134]."

"Mobile phones are used to improve point of service data collection, care delivery, patient communication, real-time medication monitoring, and adherence support."

"The use of sensors and mobile apps to collect medical, physiological, lifestyle, daily activity, and environmental data could serve as a basis for evidence-driven care practice and research activities, while allowing patients access to their health information at any given time or place."

"In a simple mHealth architecture, raw patient data can be collected from body sensors then sent to a smartphone app which constantly transmits the raw data to a care centre for analysis by health care professionals [135]."

Privacy

"The framework is made up of eight key principles [136]. (1) Individual access: Refers to the need for Individuals to be provided with simple and appropriate access to obtain their individually identifiable health information in a readable form and format. (2) Correction: States that an individual should be provided with an appropriate method to dispute the accuracy or integrity of their information, and allow for the correction or to have a dispute documented if their appeals are refused. (3) Openness and transparency: Discusses the need for openness and transparency about policies, procedures, and technologies that may have a direct impact on individuals and their personal data. (4) Individual choice: Refers to how individuals should be allowed to make informed decisions about the collection, use and disclosure of their personal data. (5) Collection, use and disclosure limitation: Discusses the extent to which an individual's information should be collected, used, or disclosed only to the extent which was previously agreed upon as well as the limit of its disclosure. (6) Data quality and integrity: Describes how individuals should take appropriate measures to ensure that their personal data are complete, accurate, and up-to-date. (7) Safeguards: refer to the necessity to ensure that there is adequate protection of personal data through appropriate safeguards to ensure its confidentiality, integrity and, availability and to prevent unauthorised or inappropriate access. (8) Accountability: Describes how the implemented principles should be monitored to ensure that appropriate enforcement is taking place as well as to disclose and mitigate any nonadherence and security breaches."

Privacy and mHealth

"Because of the limitation of some users regarding technological literacy, it is necessary to develop mHealth systems that allow patients added control over their data such as, what data are collected and who has permission to access it [137]."

"Case study, it was possible to identify the following privacy concerns: Continuous monitoring—from a wearable device that tracks activity level; Lack of encryption for data transmission—e.g., the communication between a wearable device and mobile phone; Data quality—the need for adequate checks to ensure manual data input is complete and has no missing fields; Profiling—e.g., use of location data to identify user habits; Data use and sharing—the system sharing data with the patients' health care practitioner as well as with an insurance company."

The Need for a Privacy Framework for mHealth and Chronic Diseases

"The collection of data from body sensors (C): Five main privacy threats/concerns are raised here namely: data quality; continuous monitoring; volume of data collected; surveillance; invisibility."

"The management of medical data on a smartphone app (M): Here, there are seven main privacy threats/concerns namely: lack of data security; lack of data encryption;

profiling; lack of anonymity; accessibility of data; potential risk of disclosure; and surveillance."

"The storage of data in the server/cloud (S): Here, five main privacy threats/concerns that are raised namely: data security; data encryption; lack of data anonymity; lack of data control; and accessibility of data."

"The use of data by various different types of users (U): There are six main privacy threats/concerns that are raised here namely: limitation of data use; sharing of data; confidentiality; lack of control; accessibility of data; and information misuse and abuse."

Conclusions

"The foregoing discussions in this paper have arguably demonstrated that existing privacy frameworks do not adequately address the privacy concerns of patients when using mHealth in the context of managing chronic diseases."

"The paper has also identified specific events in an mHealth scenario where users have privacy concerns and how each of these concerns is addressed by the existing privacy frameworks."

"This paper asserts that there is a need to design a suitable privacy framework for the use of mHealth to manage chronic diseases."

"The design of any new privacy framework for mHealth in that context must address the privacy threats at each of the mHealth events identified in the discussions above."

5.2.21 So You Want to Build Your Disease's First Online Patient Registry: An Educational Guide for Patient Organizations Based on US and European Experience [138]

This is a machine-generated summary of:

Wicks, Paul; Wahlstrom-Edwards, Lindsey; Fillingham, Sam; Downing, Andrea; Davies, Elin Haf: So You Want to Build Your Disease's First Online Patient Registry: An Educational Guide for Patient Organizations Based on US and European Experience [138]

Published in: The Patient—Patient-Centered Outcomes Research (2023)

Link to original: https://doi.org/10.1007/s40271-023-00619-w

If you want to cite the papers, please refer to the original.

For technical reasons we could not place the page where the original quote is coming from.

Abstract-Summary

"Patient registries fulfill a number of key roles for clinicians, researchers, non-profit organizations, payers, and policy makers."

"The complex tasks required to develop a modern registry can seem overwhelming, particularly for those who are not from a technical background."

"In this Education article, a team of co-authors from across patient advocacy, technology, privacy, and commercial perspectives who have worked on a number of such projects offer a "Registry 101" primer to help get started."

"We will outline the promise and potential of patient registries with worked case examples, identify some of the key technical considerations you will need to consider, describe the type of data you might want to collect, consider privacy risks to protect your users, sketch out some of the paths towards long-term financial sustainability we have observed, and conclude with plans to mitigate some of the challenges that can occur and signpost interested readers to further resources."

Background

"Patient registries are a relatively modern invention, dating back some 50 years and defined as "an organized system that uses observational study methods to collect uniform data (clinical and other) to evaluate specified outcomes for a population defined by a particular disease, condition, or exposure, and that serves one or more predetermined scientific, clinical, or policy purposes" [139]."

"A well-designed registry has much broader functions, in that it can potentially serve as the foundation for multiple studies from different disciplines that serve many stakeholders [139]."

"Such progress includes increasing patient and caregiver involvement in governing a registry, connecting registry data to clinical care, and supporting advocacy for non-profit organizations looking to generate evidence [140]."

"We are a group of authors specializing in digital health research (PW), registry partnerships with patient organizations (LWE), setting up a patient organization's first registry (SF), data governance and privacy issues for patients (AD), and regulations, registry development, and commercialization (EHD)."

The Promise and Potential of a Well-Designed Patient Registry

"Patients and family members seek a variety of sources to answer these questions including their healthcare providers, the medical/scientific community, and the experiences of other peers like them who have been down a similar path."

"At an organizational level, registries give authority and credibility for non-profit organizations to present data-driven insight and business cases."

"In some fields, a higher order harmonization group can come together to ensure that similar core data elements are collected by multiple registries and to offer an integrated view [141–143]."

"Regulators are a key stakeholder and harmonizer of standards in the health field, and groups such as the European Medicines Agency Cross-Committee Task Force on Patient Registries provide guidance and advice on how to best structure registries [144]."

"Between 2005 and 2013, the European Medicines Agency requested that over 30 drug manufacturers develop a registry to inform long-term safety and risk-management profiles, particularly in rare or "orphan" diseases [144]."

Starting with the End in Mind

"The point of a registry is to answer questions."

"The types of questions we can answer with a registry might include scientific, clinical, and policy concerns."

"Scientific questions might include: What sort of people have this disease?"

"Clinical questions might include: How are the outcomes of people with this disease changing over time?"

"As you start thinking through your objectives for a patient registry, it can be useful to keep a list of these questions, because this will shape what data you collect, how large a sample you need, and how burdensome it might be to take on this endeavor."

"As you meet with other stakeholders, it is also important to interview them to understand what sort of questions they are hoping to answer with a registry."

Technical Considerations

"The non-profit running the registry may need a series of administrative tools that allow maintenance, software upgrades, or to set the permissions for other categories of user."

"Each of these users has a slightly different set of requirements and permissions to access or modify the registry data, but they also share some common requirements."

"The most important users are the registry participants themselves and they will have their own needs and expectations."

"While there is an increasing expectation that "there's an app for that", smartphone users risk being overwhelmed by the range of notifications, permissions, and settings that need to be managed when controlling their health data."

"Adding features like a mobile app or clinical trial modules could be too significant an undertaking for the team that originally built a simple web-based data collection tool, and there is a risk of the code becoming out of date as browsers and mobile devices evolve."

Data Collection Considerations

"The harder you make it to enter the right answer the more likely you will encounter errors in the data, or your participants will simply give up."

"If you have a national registry then at some point you might want to compare your data with that from another country."

"You do not have to invent these yourself, there are existing standards like Logical Observational Identifier Names and Codes, which describe laboratory tests and their results, or International Classification of Diseases codes; however, these standards do not apply across all types of data that might be collected as part of a registry, and many rare diseases do not yet have an International Classification of Diseases code."

"If your registry has plenty of explanatory text on web pages or uses branching logic, it is best practice to build a "codebook" showing how the data are entered, validated, and what the user sees, preferably with screenshots, and to keep it updated as you make changes to data entry screens."

Data Governance, Privacy, and Security

"While the Health Insurance Portability and Accountability Act does not typically apply to a registry or a platform for which an individual shares their data, users may still have specific expectations with regard to the privacy and sharing rights applicable to their data."

"Truly informed consent should ensure that each individual patient understands who the guardian of the data is and how decisions are made about access and use."

"Beyond focus on informed consent, running a patient registry will also require you follow laws and standard practices to ensure your data are stored securely."

"You will need to learn the basics of how to properly encrypt the database of your registry, and ensure there are clearly defined internal control policies to access the data in ways that honor the consent model entrusted by your registrants [145]."

Building a Research Agenda

"Based on a famous quote [146], the questions that are "knowns knowns" describe your sample and assess how it compares to the scientific literature."

"If you have a large and relatively unbiased method of recruiting participants, then we might find that the participants in your registry are very similar to those reported in the published scientific literature."

"In an analysis comparing 10,255 members of the PatientsLikeMe multiple sclerosis registry with 4039 members of a specialist academic center database, the two samples were similar on age, age at onset, disease duration, gender ratio, family history, race, MS subtype, and even education level."

"Once you have established the representativeness of your sample, the next step is to study "known unknowns" [146]."

"A patient registry lowers the barriers to answer research questions that are obscure, under-researched, would be challenging to fund, or represent a long-standing gap in the literature."

Commercialization and Sustainability

"When the grant comes to an end, then either the data capture or maintenance of the registry comes to an end, or the work becomes reliant on the unpaid (and finite) good will of the host organization."

"Registry costs to consider include staff time (i.e., management, product management, front end user interface, back end database development, quality assurance testing, infrastructure operations), platform maintenance and upkeep, recruitment support and initiatives, community moderation, marketing, and web hosting."

"Some provide a free platform for advocacy organizations and use shared data ownership to help them recover the costs of maintaining the platform."

"Others charge organizations a licensing fee for use of the platform, but do not require shared data ownership."

"Another method organizations use to secure support for their work is to have a "corporate circle" or other membership scheme through which organizations solicit sponsorship from several relevant for-profit partners through which they maintain the registry."

Privacy Risks

"We discuss how to mitigate the risk of community data misuse."

"Many advocacy organizations created Facebook Groups to provide social support, reach an audience, and to grow their communities [147]."

"Many new or emerging registry tools may appear to be an "easy fix," to reach your community."

"The lack of rights to your collective data may quickly sow mistrust and even cause real harm, as shown by patient groups who tried to emerge from using Facebook as a lightweight registry [148]."

"Consider how a third-party platform may target ads to your community or resell your community's data [149], and consider how sensitive data about your community can be leaked to third parties or data brokers [150]."

"Consider your rights to the data as you adopt any new technology, and if you are uncertain, it is usually worth investing in legal counsel to help you understand your country's health data privacy and compliance laws."

Conclusions

"For someone just starting their journey on developing a patient registry, this brief (but broad) education article may appear daunting."

"This is now a well-worn path and you do not have to create every aspect of your registry from a blank slate."

"There are now a range of developers and software providers who can get you started much quicker than in the past, though continuity and interoperability must always be primary considerations when partnering with another organization."

"It might be another pediatric indication like yours, or a different form of cancer, or someone who has developed a registry around a similar drug or device that might apply in your condition but for a different indication."

5.2.22 PRISED Tangle: A Privacy-Aware Framework for Smart Healthcare Data Sharing Using IOTA Tangle [151]

This is a machine-generated summary of:

Abdullah, Sidrah; Arshad, Junaid; Khan, Muhammad Mubashir; Alazab, Mamoun; Salah, Khaled: PRISED tangle: a privacy-aware framework for smart healthcare data sharing using IOTA tangle [151]

Published in: Complex and Intelligent Systems (2022)

Link to original: https://doi.org/10.1007/s40747-021-00610-8

If you want to cite the papers, please refer to the original.

For technical reasons we could not place the page where the original quote is coming from.

Abstract-Summary

"Healthcare has evolved significantly in recent years primarily due to the advancements in and increasing adoption of technology in healthcare processes such as data collection, storage, diagnostics, and treatment."

"The emergence of the industrial internet of things (IIoT) has further evolved e-Health by facilitating the development of connected healthcare systems which can significantly improve data connectivity, visibility, and interoperability leading to improved quality of service delivered to patients."

"Such technological advancements come with their perils—there are growing concerns with regards to the security and privacy of healthcare data especially when collected, shared, and processed using cutting-edge connected sensor devices affecting the adoption of next-generation e-healthcare systems."

"We use masked authenticated messaging (MAM) over the Tangle to achieve secure data sharing within a healthcare system and provide a proof-of-concept of applying the proposed approach for securing healthcare data in a connected IIoT environment."

Introduction

"We use masked authenticated messaging (MAM) over the Tangle to achieve secure data sharing within a healthcare system and provide a proof-of-concept of applying the proposed approach for securing healthcare data in a connected IIoT environment."

"The objective of the research is to provide a decentralized framework that enables sharing and transport of health data in a secure and private environment by integrating the Tangle [152] which could support secure, fee-less, highly-scalable and granular medical data exchange."

"The main contributions of the paper include: (1) Discussion and demonstration on how the Tangle Version 1 (before Coordicide) provides improved security and privacy to the data sent by IoT devices. (2) Design of a scheme for secure transmission of medical or health data based transferred through IoT sensors to the Tangle using masked authenticated messaging (MAM) communication protocol. (3) Working "Proof-of-Concept" for transport of health data to a distributed ledger using masked authenticated messaging (MAM)."

Design Goals

"E-Health is defined as an amalgamation of information and communication of that information in terms of sharing and storage of healthcare data."

"With the advancement in e-Health solutions, more focus has been positioned on accomplishing the security and privacy of the health data as digitizing them will lead to an array of attacks such as denial of service, insider attacks, and information leak [153]."

"It deals with the protection of data in terms of authorized access to the health information [154]."

"The security of data emphasizes confidentiality, integrity, and availability of medical data which is managed by e-health system [155]."

"Confidentiality of data is also a prime factor in e-health."

"Authorization Another critical factor in a e-health system is the authorization of the data."

"The patient has the access to their health data and they have complete authority over it."

Related Work

"Similar blockchain architecture for storing health record was proposed by the authors in [156], in which they have discussed the dilemma between privacy and accessibility of patients' records."

"The proposed architecture to access electronic health records (EHRs) ensures data privacy and accessibility [156]."

"The Griggs and others have taken a more advanced perspective on patient data by analyzing security concerns of IoT devices used in health data."

"From the previous discussion on existing work done on maintenance of electronic health records (EHRs), electronic medical records (EMRs) and personal health records (PHRs) on distributed ledger technology, it can be seen that many experts have presented their solutions, protocols, frameworks and architecture to tackle two prime concerns, i.e., privacy and security of stored data."

"The authors, in this research, have proposed a framework in which IOTA Tangle is used as a ledger for storing medical data."

Background

"This concept in blockchain that a node needs to pay some fees for the transaction is also a notable drawback."

"In the Tangle, the transactions are also capable of carrying the message."

"Masked authenticated messaging (MAM), a second layer data communication protocol and an experimenting module, enables a node to publish or fetch data over the Tangle [157, 158]."

"Since MAM transactions are stored in the Tangle, it contributes to the hashing power in the ledger and maintains data integrity."

"Since the data in IOTA are decentralized (any node can send a message to any address), there is a risk that a message transaction might contain malicious data sent by any node to hijack the channel."

"They are: Public In this mode, the address transaction is the Merkle tree's root [159] and that address serves as both the channel ID and channel key [160]."

"Any node can fetch this message and decode it using the transaction's address."

The Proposed Scheme

"The proposed framework is a simple design at the initial level to demonstrate the deployment of IOTA Tangle and MAM while collecting real-time health data from the application."

"The implementation of the proposed framework presents a proof of concept of sending data securely from a smart device to IOTA Tangle, which is done via gateway."

"Research questions The system was designed with the following questions in mind: (1) Do the users have control of their health data in the existing healthcare systems? (2) Do the existing healthcare systems provide GDPR compliance over the health data? (3) What are the optimal solutions for providing privacy and security in medical and e-Health applications (4) Is the health data that is shared over secure and private? (5) Can the concerned entities view the health data with a user's permission?"

"For security, the user's health data is published using MAM restricted mode."

Method

"This test was carried out on 5 android smartphones with different android versions."

"They will have to publish the data to Firebase first."

"This data is then published to IOTA Tangle using a Web API."

"To check whether the data has been published to IOTA Tangle or not, IOTA MAM Explorer is used."

"That patient can share the QR code with their physician to give them access to view their data."

"Published data can be fetched from IOTA tangle using the developed web API."

"IOTA Devnet [161] was used to publish and fetch data from IOTA Tangle."

Evaluation

"For the same reason, i.e. the size of trytes, there is less median delay for the message of size m. It can be seen that the private node, node B, has exhibited a clear improved performance timing for both the messages, m and n. This is probably due to the reduced load in transaction requests that the private node, node B, experiences."

"It can be seen that the major delay caused when appending transaction to the Tangle is due to the "attach to tangle" operation corresponding to the "proof of work" which is carried out in each transaction."

"In a MAM transaction, the data is encrypted before attaching it to the Tangle."

"The tests were carried out on the current MAM library and it was observed that the overall delay of attaching the transaction is probably influenced by a small amount of "proof-of-work", which has randomness associated with it."

Security Analysis

"To further improve the security and encryption mechanism, MAM [157] protocol is used to encrypt the data stream on the Tangle."

"It offers tamper-proof solutions and helps in attaining the security requirements such as confidentiality, authorization, availability, and integrity."

"This section discusses the security requirements and how these requirements are achieved by the Tangle."

"Confidentiality is one of the most important requirements of security."

"The Tangle offers confidentiality by providing encryption to the data."

"For the current scenario of the proposed framework, MAM protocol is used in restricted mode, thus only enabling the authorized users to communicate with the data."

"Data confidentiality is well maintained using the Tangle."

"Availability is another important security requirement."

"Integrity The Tangle also offers this security feature in addition to the features mentioned above."

Conclusion and Future Work

"This research has evaluated and discussed the chief performance parameters of the transaction on the Tangle as well as masked authenticated messaging (MAM) Protocol."

"Another main factor of this research is masked authenticated messaging (MAM) protocol which allows for smooth and secure transactions over the Tangle."

"MAM module has also enabled the user to exercise authority over access control of data."

"One more similar work that can be carried out is addressing the patients in emergency and critical conditions where patients are not in a position to share their QR Code or publish their recent data to the Tangle."

"These different data can be published to the Tangle with different MAM modes."

"Future work can also be focused on extending the trial count from 100 to a larger set to verify, authenticate and endorse the performance parameters of the Tangle [162]."

5.2.23 Share and Protect Our Health Data: An Evidence Based Approach to Rare Disease Patients' Perspectives on Data Sharing and Data Protection—Quantitative Survey and Recommendations [163]

This is a machine-generated summary of:

Courbier, Sandra; Dimond, Rebecca; Bros-Facer, Virginie: Share and protect our health data: an evidence based approach to rare disease patients' perspectives on data sharing and data protection—quantitative survey and recommendations [163]

Published in: Orphanet Journal of Rare Diseases (2019)

Link to original: https://doi.org/10.1186/s13023-019-1123-4

If you want to cite the papers, please refer to the original.

For technical reasons we could not place the page where the original quote is coming from.

Abstract-Summary

"The needs and benefits of sharing health data to advance scientific research and improve clinical benefits have been well documented in recent years, specifically in the field of rare diseases where knowledge and expertise are limited and patient populations are geographically dispersed."

"Understanding what patients want and need from rare disease research and data sharing is important to ensure their participation and engagement in the process, and to ensure that these wishes and needs are embedded within research design."

"The aim of this survey was to explore patient and family perspectives on data sharing and data protection in research and healthcare settings and develop relevant recommendations to support shaping of future data sharing initiatives in rare disease research."

"Rare disease patients, regardless of the severity of their disease and their socio-demographic profile, are clearly supportive of data sharing to foster research and improve healthcare."

"Rare disease patients' willingness to share their data does come with specific requirements in order to respect their privacy, choices and needs for information regarding the use of their data."

"To ensure sustainability and success of international data sharing initiatives in health and research for rare diseases, appropriate legislations need to be implemented and multi-stakeholder efforts need to be pursued to foster cultural and technological changes enabling the systematic integration of patients' preferences regarding sharing of their own health data."

Introduction

"Data sharing within, and across rare disease groups can produce knowledge more efficiently by minimising the potential for duplicated studies, but also reducing the burden of research on small yet willing patient communities [164]."

"Understanding what patients want from rare disease research and data sharing is important to ensure their participation and engagement in the process, and also to ensure that consideration of their needs are embedded within research design."

"Rare disease patients' perspectives are needed to contribute to the debate on the management, sharing and protection of data, in order to reconcile tensions within the research process with what matters most to patients."

"Trust is a key issue for patients involved in rare disease research, and it could be argued that this becomes even more evident in data sharing, with the onus on researchers, institutions and collaborations to recognise this as a responsibility."

Materials and Methods

"Particular questions were also influenced by previous quantitative surveys which have specifically focused on data sharing and data protection, including for example the European Commission survey programme (Special Eurobarometer [165, 166]), which allows comparison between rare disease patients' perspectives and those of the general public."

"The questionnaire was mainly composed of close-ended questions with defined response categories, addressing the following areas: The degree of sensitivity attributed to different types of health-related information; The trust placed in different stakeholders that could be involved in data sharing initiatives; The different purposes for which they would be willing to share their data; The type of information rare disease patients would need to receive to engage in a data sharing project; The ways and the frequency to which they would want to receive information of the related projects; The degree of control they would want to have over their data; and."

Results

"Following the importance of gaining more information on the rare disease; options that would give patients and carers the possibility to have more control over the data are selected by about one quarter of the respondents: 27% declare that having the possibility to access their health information would encourage them to participate, 25% chose the possibility to withdraw their data at any time of the project—this option being more important for patients with higher level of education (27%) than those with lower level of education (18%)—and 24% opted for the option to be contacted if their information has been misused."

"Looking at the results of this question as a dependant or explanatory variable also shows that the willingness to share data is very lightly affected by the perceived sensitivity of data: respondents who perceive all type of the above health-information as sensitive are more than 90% to be willing to share their data for each health-related purposes."

Discussion

"Compared to the general population, rare disease patients seem to be more inclined to share their data."

"The support for data sharing in the context of rare disease research is aligned with the work undertaken by the European Reference Networks to establish a dedicated data-sharing platform enabling information exchange and mutual learning to improve rare disease patients' diagnosis and care, while also contributing to the development of research and innovation."

"The results clearly demonstrate that rare disease patients show various levels of trust in different actors and stakeholders who could be involved in data sharing platforms and initiatives."

"The high level of trust towards healthcare professionals involved in rare disease patient's daily care is important: data sharing initiatives would certainly benefit from involving general practitioners and other healthcare professionals in the management and communication of data sharing initiatives."

Conclusion

"The findings of this project add to the emerging literature about patient engagement in rare disease research and the value of, and barriers to, sharing data."

"Policy makers should ensure implementation of appropriate legislations at European and national levels and pursue efforts to foster cultural, technological and infrastructural changes to further develop international data sharing initiatives in health and research for rare diseases."

"All stakeholders involved in data sharing initiatives need to promote the development of, and implement, dynamic systems enabling: (i) the possibility to express different attitudes and preferences and (ii) access to updated information on research outcomes to increase patient participation in research and stimulate data sharing whilst respecting patients' preferences."

"Funders, clinicians and researchers need to emphasise potential health benefits of research studies and healthcare initiatives on future generations and other disease areas, as an incentive for wider participation in data sharing initiatives."

5.2.24 The Italian Neuromuscular Registry: A Coordinated Platform Where Patient Organizations and Clinicians Collaborate for Data Collection and Multiple Usage [167]

This is a machine-generated summary of:

Ambrosini, Anna; Calabrese, Daniela; Avato, Francesco Maria; Catania, Felice; Cavaletti, Guido; Pera, Maria Carmela; Toscano, Antonio; Vita, Giuseppe; Monaco, Lucia; Pareyson, Davide: The Italian neuromuscular registry: a coordinated platform where patient organizations and clinicians collaborate for data collection and multiple usage [167]

Published in: Orphanet Journal of Rare Diseases (2018)

Link to original: https://doi.org/10.1186/s13023-018-0918-z

If you want to cite the papers, please refer to the original.

For technical reasons we could not place the page where the original quote is coming from.

Abstract-Summary

"The worldwide landscape of patient registries in the neuromuscular disease (NMD) field has significantly changed in the last 10 years, with the international TREAT-NMD network acting as strong driver."

"In Italy, several NMD POs and Fondazione Telethon undertook the development of a TREAT-NMD linked patient registry in 2009, with the referring clinical network providing input and support to this initiative through the years."

"The Italian NMD registry is based on an informatics technology platform, structured according to the most rigorous legal national and European requirements for management of patient sensitive data."

"The platform's design permits expansion to incorporate new modules and new registries, and is suitable of interoperability with other international efforts."

"When the Italian NMD Registry was initiated, an ad hoc legal entity (NMD Registry Association) was devised to manage registries' data."

"These disease-specific registries are at different stage of development, and the NMD Registry itself has gone through several implementation steps to fulfil different technical and governance needs."

"The new governance model is based on the agreement between the NMD Registry Association and the professional societies representing the Italian NMD clinical network."

"Up to now the NMD registry has collected data on more than 2000 individuals living with a NMD condition."

"The Italian NMD Registry is a flexible platform that manages several condition-specific databases and is suitable to upgrade."

"It favored patient empowerment and their direct participation in research, while also engaging the expert clinicians of the Italian network in the collection of accurate clinical data according to the best clinical practices."

Background

"TREAT-NMD engaged all partners in the development of DMD and SMA patient registries linked to its international Global Registry, which is part of a research infrastructure acknowledged as an "IRDiRC (International Rare Diseases Research Consortium) recognised resources" [168]."

"Fondazione Telethon (Telethon) was the Italian partner of the TREAT-NMD EC project and, in this role, it engaged the Italian neuromuscular POs to develop NMD patient registries."

"One of the main goals of the NMD Alliance is to consolidate and expand the Italian NMD Registry by incorporating new disease registries and map the activities of the clinical centres."

"In line with the TREAT-NMD priorities, the first two registries developed by ADR addressed DMD and SMA and were, consequently, linked to the TREAT-NMD Global Registry; they envisage the direct involvement of patients, or parents in case of children, in the data collection."

Methods

"The informatics technology (IT) platform is a software infrastructure that provides basic functionalities such as: disease-specific forms for personal and clinical data collections (connected via pseudonymised codes); a module for the registry configuration of the participating clinical centres and their staff; a module for operator authentication; safety and protection solutions aligned with local regulations; storage solutions and data recovery."

"A web page specific for each registry illustrates the type of data that are collected, the composition of the Scientific Committee, and the operating procedures, with a link to the information leaflet and Patient Consent form."

"The NMD Registry has developed Standard Operating Procedures (SOPs), addressing (among others): how to create a new registry, role and responsibilities of all actors, modalities of data registration (clinical reported forms and support by clinicians to patients for the patients' reported forms), and data access."

Results

"At the start of the initiative, the ELAC provided ADR with advice on the Italian and international legal requirements on data protection, validated the privacy design of the IT platform, the data management procedures and flowchart, and approved the Italian version of the Informed Consent for DMD and SMA registries."

"The ELAC also provided the Scientific Coordinator of the CMT Registry with advice on the Clinical Protocol and the patient Informed Consent before the submission of the documents for approval to the Institutional Review Board (IRB) of the coordinator and partners' centres."

"By June 2018, 721 CMT patients (mean age 46 years, range 5–88 years) had registered and chosen one of the nine reference centres; up to now clinical information has been entered in the Registry for 584 of them."

Discussion

"The Registry is based on an IT platform hosting several NMD disease-specific databases, each of them having peculiarities in terms of modality of data collection (patient- or clinician-reported forms) and purposes."

"Being the POs involved in the Registry governance, the fulfilment of the patients' willingness to make their clinical data available to the broad scientific community is guaranteed."

"The NMD Registry platform is expected to become the referring structure for future registry initiatives, engaging the young generation of clinicians in data collection through an educational process that has its foundation in the FAIR concepts and ontologies-based classification [169–172]."

"They guarantee both direct input of patients in the registry and the collection of accurate and validated clinical data that are recorded in a systematic and scientifically sound manner and can be rapidly analysed."

Conclusions

"CMT Registry: Davide Pareyson and Daniela Calabrese (Milan, authors); Isabella Moroni, Emanuela Pagliano, Chiara Pisciotta, Giuseppe Piscosquito, and Stefano Carlo Previtali (Milan); Franco Gemignani and Isabella Allegri (Parma); Gian Maria Fabrizi and Tiziana Cavallaro (Verona); Angelo Schenone, Marina Grandis, and Chiara Gemelli (Genoa); Luca Padua and Costanza Pazzaglia (Rome); Lucio Santoro and Fiore Manganelli (Naples); Aldo Quattrone and Paola Valentino (Catanzaro); Giuseppe Vita and Anna Mazzeo (Messina)."

"MGSD Registry: Antonio Toscano (Messina, author); Corrado Angelini (Venice); Bruno Bembi (Udine); Andrea Martinuzzi (Conegliano); Paola Tonin (Verona); Massimiliano Filosto (Brescia); Lorenzo Maggi (Milan); Tiziana Mongini (Turin); Claudio Bruno (Genoa); Maria Alice Donati (Florence); Gabriele Siciliano (Pisa); Serenella Servidei (Rome)."

"TTR-FAP Registry: Giuseppe Vita (Messina, author); Gian Maria Fabrizi and Tiziana Cavallaro (Verona); Davide Pareyson and Silvia Fenu (Milan); Giampaolo Merlini and Laura Obici (Pavia); Alessandro Mauro (Turin); Marina Grandis and Chiara Gemelli (Genoa); Claudio Rapezzi (Bologna); Mario Sabatelli (Rome); Lucio Santoro and Fiore Manganelli (Naples); Lorenza Magliano (Caserta); Costanza Barcellona (Messina)."

5.2.25 Protection of Visual Privacy in Videos Acquired with RGB Cameras for Active and Assisted Living Applications [173]

This is a machine-generated summary of:

Climent-Pérez, Pau; Florez-Revuelta, Francisco: Protection of visual privacy in videos acquired with RGB cameras for active and assisted living applications [173]

Published in: Multimedia Tools and Applications (2021)

Link to original: https://doi.org/10.1007/s11042-020-10249-1

If you want to cite the papers, please refer to the original.

For technical reasons we could not place the page where the original quote is coming from.

Abstract-Summary

"Active and assisted living technologies are much needed, but some aspects of them cause user rejection due to concerns on privacy."

"This is even more concerning to users when visual information is used, processed, and transmitted."

"To respond to these concerns, and maximise user acceptance, visual privacy protection measures have to be put in place."

"This paper introduces a proposal for an RGB-only based visual privacy preservation filter, which capitalises on 'deep learning'-based segmentation and pose detectors."

"A background update scheme is presented, which limits leakage of sensitive information when detection fails."

"Dilation of the mask can further prevent information leakage, but a trade-off is necessary to correctly update background information."

Introduction

"Regarding supporting technologies, there are several important landmark components that enable the provision services for AAL based on computer vision, specifically for some of the more complex applications such as human action recognition and human behaviour understanding: namely, machine learning, which encompasses the current trend of pervasive use of neural networks (and 'deep learning'); as well as the development and availability of new camera sensors, modalities (IR, Depth, stereo), and increasing affordability."

"Along with [174], this work is part of the ongoing efforts in the EU-funded project on privacy-aware and acceptable lifelogging services for older and frail people (the PAAL project), which: aims to understand users' requirements in terms of perceived benefits and barriers and to increase the awareness of issues associated with lifelogging, establishing guidelines for responsible research in these technologies.—"

Visual Privacy Preservation

"This has led to an increase in research aimed at visual privacy preservation methods in different contexts: from the social media advisor proposed by Orekondy et al. [175], which advises users on the suitability for upload of certain pictures based on their contents (faces, credit cards being shown, passersby, etc.); to systems on board unmanned aircraft systems or vehicles (UAS/UAVs), such as the one by Babiceanu et al. [176], to limit the capture of video to a certain area of interest, as well as blurring non-relevant faces, car plates, or other sensitive information that might be recorded accidentally."

"To privacy preservation is related with retaining the capability to recognise the activity while the video is degraded or transformed heavily in such way that is either fully unrecognisable by humans, or alternatively, the identities of the people appearing in the video are [177, 178] (i.e. related to the concepts of secure processing and blind vision)."

"At integrating context in visual privacy preservation is by considering not just the context of the physical environment, but also the social context, that is, the relationships between the end user of the technology and the people having access to the user's information."

Privacy Preservation in RGB Videos

"To Padilla-López and others [179], we propose a series of different filters or masks of sensitive information on the image: transparency (invisibility), replacement by an avatar, pixelation, blurring and embossing; but this time with RGB information only."

"To demonstrate the current possibilities for visual privacy preservation using person detectors directly on RGB data (instead of RGB-D), available implementations of DensePose and Mask R-CNN were used."

"To further protect the model from non-moving foreground objects (e.g. a sitting person), the background model is only updated with the information outside the detection mask."

"This is a rare case (e.g. only 0.05% of frames in MuHAVi-MAS generated no mask at all), and the background quickly recovers, as usually only one frame in the whole queue is affected, and the average value is closer to the rest, therefore mitigating the leakage of sensitive information on the person's identity."

Conclusion

"It has been shown how to use off-the-shelf deep neural network implementations for the purpose of visual privacy preservation."

"It has been shown that moderate dilation around the person detection is useful to further preserve the privacy of imperfectly detected limbs or identifying pieces of clothing."

"As part of future work, we intend to use omnidirectional cameras mounted on the ceiling to detect and recognise human actions."

"To solve this issue, we plan to use side view cameras and existing detectors to train a human pose detector which works directly on the top view, taking advantage of camera topology and calibration to automatically collect ground truth from the side views which would then be used for training on the top view."

5.2.26 Privacy-Preserving Data Sharing Infrastructures for Medical Research: Systematization and Comparison [180]

This is a machine-generated summary of:

Wirth, Felix Nikolaus; Meurers, Thierry; Johns, Marco; Prasser, Fabian: Privacy-preserving data sharing infrastructures for medical research: systematization and comparison [180]

Published in: BMC Medical Informatics and Decision Making (2021)

Link to original: https://doi.org/10.1186/s12911-021-01602-x

If you want to cite the papers, please refer to the original.

For technical reasons we could not place the page where the original quote is coming from.

Abstract-Summary

"Various approaches and infrastructures have been developed that aim to ensure that patients and research participants remain anonymous when data is shared."

"The aim of the work described in this paper was to develop a systematization for the degree of privacy protection provided and the trade-offs taken by different data sharing methods."

"Contribution, we categorized popular data sharing approaches and identified research gaps by analyzing combinations of promising properties and features that are not yet supported by existing approaches."

"Three axes relate to privacy protection aspects and were adopted from the popular Five Safes Framework: (1) safe data, addressing privacy at the input level, (2)

safe settings, addressing privacy during shared processing, and (3) safe outputs, addressing privacy protection of analysis results."

"Three additional axes address the usefulness of approaches: (4) support for de-duplication, to enable the reconciliation of data belonging to the same individuals, (5) flexibility, to be able to adapt to different data analysis requirements, and (6) scalability, to maintain performance with increasing complexity of shared data or common analysis processes."

"Using the systematization, we identified three different categories of approaches: distributed data analyses, which exchange anonymous aggregated data, secure multi-party computation protocols, which exchange encrypted data, and data enclaves, which store pooled individual-level data in secure environments for access for analysis purposes."

"There are fundamental differences between different data sharing approaches and several gaps in their functionality that may be interesting to investigate in future work."

"Our systematization can make the properties of privacy-preserving data sharing infrastructures more transparent and support decision makers and regulatory authorities with a better understanding of the trade-offs taken."

Background

"Despite the fact that privacy-preserving infrastructures are often considered to be the most important enabler for comprehensive data sharing in the medical domain and despite the multitude of technological approaches available and studies that have successfully utilized such technologies (see above), these infrastructures are only rarely used for sharing healthcare and medical research data today."

"We also believe that the systematization can contribute to developing instruments for assessing the usefulness of data sharing infrastructures, i.e. the impact that their protection mechanisms have on options to analyze data compared to the simple (but often not feasible) approach of pooling all data in a common database."

"Our work provides the following contributions: (1) We present a high-level and technology-agnostic framework consisting of three axes describing the degree of protection and three axes describing the degree of usefulness provided by data sharing infrastructures. (2) We use this framework to analyze and compare ten different real-world data sharing platforms."

Methods

"An important example is data anonymization, where, as also mentioned above, the quality of output data often must be traded off against the degree of privacy protection achieved (see e.g. [181])."

"We suggest to assess the usefulness of infrastructures for sharing medical data in terms of three different axes that reflect important requirements in multi-institutional medical research: (1) De-duplication/record-linkage, which refers to the ability to combine data from different sources while taking into account that some records might relate to one another (e.g. to the same patient), (2) Flexibility, which reflects the degree to which a solution is able to support different types of statistical analyses and use cases as well as adapt to different analytical requirements as they can change over time and (3) Scalability, that refers to how an infrastructure performs when the amount of data or the complexity of an analysis increases."

Results

"It is a generic solution conceptualizing a container-based data sharing infrastructure that can be used to implement wide a range of meta-analysis approaches (Flexibility, axis 2.2)."

"Approaches using cryptography-based secure multi-party computation protocols to ensure that only encrypted individual-level data leaves the participating sites form an important additional category of data sharing infrastructures."

"Important examples of approaches from this category are: MedCo The open source software MedCo uses additively homomorphic encryption to enable researchers to perform analyses on encrypted data across sites [182]."

"The solution is somewhat generic, as typical data analysis methods are supported, but extensibility is limited as additional software, packages and functionalities can only be implemented by the enclave (Flexibility, axis 2.2)."

"The solution is specific, since its software and functionalities focus on integration and analysis of claims data (Flexibility, axis 2.2)."

Discussion

"Our work is the first to apply the framework to common biomedical data sharing infrastructures, however."

"We have complemented the Five Safes framework for modeling privacy protection with additional axes for systematizing the usefulness of data sharing technologies, considering common requirements from biomedical research."

"Other frameworks for data sharing in biomedical research have been proposed, which can also be used to analyze different technical approaches."

"Some data sharing infrastructures and approaches support different privacy models to provide Safe Data and Safe Outputs, e.g. Differential Privacy [183] or solutions limiting the uniqueness of disclosed data, such as cell suppression [184] or k-anonymity [185]."

"A challenge with privacy-preserving data sharing infrastructures is that access to individual-level data in some cases cannot be provided at all, although access to data from at least one site is often needed to develop analysis algorithms that can then be executed in the distributed network."

Conclusion

"We proposed a high-level framework for analyzing and comparing privacy-preserving data sharing infrastructures for medical research."

"We believe that our framework makes the properties of data sharing approaches more transparent and can serve as a starting point for developing more comprehensive systematizations, ultimately supporting decision makers and regulatory authorities in gaining a better understanding of the trade-offs taken."

"We have shown that our systematization is of value, by using it to analyze existing solutions, showing that there are fundamental differences between them."

"Our results also provide insights into gaps, regarding the systematization itself as well as the current landscape of data sharing infrastructures, that may be worth exploring in the future."

5.2.27 Health Data Privacy Through Homomorphic Encryption and Distributed Ledger Computing: An Ethical-Legal Qualitative Expert Assessment Study [186]

This is a machine-generated summary of:

Scheibner, James; Ienca, Marcello; Vayena, Effy: Health data privacy through homomorphic encryption and distributed ledger computing: an ethical-legal qualitative expert assessment study [186]

Published in: BMC Medical Ethics (2022)

Link to original: https://doi.org/10.1186/s12910-022-00852-2

If you want to cite the papers, please refer to the original.

For technical reasons we could not place the page where the original quote is coming from.

Abstract-Summary

"Hospitals and research institutes are developing technical solutions for sharing patient data in a privacy preserving manner."

"Homomorphic encryption represents a potential solution for conducting feasibility studies on cohorts of sensitive patient data stored in distributed locations."

"Distributed ledger technology provides a permanent record on all transfers and processing of patient data, allowing data custodians to audit access."

"A significant portion of the current literature has examined how these technologies might comply with data protection and research ethics frameworks."

"In the Swiss context, these instruments include the Federal Act on Data Protection and the Human Research Act."

"These included legal and clinical data management staff, as well as clinical and research ethics experts."

"These interviews were carried out with two series of vignettes that focused on data discovery using homomorphic encryption and data erasure from a distributed ledger platform."

"For our first set of vignettes, interviewees were prepared to allow data discovery requests if patients had provided general consent or ethics committee approval, depending on the types of data made available."

"For our second set, there was disagreement amongst interviewees on whether they would delete patient data locally, or delete data linked to a ledger with cryptographic hashes."

"Our interviewees were also willing to delete data locally or on the ledger, subject to local legislation."

Introduction

"The GDPR restricts the use and transfer of sensitive personal data, including genetic, medical and health related data."

"As we discuss in this paper, HE can be useful for researchers who wish to conduct data discovery or feasibility studies on patient records [187]."

"Some scholars have proposed using DLT style implementations to create an auditable record of access to patient data [188, 189]."

"Recent studies indicate that whilst most health research projects in Switzerland use coded data [190]."

"Our paper is the product of an ethico-legal assessment of HE and DLT conducted as part of the Data Protection and Personalised Health (DPPH) project."

"The DPPH project is an SPHN-Personalized Health and Related Technologies (PHRT) driver project designed to assess the use of these technologies for multisite data sharing."

Methods

"Our interviewees were drawn from legal practitioners, ethics practitioners and clinical data managers working at Swiss hospitals, research institutes and governance centres."

"Part A of our vignette contained three scenarios concerning a data discovery or feasibility request on patient data."

"Unlike the other two scenarios, this scenario concerned a data discovery request from non-SPHN affiliated commercial research institute, for conducting drug discovery on an anti-melanoma drug."

"If the interviewee wanted an ethics committee to review a data discovery request due to existing policy, this answer was labelled as 'policy consistency'."

"These included whether the interviewee would release the data without further consent, seek informed consent from patients or refer the matter to an ethics committee."

"This qualitative data analysis was complemented with numerical data analysis to measure how many interviewees were willing to accept data transfer requests under the scenarios depicted in our vignettes."

Results

"Another interviewee who would have permitted the request on non-genetic data with general consent would not have permitted access to this data without specific consent from the patient."

"Willingness, these interviewees argued that if patients were not made aware that their data might be used for commercial research, it was questionable whether informed consent had been obtained in the circumstances."

"A consistent theme that emerged across all three scenarios amongst interviewees was guaranteeing that the requesting researcher or institution needed the data for specified, explicit and legitimate purposes."

"If patients were only allowed to opt out or refuse to consent to having their encoded data, but not their anonymised data used for research purposes, these interviewees reasoned that this would impinge on patient rights."

"One interviewee pointed out that the general consent form in the scenario guaranteed that patient data would not be available for new research projects if the patient withdrew their consent."

Discussion

"Further, one interviewee mentioned that a general consent form should not only allow a patient to opt out of having their data encoded, but also having their data anonymised."

"Another interviewee from a university hospital noted their institution had developed a general consent form that allowed opt outing of further use for both encoded and anonymised data."

"Although this consent form went beyond the legal requirements, it nevertheless offered the patient more control over their data compared to other general consent forms."

"Another interviewee mentioned this public trust could be accentuated with a general consent form that would allow the patient to seek further information about the research projects their data is used for."

"This general consent form should highlight whether a patient's data might be used for commercial research purposes."

"One interviewee questioned whether a patient could give explicit and informed consent to having their data processed using this technology."

Conclusions

"We described an interview study on the use of HE and DLT for processing patient data."

"With respect to the first set of requests, most of our interviewees were prepared to permit processing, provided that general consent had been obtained from patients to do so."

"Advanced privacy enhancing technologies have the potential to fill the regulatory gaps that exist under current data protection laws in Switzerland."

"General consent forms should be amended to offer patients the opportunity to opt out to having their data anonymised using advanced privacy enhancing technologies."

"A public education campaign should be targeted at explaining how these technologies work to give patients the opportunity to understand how their data will be processed."

"Future research should address how patients might respond to the use of these technologies to process their data."

5.2.28 How to Integrate Legal Requirements into a Requirements Engineering Methodology for the Development of Security and Privacy Patterns [191]

This is a machine-generated summary of:

Compagna, Luca; El Khoury, Paul; Krausová, Alžběta; Massacci, Fabio; Zannone, Nicola: How to integrate legal requirements into a requirements engineering methodology for the development of security and privacy patterns [191]

Published in: Artificial Intelligence and Law (2008)

Link to original: https://doi.org/10.1007/s10506-008-9067-3

If you want to cite the papers, please refer to the original.

For technical reasons we could not place the page where the original quote is coming from.

Abstract-Summary

"Several IT solutions (e.g., Privacy Enhancing Technologies, Access Control Infrastructure, etc.) have been proposed to address security and privacy issues."

"Security engineers and legal experts should analyze the business goals of a company and its organizational structure and derive from there the points where security and privacy problems may arise and which solutions best fit such (legal) problems."

"The paper investigates the methodological support for capturing security and privacy requirements of a concrete health care provider."

Introduction

"Security and legal experts need support for the design of security patterns in the form of automated and easy-to-use tools that, on the one side, assist experts in the specification of their solutions and, on the other side, provide formal techniques for a proof-of-concept analysis of these solutions."

"We show how we can combine logic, agent-oriented methodologies, practical security engineering with legal reasoning to capture, model, and reason about practical legal patterns in a concrete industrial health care domain."

"We propose a methodological approach that intends to support security and legal experts in the definition of security patterns."

"We start from a graphical and easy-to-use modeling framework, specifically the SI* modeling framework (Massacci et al. [192]), to support system designers lacking deep security and legal knowledge to design and deploy systems where security and privacy solutions are enforced in accordance to regulations."

Primer on Security Patterns

"Security patterns have been proposed to assist designers in the identification and formulation of security solutions to recurrent security problems."

"They are defined as quadruples ⟨Context, Requirement, Solution, Consequence⟩, where Context defines the situation and conditions of applicability of the pattern, Requirement is the expression of the need for a system to enhance its privacy or security level, Solution is given as an abstract formal model whose application in the system-to-be guarantees the achievement of the requirement, and Consequence describes the potential effects of implementing the pattern."

"Fernández and Pan [193] design patterns for the most common security models such as authorization, role-based access control, and multilevel security."

"Fernández et al. [194] propose security patterns for physical access control."

"They propose to select security patterns at design time while still focusing on their efficient implementation."

Smart Items for Health Care

"The HCC equips Bob with a smart T-shirt that incorporates a motion sensor providing an alert as soon as Bob becomes passive for 2 min and monitoring devices that regularly measure his heart rate, blood pressure, body temperature, etc. The smart T-shirt conveys all measured data to an e-health terminal that allows Bob to promptly communicate them together with medical data to his doctor via the Monitoring and Emergency Response Centre (MERC), the department of the HCC responsible for receiving and handling patient requests for assistance."

"Dr. Charlie interrogates the MERC via his e-health terminal to receive Bob's medical data from the smart T-shirt and medical history."

"A social worker, Alison, receives a message from the MERC on her e-health terminal to go to the pharmacy and get the medicine to be delivered to Bob as an electronic credential."

Security and Privacy Requirements from a Legal Perspective

"Among such conditions, we mention two that are relevant for our purposes: (c1) "processing is necessary to protect the vital interests of the data subject or of another person where the data subject is physically or legally incapable of giving his consent" (Article 8, para. 2(c)); (c2) "processing of the data is required for the purposes of preventive medicine, medical diagnosis, the provision of care or treatment or the management of health-care services, and where those data are processed by a health professional subject under national law or rules established by national competent bodies to the obligation of professional secrecy or by another person also subject to an equivalent obligation of secrecy" (Article 8, para. 3)."

"If a data controller intends to process sensitive data for purposes that in accordance with a national law cannot be interpreted as preventive medicine, medical diagnosis, provision of care or treatment, or the management of health care services, the patient must give his explicit consent with such processing."

The SI* Modeling Framework

"To understand why and where solutions to security and privacy problems have to be deployed, system designers must model the goals and assets of the stakeholders within the socio-technical system, the assignments of permissions and responsibilities among them, and the trust network."

"We have adopted Secure Tropos (Giorgini et al. [195]), an agent- and goal-oriented security requirements engineering methodology designed to model and analyze security aspects of socio-technical systems."

"The language adopts the notions of trust of permission and trust of execution to model the expectation of one actor (the trustor) about the behavior of another actor (the trustee) on a goal or resource (the trustum)."

"Execution Dependency Diagram identifying the dependencies among actors and in particular to which actor the achievement of which goals or delivery of which resources has been assigned by which actor."

Making Security Patterns Aware of Legal Issues

"We present the pattern used to enforce the need-to-know principle when the data controller grants access to information that are not strictly necessary to perform the data processing."

"An important feature of the need-to-know pattern lies in making visible to the business application designer the significance of clearly separating those data that are strictly necessary to accomplish a certain task from those that are not."

"Need-to-Know and NDA should be jointly used to provide that high level of protection required by applications processing critical sensible data."

"The outsourcing pattern is captured only for cases in which data controllers and data processors are subjects established under national laws of EU Member States."

"When soliciting consent from the data subject, the data controller must provide them with information specified in a relevant law On the other hand, while planning later to outsource processing of personal data, the data controller has to look for a supplier providing sufficient guarantees in respect of the organizational and technical security measures."

Formal Analysis and Validation

"We want assurance that the application of each pattern makes the system compliant with the requirements for which it has been designed."

"The selection of countermeasures depends on several factors such as context, risk, cost of the solution, compliance with legislation, etc. The patterns proposed in this paper are intended to assist designers, including those that are not security expert, in this decision making."

"Once the violation of a property has been spotted by the inference engine, the designer should analyze the context and the requirements demanded by stakeholders and apply the appropriate pattern."

"This allows them to use the formal framework underlying SI* for verifying if the solution meets the requirement addressed by the pattern."

"The framework verifies if the properties encoding requirements are not satisfied by the solution."

Conclusions

"We address this issue by presenting an approach based on security and privacy patterns."

"Our focus is the definition of security patterns at the organizational level that intend to make the system compliant with legal requirements."

"With patterns, the paper also presents a formal framework that assists pattern designers in the validation of patterns and system designers in their application by supporting the analysis of organizational security and privacy requirements."

"In this setting, every security pattern (i.e., context, solution, requirement, and consequences) together with the union and intersection between their solutions and contexts are transformed into rules, and consistency among patterns are automatically verified before integration."

"The security patterns defined in this paper intend to understand how the organizational setting should be modified to meet the legal requirements."

5.2.29 *Pharmacogenetic Research and Data Protection—Challenges and Solutions [196]*

This is a machine-generated summary of:

Reischl, J; Schröder, M; Luttenberger, N; Petrov, D; Schümann, B; Ternes, R; Stürzebecher, S: Pharmacogenetic research and data protection—challenges and solutions [196]

Published in: The Pharmacogenomics Journal (2006)

Link to original: https://doi.org/10.1038/sj.tpj.6500372

If you want to cite the papers, please refer to the original.

For technical reasons we could not place the page where the original quote is coming from.

Introduction

"In this ongoing debate, it is our position that both the concerns of the public—including the view of many ethics committee members—have to be acknowledged by an elaborate practice for handling genetic samples and data, and at the same time, an understanding of the benefits and risks of pharmacogenetics has to be promoted."

"Our data protection concept (DPC) for pharmacogenetic studies in clinical research is not meant to indicate that we support 'genetic exceptionalism', but rather that we provide the necessary tools to flexibly react to and be 'compliant' with the public perception and legal policies, respectively."

"The scope of this paper will not include a detailed discussion of ethical and societal questions that are being raised in the context of genetic and pharmacogenetic research, but will focus on data protection highlighting the principles and the most important details of our concept."

Concept

"It has been Schering AG's approach for the last 3 years that clinical end points determine study design and sample size and to let clinical results drive the pharmacogenetic questions and related analyses."

"To handle, process, and analyze pharmacogenetic samples and data, an IT-based management system called GENOMatch was set up at Schering AG."

"GENOMatch is designed to achieve an optimized balance between confidentiality of genetic information and the informational risk for volunteers participating in pharmacogenetic trials on the one hand and the requirements of the researcher to accurately link clinical and genetic data on the other."

"The GENOMatch design aims at separation of information and responsibilities via a restricted access right approach based on the respective role and institution and was determined by the following key requirements: GENOMatch must support the smooth integration of pharmacogenetic research processes into already existing clinical development processes."

GENOMatch Implementation

"The complete organization of handling pharmacogenetic samples and data are represented and safeguarded by the GENOMatch process either by the application itself or via Standard Operation Procedures (SOPs)."

"The following sections describe the major functions/institutions of GENOMatch necessary for 'tracking' a sample and its associated data from sample collection in the clinic (study center) to the aggregated result of pharmacogenetic analyses."

"To allow the generation of a keycode list for any given study, two kinds of sample identifier pairs per sample are transmitted to the Secure Identity Management (SIM) center by different persons in the CSR: (1) PN–BC1, and (2) BC1–BC2."

"The 'tracking database' stores information on sample tracking, recruitment and user related data."

"GENOMatch uses external sources to upload clinical data and study related information (e.g. trial site information)."

"Match IDs are replaced by the corresponding BC2s and all patient data sets with no corresponding genetic sample will be dropped."

Feedback of Individual Genetic Data to Patients

"It is for these reasons that we have designed a feedback process that allows the return of individual genetic information without interfering with the data protection policy of GENOMatch."

"Before any such feedback of genetic information is given, a complete re-run of genetic analyses according to the procedure described for the primary sample will be performed, that is, only confirmed results will be forwarded to the physician/the patient."

"The physician chosen by the patient authorizes himself with the PWD1 and a second informed consent where the patient confirms his or her interest in individual feedback at the FHC."

"Only after replication of findings, an individual feedback report will be generated by the PBE and encrypted in a way that both passwords, the password of the physician (PWD1) and the password of the patient (PWD2), are necessary to access the information."

Conclusion

"Although we consider many of the special measures of pharmacogenetic sample and data handling as a 'necessary compromise' that will most likely be overcome by the future development of pharmacogenetic research and the better understanding and perception of its benefits and risks in the public, a number of principles will most likely be maintained that may—in the long run—have repercussions on the handling of clinical research data as well."

"Among the special issues of pharmacogenetics, the option of feedback of relevant and validated information to the patient is debated controversially with regard to its compatibility with data protection measures, the potential uncertainty it may cause for the patient and the appropriate provision of counseling."

"We have designed the GENOMatch system to allow feedback to patients without jeopardizing the established DPC and are able to accommodate the requests of patients and doctors as well as the development of guidelines and legislation."

References

1. Massey, Aaron K.; Otto, Paul N.; Hayward, Lauren J.; Antón, Annie I. Evaluating existing security and privacy requirements for legal compliance. Requirements Engineering (2009). https://doi.org/10.1007/s00766-009-0089-5
2. Otto PN, Antón AI (2007) Addressing legal requirements in requirements engineering. In: Proceedings of the 15th IEEE international requirements engineering conference, pp 5–14
3. Breaux TD, Antón AI (2008) Analyzing regulatory rules for privacy and security requirements. IEEE Trans Softw Eng 34(1):5–20
4. Breaux TD, Vail MW, Antón AI (2006) Towards regulatory compliance: extracting rights and obligations to align requirements with regulations. In: RE'06: Proceedings of the 14th IEEE international requirements engineering conference, pp 49–58
5. Barth A, Datta A, Mitchell JC, Nissenbaum H (2006) Privacy and contextual integrity: framework and applications. In: Proceedings of the 2006 IEEE symposium on security and privacy, pp 184–198
6. Breaux TD, Antón AI (2005) Mining rule semantics to understand legislative compliance. In: WPES '05: proceedings of the 2005 ACM workshop on privacy in the electronic society, pp 51–54
7. Baldassarre, Maria Teresa; Barletta, Vita Santa; Caivano, Danilo; Scalera, Michele Integrating security and privacy in software development. Software Quality Journal (2020). https://doi.org/10.1007/s11219-020-09501-6
8. Baldassarre, M. T., Caivano, D., & Visaggio, G. (2013). Empirical studies for innovation dissemination: Ten years of experience. In: 17th International Conference on Evaluation and Assessment in Software Engineering, EASE 2013. ACM International Conference Proceedings Series, ACM Press, ISBN: 978-145031848-8, https://doi.org/10.1145/2460999.2461020
9. Ardimento, P., Caivano, D., Cimitile, M., & Visaggio, G. (2008). Empirical investigation of the efficacy and efficiency of tools for transferring software engineering knowledge. Journal of Information & Knowledge Management, 7(3), 197–207 ISSN: 0219-6492
10. Hilbrich, M., & Frank, M. (2017). Enforcing security and privacy via a cooperation of security experts and software engineers: a model-based vision. IEEE 7th International Symposium on Cloud and Service Computing (SC2), Kanazawa, pp. 237–240
11. Cavoukian, A. (2012). Operationalizing privacy by design: a guide to implementing strong privacy practices. Resource document. Global Privacy and Security by Design. http://www.ontla.on.ca/library/repository/mon/26012/320221.pdf. Accessed 17 Oct 2019
12. Van Blarkom, G. W., Borking, J. J., & Olk, J. G. E. (2003). Handbook of privacy and privacy-enhancing technologies. The Case of Intelligent Software Agents. College Bescherming Bersoonsgegevens, ISBN 90-74087-33-7
13. Onesimu, J. Andrew; Karthikeyan, J.; Sei, Yuichi An efficient clustering-based anonymization scheme for privacy-preserving data collection in IoT based healthcare services. Peer-to-Peer Networking and Applications (2021). https://doi.org/10.1007/s12083-021-01077-7

14. Mohana S, Mary SASA (2016) Preserving privacy in health care information: a memetic approach. J Med Imaging Heal Informatics 6(3):779–783
15. Guan Z, Zhang Y, Wu L, Wu J, Li J, Ma Y, Hu J (2019) APPA: an anonymous and privacy preserving data aggregation scheme for fog-enhanced IoT. J Netw Comput Appl 125:82–92
16. Lu R, Heung K, Lashkari AH, Ghorbani AA (2017) A lightweight privacy-preserving data aggregation scheme for fog computing-enhanced IoT. IEEE Access 5:3302–3312
17. Song T, Li R, Mei B, Yu J, Xing X, Cheng X (2017) A privacy preserving communication protocol for IoT applications in smart homes. IEEE Internet Things J 4(6):1844–1852
18. Jayaraman PP, Yang X, Yavari A, Georgakopoulos D, Yi X (2017) Privacy preserving internet of things: from privacy techniques to a blueprint architecture and efficient implementation. Futur Gener Comput Syst
19. Sharma S, Chen K, Sheth A (2018) Toward practical privacy-preserving analytics for IoT and cloud-based healthcare systems. IEEE Internet Comput 22(2):42–51
20. Andrew J, Mathew SS, Mohit B (2019) "A comprehensive analysis of privacy-preserving techniques in deep learning based disease prediction systems," pp. 0–9
21. Ge C, Susilo W, Liu Z, Xia J, Szalachowski P, Liming F (2020) Secure keyword search and data sharing mechanism for cloud computing. IEEE Trans Dependable Secur Comput pp. 1–1
22. Ren Y et al (2020) Data query mechanism based on hash computing power of blockchain in Internet of Things. Sensors 20(1):207
23. Sweeney L (2002) k-anonymity: a model for protecting privacy. Int J Uncertain Fuzziness Knowl-Based Syst 10(05):557–570
24. Machanavajjhala A, Gehrke J, Kifer D, Venkitasubramaniam M (2006) L-diversity: privacy beyond k-anonymity. In: 22nd international conference on data engineering (ICDE'06), pp. 24–24
25. Ninghui L, Tiancheng, L, Venkatasubramanian S (2007), t-Closeness: privacy beyond k-anonymity and ℓ-diversity. In: Proceedings – international conference on data engineering, pp. 106–115
26. Prakash M, Singaravel G (2012) A new model for privacy preserving sensitive Data Mining. In: 2012 3rd international conference on computing, communication and networking technologies, ICCCNT 2012
27. Li N, Li T, Venkatasubramanian S (2010) Closeness: a new privacy measure for data publishing. IEEE Trans Knowl Data Eng 22(7):943–956
28. Rana ME, Jayabalan M, Aasif MA (2016), Privacy preserving anonymization techniques for patient data: an overview. In: Third international congress on technology, communication and knowledge (ICTCK 2016)
29. He X, Chen HH, Chen Y, Dong Y, Wang P, Huang Z (2012) Clustering-based k-anonymity. In: Lecture notes in computer science (including subseries lecture notes in artificial intelligence and lecture notes in bioinformatics), 7301 LNAI (1): 405–417
30. Li H, Guo F, Zhang W, Wang J, Xing J (2018) (a,k)-anonymous scheme for privacy-preserving data collection in IoT-based healthcare services systems. J Med Syst 42(3):56
31. Kim JW, Jang B, Yoo H (2018) Privacy-preserving aggregation of personal health data streams. PLoS One 13(11):e0207639
32. Sajjad H, Kanwal T, Anjum A, Malik SR, Khan A, Khan A, Manzoor U (2019) An efficient privacy preserving protocol for dynamic continuous data collection. Comput Secur 86:358–371
33. A mathematical theory of communication – Shannon – 1948 – bell system technical Journal – Wiley Online Library. [Online]. Available: https://doi.org/10.1002/j.1538-7305.1948.tb01338.x. Accessed: 12-Mar-2020
34. Hadar, Irit; Hasson, Tomer; Ayalon, Oshrat; Toch, Eran; Birnhack, Michael; Sherman, Sofia; Balissa, Arod Privacy by designers: software developers' privacy mindset. Empirical Software Engineering (2017). https://doi.org/10.1007/s10664-017-9517-1
35. Bandura A (1986) Social foundations of thought and action: a social cognitive theory. Prentice-Hall, Englewood Cliffs
36. Walsham G (2006) Doing interpretive research. Eur J Inf Syst 15(3):320–330

37. Strauss A, Corbin J (1994) Grounded theory methodology: an overview. In: Denzin NK, Lincoln YS (eds) Handbook of qualitative research. Sage, Thousand Oaks, pp 273–285
38. Strauss A, Corbin J (1998) Basics of qualitative research: techniques and procedures for developing grounded theory. Sage Publications, Thousand Oaks
39. Arizon-Peretz, Renana; Hadar, Irit; Luria, Gil; Sherman, Sofia Understanding developers' privacy and security mindsets via climate theory. Empirical Software Engineering (2021). https://doi.org/10.1007/s10664-021-09995-z
40. Spiekermann S, Korunovska J, Langheinrich M (2019) Inside the organization: why privacy and security engineering is a challenge for engineers. Proc IEEE 107(3):600–615
41. Hadar I, Hasson T, Ayalon O, Toch E, Birnhack M, Sherman S, Balissa A (2018) Privacy by designers: software developers' privacy mindset. Empir Softw Eng 23(1):259–289. https://doi.org/10.1007/s10664-017-9517-1
42. Zohar D, Polachek T (2014) Discourse-based intervention for modifying supervisory communication as leverage for safety climate and performance improvement: A randomized field study. J Appl Psychol 99(1):113–124
43. Christian MS, Bradley JC, Wallace JC, Burke MJ (2009) Workplace safety: A meta-analysis of the roles of person and situation factors. J Appl Psychol 94(5):1103–1127
44. Neal A, Griffin MA (2004) Safety climate and safety at work. In: Barling J, Frone MR (eds) The Psychology of Workplace Safety. American Psychological Association, Washington, DC, pp 15–34
45. Miles MB, Huberman AM, Saldaña J (2014) Qualitative data analysis: A methods sourcebook. 3rd Edition. Sage publications
46. Guba EG (1981) Criteria for assessing the trustworthiness of naturalistic inquiries. Ectj 29(2):75–91
47. Wilkowska, Wiktoria; Offermann, Julia; Colonna, Liane; Florez-Revuelta, Francisco; Climent-Pérez, Pau; Mihailidis, Alex; Poli, Angelica; Spinsante, Susanna; Ziefle, Martina Interdisciplinary perspectives on privacy awareness in lifelogging technology development. Journal of Ambient Intelligence and Humanized Computing (2022). https://doi.org/10.1007/s12652-022-04486-5
48. Rashidi P, Mihailidis A (2013) A survey on ambient-assisted living tools for older adults. IEEE J Biomed Health Inform 17(3):579–590. https://doi.org/10.1109/JBHI.2012.2234129
49. Poli A, Muñoz-Antón AM, Spinsante S, Florez-Revuelta F (2021) Balancing activity recognition and privacy preservation with a multi-objective evolutionary algorithm. In: Pires IM, Spinsante S, Zdravevski E, Lameski P (eds) Smart Objects and Technologies for Social Good. GOODTECHS 2021. Lecture Notes of the Institute for Computer Sciences, Social Informatics and Telecommunications Engineering, vol 401. Springer, Cham. https://doi.org/10.1007/978-3-030-91421-9_1
50. Diaz, Claudia; Kosta, Eleni; Dekeyser, Hannelore; Kohlweiss, Markulf; Nigusse, Girma Privacy preserving electronic petitions. Identity in the Information Society (2008). https://doi.org/10.1007/s12394-009-0012-8
51. De Cock D. Non-official information on the Belgian electronic personal identification card. 2009. https://www.cosic.esat.kuleuven.be/belpic/
52. Chaum D. Security without identification: transaction systems to make big brother obsolete. Commun ACM 1985;28(10):1030–1044
53. Camenisch J, Lysyanskaya A. An efficient system for non-transferable anonymous credentials with optional anonymity revocation. In: Advances in cryptology—EUROCRYPT 2001. New York: Springer; 2001. LNCS. vol. 2045, p. 93–118
54. Camenisch J, Hohenberger S, Kohlweiss M, Lysyanskaya A, Meyerovich M. How to win the clonewars: efficient periodic n-times anonymous authentication. In: Juels A, Wright RN, De Capitani di Vimercati S, editors. ACM conference on computer and communications security. New York: ACM; 2006. p. 201–10
55. Goldwasser S, Micali S, Rackoff C. The knowledge complexity of interactive proof-systems (extended abstract). In: STOC. New York: ACM; 1985. p. 291–304

56. Brickell E, Camenisch J, Chen L. Direct anonymous attestation. In: CCS '04: proceedings of the 11th ACM conference on computer and communications security, New York, NY, USA. New York: ACM; 2004. p. 132–145. ISBN 1-58113-961-6. https://doi.org/10.1145/1030083.1030103
57. Teranishi I, Furukawa J, Sako K. k-times anonymous authentication (extended abstract). In: Lee PJ, editor. ASIACRYPT. Lecture notes in computer science. New York: Springer; 2004. vol. 3329, p. 308–22. ISBN 3-540-23975-8
58. Damgård I, Dupont K, Østergaard Pedersen M. Unclonable group identification. In: EUROCRYPT. 2006. p. 555–72
59. Fiat A, Shamir A. How to prove yourself: practical solutions to identification and signature problems. In: Odlyzko AM, editor. CRYPTO '86. Lecture notes in computer science. New York: Springer; 1987. vol. 263, p. 186–94
60. Dingledine R, Mathewson N, Syverson P. Tor: the second-generation onion router. In: Proceedings of the 13th USENIX security symposium; 2004
61. Dumas, Marlon; García-Bañuelos, Luciano; Jääger, Joosep; Laud, Peeter; Matulevičius, Raimundas; Pankova, Alisa; Pettai, Martin; Pullonen-Raudvere, Pille; Toots, Aivo; Tuuling, Reedik; Yerokhin, Maksym Multi-level privacy analysis of business processes: the Pleak toolset. International Journal on Software Tools for Technology Transfer (2021). https://doi.org/10.1007/s10009-021-00636-w
62. Pullonen, P., Matulevičius, R., Bogdanov, D.: PE-BPMN: privacy-enhanced business process model and notation. In: Proceedings of the 15th International Conference on Business Process Management (BPM), pp. 40–56. Springer (2017)
63. Pullonen, P., Tom, J., Matulevicius, R., Toots, A.: Privacy-enhanced BPMN: enabling data privacy analysis in business processes models. Softw. Syst. Model. 18(6), 3235–3264 (2019)
64. Bhowmick, A., Duchi, J., Freudiger, J., Kapoor, G., Rogers, R.: Protection against reconstruction and its applications in private federated learning (2019). arXiv preprint arXiv:1812.00984
65. Vidhani, Kumar; Banahatti, Vijayanand; Lodha, Sachin Challenges in enabling privacy self management. CSI Transactions on ICT (2021). https://doi.org/10.1007/s40012-021-00335-6
66. Eu data protection directive. https://epic.org/privacy/intl/eu_data_protection_directive.html. Last accessed: 16-August-2021
67. Matuszewska K (2021) When design goes awry - how dark patterns conflict with GDPR and CCPA. https://piwik.pro/blog/how-dark-patterns-conflict-with-gdpr-ccpa/. Last accessed: 16-August
68. Koebe, Till; Arias-Salazar, Alejandra; Schmid, Timo Releasing survey microdata with exact cluster locations and additional privacy safeguards. Humanities and Social Sciences Communications (2023). https://doi.org/10.1057/s41599-023-01694-y
69. Alkire S, Kanagaratnam U, Suppa N (2019) The Global Multidimensional Poverty Index (MPI) 2019. OPHI MPI Methodological Note 47. Tech. Rep., Oxford Poverty and Human Development Initiative, University of Oxford. https://www.ophi.org.uk/wp-content/uploads/OPHI_MPI_MN_47_2019_vs2.pdf
70. Méndez F, Bravo, O (2011) Costa Rica Mapas de Pobreza 2011. Tech. Rep., INEC Costa Rica, San José, Costa Rica. https://www.inec.cr/sites/default/files/documentos/pobreza_y_presupuesto_de_hogares/pobreza/metodologias/documentos_metodologicos/mepobrezacenso2011-01.pdf.pdf
71. Zhang, Guangsheng; Liu, Bo; Zhu, Tianqing; Zhou, Andi; Zhou, Wanlei Visual privacy attacks and defenses in deep learning: a survey. Artificial Intelligence Review (2022). https://doi.org/10.1007/s10462-021-10123-y
72. Oh SJ, Augustin M, Fritz M, Schiele B (2018) Towards reverse-engineering black-box neural networks. In: 6th international conference on learning representations, ICLR 2018. OpenReview.net
73. Tanuwidjaja HC, Choi R, Kim K (2019) A survey on deep learning techniques for privacy-preserving. Machine learning for cyber security. Springer, Cham, pp 29–46

74. Boulemtafes A, Derhab A, Challal Y (2020) A review of privacy-preserving techniques for deep learning. Neurocomputing 384:21–45
75. Tanuwidjaja HC, Choi R, Baek S, Kim K (2020) Privacy-preserving deep learning on machine learning as a service-a comprehensive survey. IEEE Access 8:167425–167447
76. Ha T, Dang TK, Le H, Truong TA (2020) Security and privacy issues in deep learning: a brief review. SN Comput Sci 1:253
77. He Y, Meng G, Chen K, Hu X, He J (2020a) Towards security threats of deep learning systems: a survey. IEEE Transactions on Software Engineering
78. Liu X, Xie L, Wang Y, Zou J, Xiong J, Ying Z, Vasilakos AV (2021d) Privacy and security issues in deep learning: a survey. IEEE Access 9:4566–4593
79. He Y, Rahimian S, Schiele B, Fritz M (2020b) Segmentations-leak: membership inference attacks and defenses in semantic image segmentation. In: Computer vision—ECCV 2020. Springer International Publishing, pp 519–535
80. Park Y, Kang M (2020) Membership inference attacks against object detection models. arXiv:200104011 [cs]
81. Salem A, Zhang Y, Humbert M, Berrang P, Fritz M, Backes M (2019) ML-leaks: model and data independent membership inference attacks and defenses on machine learning models. In: 26th Annual network and distributed system security symposium, NDSS 2019. The Internet Society
82. Yeom S, Giacomelli I, Fredrikson M, Jha S (2018) Privacy risk in machine learning: analyzing the connection to overfitting. In: 2018 IEEE 31st computer security foundations symposium (CSF), pp 268–282
83. Chen Z, Zhu T, Xiong P, Wang C, Ren W (2021) Privacy preservation for image data: a GAN-based method. Int J Intell Syst 36:1668–1685
84. Liu B, Ding M, Xue H, Zhu T, Ye D, Song L, Zhou W (2021b) DP-Image: differential privacy for image data in feature space. arXiv:210307073 [cs]
85. Phan N, Wu X, Hu H, Dou D (2017b) Adaptive laplace mechanism: differential privacy preservation in deep learning. In: 2017 IEEE international conference on data mining (ICDM), pp 385–394
86. Rahman MA, Rahman T, Laganière R, Mohammed N (2018) Membership inference attack against differentially private deep learning model. Trans Data Priv 11:61–79
87. Truex S, Liu L, Gursoy ME, Yu L, Wei W (2019) Demystifying membership inference attacks in machine learning as a service. IEEE Transactions on Services Computing
88. Chen D, Yu N, Zhang Y, Fritz M (2020b) GAN-leaks: a taxonomy of membership inference attacks against generative models. In: Proceedings of the 2020 ACM SIGSAC conference on computer and communications security. Association for Computing Machinery, pp 343–362
89. Leino K, Fredrikson M (2020) Stolen memories: leveraging model memorization for calibrated white-box membership inference. In: 29th USENIX security symposium (USENIX Security 20), pp 1605–1622
90. Choquette-Choo CA, Tramer F, Carlini N, Papernot N (2021) Label-only membership inference attacks. In: International conference on machine learning
91. Nasr M, Shokri R, Houmansadr A (2018) Machine learning with membership privacy using adversarial regularization. In: Proceedings of the 2018 ACM SIGSAC conference on computer and communications security. Association for Computing Machinery, pp 634–646
92. Attili, V. S. Prakash; Mathew, Saji K.; Sugumaran, Vijayan Information Privacy Assimilation in IT Organizations. Information Systems Frontiers (2021). https://doi.org/10.1007/s10796-021-10158-0
93. Attili, V. S. P., Mathew, S. K., & Sugumaran, V. (2018). Understanding information privacy assimilation in IT organizations using multi-site case studies. Communications of the Association for Information Systems, 42(4), 66–94
94. Conger, S., Pratt, J. H., & Loch, K. D. (2013). Personal information privacy and emerging technologies. Information Systems Journal, 23(5), 401–417. https://doi.org/10.1111/j.1365-2575.2012.00402.x

95. Kim, D. J., Yim, M. S., Sugumaran, V., & Rao, H. R. (2016). Web assurance seal services, trust and consumers' concerns: An investigation of e-commerce transaction intentions across two nations. European Journal of Information Systems, 25(3), 252–273
96. Johnson, E. C. (2011). AICPA/CICA Privacy Maturity Model. Retrieved from https://www.kscpa.org/writable/files/AICPADocuments/10-229_aicpa_cica_privacy_maturity_model_finalebook.pdf. Accessed 1 Feb 2021
97. Belanger, F., & James T. L. (2020). A theory of multilevel information privacy management for the digital era, Information Systems Research, 31(2), 510–536
98. Henseler, J., Ringle, M. C., & Sinkovics, R. R. (2009). The use of partial least squares path modeling in international marketing. Advances in International Marketing, 20(1), 277–319
99. Hair, J. F. J., Hult, G. T. M., Ringle, C. M., & Sarstedt, M. (2016). A primer on partial least squares structural equation modeling (PLS-SEM). Sage Publications
100. Ringle, C. M., Sarstedt, M., & Straub, D. W. (2012). A critical look at the use of PLS-SEM. MIS Quarterly, 36(1), 3–14
101. Devriendt, Thijs; Shabani, Mahsa; Lekadir, Karim; Borry, Pascal Data sharing platforms: instruments to inform and shape science policy on data sharing?. Scientometrics (2022). https://doi.org/10.1007/s11192-022-04361-2
102. Oliveira, J. L., Trifan, A., & Bastião Silva, L. A. (2019). EMIF catalogue: A collaborative platform for sharing and reusing biomedical data. International Journal of Medical Informatics, 126, 35–45
103. Altmetrics Working Group of the Open Science Policy Platform (2017). Recommendations of the OSPP on next-generation metrics. Retrieved from https://ec.europa.eu/research/openscience/pdf/ospp_metrics_wg_recommendations_final.pdf#view=fit&pagemode=none
104. Leonelli, S. (2017). Mutual learning exercise : Open science—altmetrics and rewards incentives and rewards to engage in open science activities. Retrieved from https://rio.jrc.ec.europa.eu/policy-support-facility/mle-open-science-altmetrics-and-rewards
105. Working Group on Rewards under Open Science. (2017). Evaluation of research careers fully acknowledging open science practices. European Commission
106. Wouters, P., Ràfols, I., Oancea, A., Caroline, S., Kamerlin, L., Britt, J., & Jacob, M. (2019). Indicator frameworks for fostering open knowledge practices in science and scholarship. European Commission
107. Allen, L., O'Connell, A., & Kiermer, V. (2019). How can we ensure visibility and diversity in research contributions? How the contributor role taxonomy (CRediT) is helping the shift from authorship to contributorship. Learned Publishing, 32(1), 71–74. https://doi.org/10.1002/leap.1210
108. Bierer, B. E., Crosas, M., & Pierce, H. H. (2017). Data authorship as an incentive to data sharing. The New England Journal of Medicine, 376(17), 1684–1687. https://doi.org/10.1056/NEJMsb1616595
109. Hemphill, Libby; Schöpke-Gonzalez, Angela; Panda, Anmol Comparative sensitivity of social media data and their acceptable use in research. Scientific Data (2022). https://doi.org/10.1038/s41597-022-01773-w
110. Hill, E. M., Turner, E. L., Martin, R. M. & Donovan, J. L. "Let's get the best quality research we can": public awareness and acceptance of consent to use existing data in health research: a systematic review and qualitative study. BMC Med. Res. Methodol. 13, 72 (2013)
111. Martin, K. E. & Nissenbaum, H. What is it about location? Berkeley Technol. Law J. 35 (2020)
112. Hemphill, L. Personal and social media data survey [data set]. University of Michigan - Deep Blue Data https://doi.org/10.7302/6vjf-av59 (2022)
113. Ishii, Kaori Comparative legal study on privacy and personal data protection for robots equipped with artificial intelligence: looking at functional and technological aspects. AI & SOCIETY (2017). https://doi.org/10.1007/s00146-017-0758-8
114. Institute for Information and Communications Policy of the Ministry of Internal Affairs and Communications (2016) AI network review meeting report: challenges for achieving wisdom network society (WINS). http://www.soumu.go.jp/menu_news/s-news/01iicp01_02000050.html. Accessed 12 July 2017 (in Japanese)

115. Office of the Privacy Commissioner of Canada (2016) Consent and privacy https://www.priv.gc.ca/media/1806/consent_201605_e.pdf. Accessed 12 July 2017
116. Fox, Grace; James, Tabitha L. Toward an Understanding of the Antecedents to Health Information Privacy Concern: A Mixed Methods Study. Information Systems Frontiers (2020). https://doi.org/10.1007/s10796-020-10053-0
117. Anderson, C. L., & Agarwal, R. (2011). The digitization of healthcare: Boundary risks, emotion, and consumer willingness to disclose personal health information. Information Systems Research, 22(3), 469–490
118. Bansal, G., Zahedi, F. M., & Gefen, D. (2016). Do context and personality matter? Trust and privacy concerns in disclosing private information online. Information & Management, 53(1), 1–21
119. Kordzadeh, N., & Warren, J. (2017). Communicating personal health information in virtual health communities: An integration of privacy calculus model and affective commitment. Journal of the Association for Information Systems, 18(1), 45–81
120. Kordzadeh, N., Warren, J., & Seifi, A. (2016). Antecedents of privacy calculus components in virtual health communities. International Journal of Information Management, 36(5), 724–734
121. Xu, H., Dinev, T., Smith, J., & Hart, P. (2011). Information privacy concerns: Linking individual perceptions with institutional privacy assurances. Journal of the Association for Information Systems, 12(12), 798–824
122. Zhang, X., Liu, S., Chen, X., Wang, L., Gao, B., & Zhu, Q. (2018). Health information privacy concerns, antecedents, and information disclosure intention in online health communities. Information & Management, 55(4), 482–493
123. Angst, C. M., & Agarwal, R. (2009). Adoption of electronic health records in the presence of privacy concerns: The elaboration likelihood model and individual persuasion. MIS Quarterly, 33(2), 339–370
124. Li, T., & Slee, T. (2014). The effects of information privacy concerns on digitizing personal health records. Journal of the Association for Information Science and Technology, 65(8), 1541–1554
125. Hong, W., & Thong, J. (2013). Internet privacy concerns: An integrated conceptualization and four empirical tests. MIS Quarterly, 37(1), 275–298
126. Bansal, G., Zahedi, F. M., & Gefen, D. (2010). The impact of personal dispositions on information sensitivity, privacy concern and trust in disclosing health information online. Decision Support Systems, 49(2), 138–150
127. Li, H., Gupta, A., Zhang, J., & Sarathy, R. (2014). Examining the decision to use standalone personal health record systems as a trust-enabled fair social contract. Decision Support Systems, 57, 376–386
128. Manion, Frank J; Robbins, Robert J; Weems, William A; Crowley, Rebecca S Security and privacy requirements for a multi-institutional cancer research data grid: an interview-based study. BMC Medical Informatics and Decision Making (2009). https://doi.org/10.1186/1472-6947-9-31
129. DeCew, Judith Wagner Alternatives for protecting privacy while respecting patient care and public health needs. Ethics and Information Technology (1999). https://doi.org/10.1023/a:1010012018213
130. Austin, Melissa A; Harding, Sarah E; McElroy, Courtney E Monitoring ethical, legal, and social issues in developing population genetic databases. Genetics in Medicine (2003). https://doi.org/10.1097/01.gim.0000093976.08649.1b
131. Austin M, Harding S, McElroy C Genebanks. A comparison of eight proposed international genetic databases. Community Genet 2003; 6: 37–45
132. Jusob, Farad Rafique; George, Carlisle; Mapp, Glenford Exploring the need for a suitable privacy framework for mHealth when managing chronic diseases. Journal of Reliable Intelligent Environments (2017). https://doi.org/10.1007/s40860-017-0049-7
133. Estrin D, Sim I (2010) Open mHealth architecture: an engine for health care innovation. Science 330:759–760

134. European Commission (2014) Green paper on mobile Health ("mHealth"). Brussels, 10 April 2014, COM(2014) 219 final
135. Varshney U (2014) A model for improving quality of decisions in mobile health. Decis Support Syst 62:66–77. https://doi.org/10.1016/j.dss.2014.03.005
136. Office of the National Coordinator for Health Information Technology (2008) Nationwide privacy and security framework for electronic exchange of individually identifiable health information. https://www.healthit.gov/sites/default/files/nationwide-ps-framework-5.pdf. Accessed 07 Apr 2017
137. Arora S, Yttri J, Nilse W (2014) Privacy and security in mobile health (mHealth) research. Alcohol Res Curr Rev 36:143–151
138. Wicks, Paul; Wahlstrom-Edwards, Lindsey; Fillingham, Sam; Downing, Andrea; Davies, Elin Haf So You Want to Build Your Disease's First Online Patient Registry: An Educational Guide for Patient Organizations Based on US and European Experience. The Patient - Patient-Centered Outcomes Research (2023). https://doi.org/10.1007/s40271-023-00619-w
139. Gliklich RE, Leavy MB, Dreyer NA. Registries for evaluating patient outcomes: a user's guide. Fourth edition. Agency for Healthcare Research and Quality (AHRQ); 2020. https://effectivehealthcare.ahrq.gov/products/registries-guide-4th-edition/users-guide. Accessed 9 Mar 2023
140. Nelson EC, Dixon-Woods M, Batalden PB, Homa K, Van Citters AD, Morgan TS, et al. Patient focused registries can improve health, care, and science. BMJ. 2016. https://doi.org/10.1136/bmj.i3319
141. European Medicines Agency. Patient Registries Workshop, 28 October 2016. 2017. https://www.ema.europa.eu/en/documents/report/report-patient-registries-workshop_en.pdf. Accessed 9 Mar 2023
142. TREAT-NMD Neuromuscular Network. TREAT-NMD diseases: data and Information. 2022. https://treat-nmd.org/what-we-do/data-and-information/treat-nmd-diseases/. Accessed 9 Mar 2023
143. Maelstrom. Multiple Sclerosis Metadata Collective: a collaborative effort of North American observational studies in multiple sclerosis. 2022. https://www.maelstrom-research.org/network/msmdc. Accessed 24 Oct 2022
144. Olmo CA, McGettigan P, Kurz X. Barriers and opportunities for use of patient registries in medicines regulation. Clin Pharmacol Ther. 2019;106:39–42
145. Stine K, Dang Q. Encryption basics. J Am Health Inform Manage Assoc. 2011;82:44–6
146. Wikipedia. There are known knowns. 2022. https://en.wikipedia.org/wiki/There_are_known_knowns. Accessed 9 Mar 2023
147. Zhang S. Facebook groups as therapy. 2018. https://www.theatlantic.com/technology/archive/2018/10/facebook-emotional-support-groups/572941/. Accessed 24 Oct 2022
148. Prior R. This breast cancer advocate says she discovered a Facebook flaw that put the health data of millions at risk. 2020. https://edition.cnn.com/2020/02/29/health/andrea-downing-facebook-data-breach-wellness-trnd/index.html. Accessed 24 Oct 2022
149. Ohlheiser A. Facebook is bombarding cancer patients with ads for unproven treatments. 2022. https://www.technologyreview.com/2022/06/27/1054784/facebook-meta-cancer-treatment-ads-misinformation/. Accessed 9 Mar 2023
150. Feathers T, Fondrie-Teitler S, Waller A, Mattu S. Facebook is receiving sensitive medical information from hospital websites. 2022. https://themarkup.org/pixel-hunt/2022/06/16/facebook-is-receiving-sensitive-medical-information-from-hospital-websites. Accessed 24 Oct 2022
151. Abdullah, Sidrah; Arshad, Junaid; Khan, Muhammad Mubashir; Alazab, Mamoun; Salah, Khaled PRISED tangle: a privacy-aware framework for smart healthcare data sharing using IOTA tangle. Complex & Intelligent Systems (2022). https://doi.org/10.1007/s40747-021-00610-8
152. Popov S (2015) The tangle
153. Coppolino L, D'Antonio S, Mazzeo G, Romano L (2017) Cloud security: emerging threats and current solutions. Comput Electr Eng 59:126–140. https://doi.org/10.1016/j.compeleceng.2016.03.004. http://www.sciencedirect.com/science/article/pii/S0045790616300544

154. Domínguez Mayo F. J, Escalona M, Mejías M, Aragón G, Garcia-Garcia J, Torres J, Enríquez J (2015) A strategic study about quality characteristics in e-health systems based on a systematic literature review. Sci World J 2015:1–11. https://doi.org/10.1155/2015/863591
155. Scholl M, Stine K, Hash J, Bowen P, Johnson A, Dancy Smith C, Steinberg DI (2013) An introductory resource guide for implementing the health insurance portability and accountability act (HIPAA) security rule. Retrieved from https://nvlpubs.nist.gov/nistpubs/Legacy/SP/nistspecialpublication800-66r1.pdf
156. da Conceição AF, da Silva FSC, Rocha V, Locoro A, Barguil JM. Eletronic health records using blockchain technology. arXiv:1804.10078
157. Handy P (2017) Introducing masked authenticated messaging, https://blog.iota.org/introducing-masked-authenticated-messaging-e55c1822d50e
158. IOTA, Masked authenticated messaging, https://github.com/iotaledger/mam.client.js/ (2020)
159. Zheng X, Sun S, Mukkamala RR, Vatrapu R, Ordieres-Meré J (2019) Accelerating health data sharing: A solution based on the internet of things and distributed ledger technologies. J Med Internet Res 21(6):e13583. https://doi.org/10.2196/13583
160. Brogan J, Baskaran I, Ramachandran N (2018) Authenticating health activity data using distributed ledger technologies, computational and structural. Biotechnol J 16:257–266. https://doi.org/10.1016/j.csbj.2018.06.004. http://www.sciencedirect.com/science/article/pii/S2001037018300345
161. Devnet, [Online; accessed March 07, 2020]. https://devnet.thetangle.org/nodes
162. Bartolomeu P, Vieira E, Ferreira J (2018) Iota feasibility and perspectives for enabling vehicular applications 1–7. https://doi.org/10.1109/GLOCOMW.2018.8644201
163. Courbier, Sandra; Dimond, Rebecca; Bros-Facer, Virginie Share and protect our health data: an evidence based approach to rare disease patients' perspectives on data sharing and data protection - quantitative survey and recommendations. Orphanet Journal of Rare Diseases (2019). https://doi.org/10.1186/s13023-019-1123-4
164. El Emam, K., Rodgers, S. & Malin, B. Anonymising and sharing individual patient data. BMJ 350, 1139 (2015)
165. European Commission. Attitudes on data protection and electronic identity in the European Union. Special Eurobarometer 359; 2010. Accessed Oct 2018
166. European Commission. Data protection report. Special Eurobarometer 431; 2015
167. Ambrosini, Anna; Calabrese, Daniela; Avato, Francesco Maria; Catania, Felice; Cavaletti, Guido; Pera, Maria Carmela; Toscano, Antonio; Vita, Giuseppe; Monaco, Lucia; Pareyson, Davide The Italian neuromuscular registry: a coordinated platform where patient organizations and clinicians collaborate for data collection and multiple usage. Orphanet Journal of Rare Diseases (2018). https://doi.org/10.1186/s13023-018-0918-z
168. Lochmüller H, Le Cam Y, Jonker AH, Lau LP, Baynam G, Kaufmann P, et al. 'IRDiRC recognized Resources': a new mechanism to support scientists to conduct efficient, high-quality research for rare diseases. Eur J Hum Genet. 2017;25(2):162–5
169. Gainotti S, Torreri P, Wang CM, Reihs R, Mueller H, Heslop E, et al. The RD-Connect Registry & Biobank Finder: a tool for sharing aggregated data and metadata among rare disease researchers. Eur J Hum Genet. 2018;26(5):631–43
170. Wilkinson, M.D., Dumontier, M., Aalbersberg, I.J., Appleton, G., Axton, M., Baak, A., Blomberg, N., Boiten, J.-W., da Silva Santos, L.B., Bourne, P.E.: The fair guiding principles for scientific data management and stewardship. Sci. Data 3(1), 1–9 (2016)
171. Thompson R, Robertson A, Lochmüller H. Natural history, trial readiness and gene discovery: advances in patient registries for neuromuscular diseases. In: de la Paz Posada M, Taruscio D, Groft S, editors. Rare diseases epidemiology: update and overview. Advances in experimental medicine and biology, vol. 1031. Cham: Springer; 2017. p. 97–124
172. Köhler S, Vasilevsky NA, Engelstad M, Foster E, McMurry J, Aymé S, et al. The human phenotype ontology in 2017. Nucleic Acids Res. 2017;45(D1):D865–76
173. Climent-Pérez, Pau; Florez-Revuelta, Francisco Protection of visual privacy in videos acquired with RGB cameras for active and assisted living applications. Multimedia Tools and Applications (2021). https://doi.org/10.1007/s11042-020-10249-1

174. Offermann-van Heek J, Arning K, Ziefle M (2019) All eyes on you! impact of location, camera type, and privacy-security-trade-off on the acceptance of surveillance technologies. In: Donnellan B, Klein C, Helfert M, Gusikhin O, Pascoal A (eds) Smart cities, green technologies, and intelligent transport systems, pp 131–149. Springer International Publishing, Cham
175. Orekondy T, Schiele B, Fritz M (2017) Towards a visual privacy advisor: Understanding and predicting privacy risks in images. In: Proceedings of the IEEE international conference on computer vision, pp 3686–3695
176. Babiceanu RF, Bojda P, Seker R, Alghumgham MA (2015) An onboard uas visual privacy guard system. In: 2015 Integrated communication, navigation and surveillance conference (ICNS), pp J1–1–J1–8
177. Chen J, Konrad J, Ishwar P (2018) Vgan-based image representation learning for privacy-preserving facial expression recognition. In: Proceedings of the IEEE conference on computer vision and pattern recognition workshops, pp 1570–1579
178. Wu Z, Wang Z, Wang Z, Jin H (2018) Towards privacy-preserving visual recognition via adversarial training: A pilot study. In: Proceedings of the European conference on computer vision (ECCV), pp 606–624
179. Padilla-López J, Chaaraoui A, Gu F, Flórez-Revuelta F (2015) Visual privacy by context: proposal and evaluation of a level-based visualisation scheme. Sensors 15(6):12959–12982
180. Wirth, Felix Nikolaus; Meurers, Thierry; Johns, Marco; Prasser, Fabian Privacy-preserving data sharing infrastructures for medical research: systematization and comparison. BMC Medical Informatics and Decision Making (2021). https://doi.org/10.1186/s12911-021-01602-x
181. Li T, Li N. On the tradeoff between privacy and utility in data publishing. In: Elder J, Soulié Fogelman F, editors. Proceedings of the 15th ACM SIGKDD International Conference on Knowledge Discovery and Data Mining; 2009; Paris. New York: Association for Computing Machinery; 2009. p. 517–26. https://doi.org/10.1145/1557019.1557079
182. Raisaro JL, Troncoso-Pastoriza JR, Misbach M, Sousa JS, Pradervand S, Missiaglia E, et al. MedCo: enabling secure and privacy-preserving exploration of distributed clinical and genomic data. IEEE/ACM Trans Comput Biol Bioinform. 2018;16:1328–41. https://doi.org/10.1109/TCBB.2018.2854776
183. Dwork C. Differential privacy: a survey of results. In: Agrawal M, Du D, Duan Z, Li A, editors. Theory and Applications of Models of Computation. Proceedings of the 5th International Conference on Theory and Applications of Models of Computation; 2008; Xi'an. Berlin: Springer; 2008, p. 1–19. https://doi.org/10.1007/978-3-540-79228-4_1
184. Ohno-Machado L, Vinterbo S, Dreiseitl S. Effects of data anonymization by cell suppression on descriptive statistics and predictive modeling performance. J Am Med Inform Assoc. 2002;9:S115–9. https://doi.org/10.1197/jamia.M1241
185. Sweeney L (2002) k-anonymity: a model for protecting privacy. Int J Uncertainty Fuzziness Knowl Based Syst 10(05):557–570
186. Scheibner, James; Ienca, Marcello; Vayena, Effy Health data privacy through homomorphic encryption and distributed ledger computing: an ethical-legal qualitative expert assessment study. BMC Medical Ethics (2022). https://doi.org/10.1186/s12910-022-00852-2
187. Raisaro JL, Troncoso-Pastoriza JR, Misbach M, Sousa JS, Pradervand S, Missiaglia E, et al. MedCo: Enabling secure and privacy-preserving exploration of distributed clinical and genomic data. IEEE/ACM Trans Comput Biol Bioinform. 2018; 1–1
188. Azaria A, Ekblaw A, Vieira T, Lippman A. MedRec: Using blockchain for medical data access and permission management. In: 2016 2nd International Conference on Open and Big Data (OBD). 2016. p. 25–30
189. Troncoso-Pastoriza JR, Raisaro JL, Gasser L, Ford B, Hubaux J-P. MedChain: accountable and auditable data sharing in distributed medical scenarios. 2019
190. Driessen S, Gervasoni P. Research projects in human genetics in Switzerland: analysis of research protocols submitted to cantonal ethics committees in 2018. Swiss Med Wkly. 2021. https://doi.org/10.4414/smw.2021.20403.

191. Compagna, Luca; El Khoury, Paul; Krausová, Alžběta; Massacci, Fabio; Zannone, Nicola How to integrate legal requirements into a requirements engineering methodology for the development of security and privacy patterns. Artificial Intelligence and Law (2008). https://doi.org/10.1007/s10506-008-9067-3
192. Massacci F, Mylopoulos J, Zannone N (2007) An ontology for secure socio-technical systems. In: Handbook of ontologies for business interaction. The IDEA Group, Hershey, pp 188–207
193. Fernández EB, Pan R (2001) A pattern language for security models. In: Proceedings of PLoP'01
194. Fernández EB, Ballesteros J, Desouza-Doucet AC, Larrondo-Petrie MM (2007) Security patterns for physical access control systems. In: Proceedings of DBSec'07, LNCS 4602. Springer-Verlag, Berlin, pp 259–274
195. Giorgini P, Massacci F, Zannone N (2005) Security and trust requirements engineering. In: FOSAD 2004/2005, LNCS 3655. Springer-Verlag, Berlin, pp 237–272
196. Reischl, J; Schröder, M; Luttenberger, N; Petrov, D; Schümann, B; Ternes, R; Stürzebecher, S Pharmacogenetic research and data protection – challenges and solutions. The Pharmacogenomics Journal (2006). https://doi.org/10.1038/sj.tpj.6500372
197. Wong R, Li J, Fu A, Wang K (2009) (A, K)-anonymous data publishing. J Intell Inf Syst 33(2):209–234
198. Eurordis-NORD-CORD Joint Declaration of 10 Key Principles for Rare Disease Patient Registries 2012. https://www.eurordis.org/content/eurordis-nord-cord-release-joint-declaration-10-key-principles-rare-disease-patient-registries, 2012, 12 November. In: EURORDIS https://www.eurordis.org/content/eurordis-nord-cord-release-joint-declaration-10-key-principles-rare-disease-patient-registries. Accessed 25 June 2018
199. EUCERD Recommendations on Rare Disease Patient Registration and Data Collection (5 June 2013). In: EUCERD Recommendations http://www.eucerd.eu/?page_id=13. Accessed 25 June 2018
200. Thompson R, et al. RD-connect: an integrated platform connecting databases, registries, biobanks and clinical bioinformatics for rare disease research. J Gen Intern Med. 2014;29(3):780–7
201. RD-Connect. https://rd-connect.eu/. Accessed 25 June 2018
202. Lochmüller H, et al. RD-connect, NeurOmics and EURenOmics: collaborative European initiative for rare diseases. Eur J Hum Genet. 2018;26(6):778–85

The manufacturer's authorised representative in the EU is Springer Nature Customer Service Centre GmbH, Europaplatz 3, 69115 Heidelberg, Germany. If you have any concerns regarding our products, please contact ProductSafety@springernature.com

Printed and bound by CPI Group (UK) Ltd, Croydon, CR0 4YY
04/08/2026
02185640-0003